The Intentional Brain

The Intentional Brain

Motion, Emotion, and the Development
of Modern Neuropsychiatry

MICHAEL R. TRIMBLE, MD

Johns Hopkins University Press
Baltimore

2 4 6 8 9 7 5 3 1

Johns Hopkins University Press
2715 North Charles Street
Baltimore, Maryland 21218-4363
www.press.jhu.edu

Library of Congress Cataloging-in-Publication Data

Trimble, Michael R., author.
The intentional brain : motion, emotion, and the development of modern
neuropsychiatry / Michael R. Trimble.
    p. ; cm.
Includes bibliographical references and index.
ISBN 978-1-4214-1949-7 (hardcover : alk. paper) — ISBN 1-4214-1949-1 (hardcover :
alk. paper) — ISBN 978-1-4214-1950-3 (electronic) — ISBN 1-4214-1950-5 (electronic)
    I. Title.
[DNLM: 1. Neuropsychiatry.   2. Behavior—physiology.   3. Creativity.
    4. Emotions—physiology.   5. Mental Disorders—physiopathology.
        6. Nervous System Diseases—psychology. WL 102]
    RC346
    616.8—dc23   2015028202

A catalog record for this book is available from the British Library.

*Special discounts are available for bulk purchases of this book. For more information,
please contact Special Sales at 410-516-6936 or
specialsales@press.jhu.edu.*

To Jackie A, for friendship, professionalism, and unwavering support at challenging times

The Tree of Knowledge in Eden provokes humankind to transgression, to lasting exile and *misère*. Prometheus is sentenced to unenduring torture for his theft of theoretical and practical sagacity from the jealous gods. The striving intellect of Faust overreaches and precipitates his soul into hell. An ineradicable crime attaches to the defining excellence of the human spirit. Measureless vengeance is visited on those who would "teach eternity" (Dante). Hunters after truth are in turn hunted as if some organic contradiction opposed the exercise of the mind and at-homeness in natural life. Yet the impulse to taste of the forbidden fruit, to steal and master fire, to pose ultimate questions as does Faust, is unquenchable. Be it at the cost of personal survival or social ostracism. . . . The magnet is the unknown and man is the animal which asks.

George Steiner, *The Poetry of Thought*

CONTENTS

Time flaps on the mast. There we stop; there we stand. Rigid, the skeleton of habit alone upholds the human frame.

Virginia Woolf, 1925/1996, 48

This book supports a phenomenological approach to understanding human experiences. This includes the importance of narrative, of telling stories, as the backbone of human culture and individual lives, and the competing desires that enchant the world that our body encounters, beckoning us forever onward. The book is not simply about neuropsychiatry as a medical discipline, but it is in many ways much more a reflection on the way the brain and its functions have been viewed over the centuries, as well as on the huge change in orientation, germinating within romanticism, which has given us an understanding of our dynamic, active, creative brain. It is about the dawning and development of artistic creativity and the conflicts between different philosophical views of our world and our position in it. Congruence not disharmony, connectivity not separation, uniting not splitting, wholeness not hollowness, the self in the world not alienated from it—all of these are embedded in the romantic view.

Looking back on a career of more than 40 years in clinical neurology and psychiatry and my involvement in experimental neuroscience, I realize that the struggles for recognition of the discipline of neuropsychiatry are still with us. In this book, my contribution, I explore some of the reasons and the intellectual contests involved. Certain disorders are relevant to the developing background of today's practice of neuropsychiatry. These include especially epilepsy, syphilis, hysteria, movement disorders, dementia, psychosis, and melancholia. But such a portrayal cannot ignore interesting ideas that held sway in Europe throughout much of the nineteenth century, such as the rise and fall of monomania and the degeneration theory.

Neuropsychiatry is concerned not only with descriptions of clinical abnormalities that relate to our understanding of brain-behavior relationships but also with the meaning of abnormal behavior. This requires consideration of content as well as form and the various life contingencies that impinge on patients, which may influence their symptoms. This also calls for consideration of the huge gulf between disease (pathology) and illness (what patients present with) and a propensity to tolerate uncertainty. William Alwyn

Lishman, in his paper "What Is Neuropsychiatry?," explained that neuro-psychiatry was not an "all-exclusive domain" embracing only the neurosciences, but that "social, developmental, psychodynamic and interpersonal forces must also be considered." He warns, "One must take issue with those who claim that the neuropsychiatric approach can account for all mental disorder. Such vaulting ambition is reminiscent of the once proud claims of psychodynamic theory."[1]

Neuropsychiatry is not simply an offshoot of psychiatry. It is a discipline that has arisen out of a clinical need for patients who have fallen badly between the cracks engendered by the developments of the clinical neurosciences in the twentieth century. Every psychiatric disorder is accompanied by abnormal movements; movement disorders such as Parkinson's disease invariably have associated mental state changes, and Gilles de la Tourette syndrome is not simply a tic disorder. Likewise, epilepsy is not just about seizures. It is a diagnosis of a continuous interruption of cerebral electro-chemical activity in the brain with intermittent eruptions of such excessive activity that a clinical seizure is observed. If the first question of the treating physician is "How many fits have you had?" then the physician needs to review his or her understanding of the disorder. Living with a fear of the next seizure, the social limitations and stigma of the diagnosis, and the need in many cases for continuous medications harbors much more than the "fit" itself. Even the word *seizure* has considerable historical and social baggage attached to it.

Such euphemisms abound in clinical neurological practice; in some cases they dance around terms such as *functional, supratentorial, non-neurological, medically unexplained*, and the like, avoiding the well-trodden, always reemerging *hysteria*, with implications of exaggeration of complaints if not frank malingering. Dementia in its various guises has become the feared scourge of the twenty-first century. Barely of interest to an earlier generation of neurologists, who left it to "geriatric" psychiatrists to pick up the pieces, it most obviously requires a neuropsychiatric approach to understanding and management.

The narrative of this book echoes the development of Western civilization, from the early city-states of the Greeks (*civitas*) to the intellectual and artistic achievements that remain part of our contemporary culture. Two great figures whose ideas reverberate through this text are Aristotle and Plato. The mathematician and philosopher Alfred North Whitehead said that all Western philosophy was a footnote to Plato, yet it was Aristotle who thought

that the philosopher should end with medicine and the physician commence with philosophy.[2] According to the historian Arthur Herman, the idea
that the rise of modern science involved a struggle between reason and religion is not merely wrong but also misleading.[3] It is true that the overwhelming pressures of the monotheistic religions on scientific advancement have
been profound, yet the real struggle has been between Aristotle and Plato.
Both were concerned with things eternal and immutable, but they also
needed to assimilate into their ideas what was transitory and changing. As
the cultural philosopher George Steiner phrased it, "Only, moreover, in
classical Greece and its European legacy, is the theoretical applied to the
practical in the guise of a universal critique of all life and of its goals. There
is a sharp distinction to be drawn between this phenomenology and the
'mythicopractical' fabric of Far Eastern or Indian models. The seminal fact
of wonder, *thaumazein*, and of theoretical-logical development is Platonic
and Aristotelian to the core."[4] For Plato the idealist, wisdom came not through
the senses, but through contemplation, reason, and wonder, while for Aristotle, knowledge was derived a posteriori, from experience.

The earliest intimation of what today we call neuropsychiatry begins with
Hippocrates, who laid the foundations of our clinical approach to medicine.
His affirmations that the brain is the seat of emotions, as well as epilepsy
and diseases of the mind, began a trail of thought and experiment leading to
our current attempts to understand mind–brain relationships. Some other
civilizations, such as the Egyptians and Babylonians, left documentary evidence that disorders we still identify were a part of their medical experience,
including descriptions of what we now call mental or psychiatric disorders, but
they left no hints as to what they thought might lie behind such illnesses, and
nothing of what they thought about the brain and its functions.

I seek in this book to explore ideas over the past 2,500 years which have
contributed to the intellectual legacy of what is now referred to as neuropsychiatry. The writings and speculations of our predecessors, without the tools
we now have for investigation and a variety of well-honed treatments, are
worthy of consideration.

The empirical view of history, preferred by the Anglo-Saxon tradition,
would be based firmly on facts, seeking natural "laws" that are discoverable.
Yet there has been the alternative view of history, as a venture seeking not so
much knowledge but understanding, linked especially to the European schools
of hermeneutics, or text interpretation. Meaning in these paradigms precedes
knowledge, the historian taking what scattered pieces of documentation or

artifacts are available and working them into a tapestry of identifiable ideas, however incomplete the picture. The historian-philosopher Wilhelm Dilthey realized that there were no natural "laws" for history, as in the physical sciences, and drew distinctions between the latter (*Naturwissenschaften*) and the human sciences (*Geisteswissenschaften*), discussed more in later chapters. While the natural sciences seek explanation by way of cause and effect, in contrast, the human sciences explore "understanding": the relations of parts to the whole and to what might be referred to as the human spirit, understanding as a manifestation of lived experience (*Erlebnis*). The German *verstehen* ("to understand") applied to several areas of inquiry, including historical texts, law, philosophy, and, as we see in chapter 14, through Karl Jaspers, psychiatry. Romantic in orientation and concerned with language as the basis for the construction of all human narratives, including science, this approach acknowledged that historical knowledge was not acquired in the same way as scientific knowledge, and that general "laws" of history are very hard to come by. As the historian Jacob Burckhardt noted, "History is the record of what one age finds worthy of note in another."[5] What emerges as having historical significance in this account is more than tinged with the writer's personal indulgence, for which I can only apologize.

One perspective on history, which is hard to shift, is that its trajectory is like the arrow of Odysseus, traveling with precision in only one direction through 12 narrow axes; time's arrow is always straight and the aim teleological. History in this view is not only a progressive science but also a science of progress. But there are many turning points in history; the question is which turn to take and how to negotiate the chosen path. In science Thomas Kuhn's *The Structure of Scientific Revolutions* (1962) shattered the idea of scientific discovery being one of continuous progress, based on nineteenth-century positivism, implying that paradigm shifts occur at times when anomalies in the data reach such a point that the theories they portend to support become untenable. The intellectual consensus breaks down, new models emerge that are incompatible with the older ones, and so it goes on. A key example that has to be acknowledged in the history of psychiatry is the rapid transition from the psychoanalytic emphasis of the first part of the twentieth century to the biological paradigms of the later decades.

This dramatic shift of the underlying basis for the discipline has a historical backdrop, which has been discussed many times by both the physicians who lived through the changes and the professional historians, often with quite competing visions of the different interests involved and their cultural and social environments. I am not a professional historian but a researcher and practitioner with considerable interest and curiosity to understand from whence my discipline came and how we got to where we are today. This book is not intended to document a linear trajectory, nor one with a teleological directive; indeed, it is not a history in any traditional sense. History belongs for me, as Nietzsche described it, "to him who preserves and reveres—to him who looks back to whence he has come, to where he came into being, with love and loyalty; with this piety he as it were gives thanks for his existence."[6]

There will undoubtedly be those who would criticize this approach. Indeed, this book is not intended to be a historical exegesis as might be written by a professional historian. If there is criticism over my use of terms such as *romanticism* or *Enlightenment* for various historical periods, this is well accepted, since all isms and epochs retrospectively allocated are bound to be controversial. The chapter headings and divisions are interlinked with certain historically bound ideas, and I hope they serve to highlight trends that relate not only to neuroscience but also to the arts and culture of the times. I make no apology for linking neuroscience perspectives to poetry, literature, or developments of musical style. Literature is an essential pathway to feeling the human situation and examining an individual's responses to the environment, with the contingencies of everyday life of those who succeed and those who suffer. If it was Samuel Taylor Coleridge who in England opened up a discussion of the creative brain and introduced us to the word *unconscious* (1818), it was Jane Austen who around the same time explored the inner motives of her characters, and Johann Wolfgang von Goethe who gave us both Werther and Faust.

Many explorers of the brain discussed in this book, such as Franz Joseph Gall, Erasmus Darwin, John Hughlings Jackson, Sigmund Freud, and Sir Charles Sherrington, considered philosophy central to their enterprise. They were interested in minds as well as brains, and they searched for new insights within the confines of the times and the ideas prevalent when they were alive.

There are some who proclaim that there is no conceptual core or stable definition of neuropsychiatry. German Berrios in his historical overview of the subject describes the "elusive nature of the social practice of neuropsychiatry [as] the most important obstacle for writing its history." Indeed, this is a perspective in which psychiatry itself "appears to be just another discipline created by society to perform certain anthropological, managerial and policing duties."[7] But surely there is a history to be explored in one way or another?

There may be difficulties in defining neuropsychiatry and finding the right words to counterbalance such criticisms. As will become clear, neuropsychiatry as explored in this book has veritable origins, but it is not until chapter 10 that the recognizable discipline of today materializes. In the late nineteenth century, psychopathologists were neuropathologists and vice versa, and their interest was with the brain and its abnormalities in those with certain behavior disorders. There was, however, a clear conceptual and practical schism by the late nineteenth century, widening in the first half of the twentieth century, such that a brain-based neurology and a psychological psychiatry that avoided flirting with neuroscience were the eventual outcomes. Over this time there were practitioners, especially in Europe, who were forerunners of today's neuropsychiatrists, but they perhaps were viewed as neither neurologists nor psychiatrists, but akin to a mythological chimera.

The term *neuropsychiatry* entered the English language in 1918, some 70 years after the word *psychiatry* (1847). Shortly thereafter emerged the neuropsychiatrist (1922). The situation has now altered considerably. There has been a growing tradition in the twentieth century for books to be published with "Neuropsychiatry" in the title. Stanley Cobb's *Foundations of Neuropsychiatry* was first published in 1936. There was a series titled *Recent Advances in Neurology and Neuropsychiatry* which began in 1946 and ran for several years under the editorship of Lord Brain. Derek Richter's *Perspectives in Neuropsychiatry* was published in 1950. He became director of the Medical Research Council's Neuropsychiatric Research Unit in Carshalton, England, in 1960.

The modern era of clinical neuropsychiatric texts perhaps began around the 1980s. My own *Neuropsychiatry* was published in 1981, and Jeff Cummings's *Clinical Neuropsychiatry* in 1985.[8] On the bookshelves in front of me are a dozen books with "Neuropsychiatry" emblazoned on the spines. The British Neuropsychiatric Association was established in 1987, the American Neuropsychiatric Association in 1988, and the Japanese Neuropsychiatric Association in 1996. The International Neuropsychiatric Association (INA) was formed in 1998—neuropsychiatry is now a well-recognized discipline in many countries.

There are several suggested definitions. Those who would narrowly limit neuropsychiatry simply with a claim that its flagship is "mental disorders are disorders of the brain" miss the wider implications of the subject and conflate it with an allied but separate enterprise, namely, behavioral neurology.[9] As explored further in chapter 13, behavioral neurology arose from a subspecialty of neurology, the neurology of behavior, which was most interested in correlating focal structural lesions with behavioral changes. Behavioral neurology received its imprimatur with the publication in 1965 of Norman Geschwind's paper "Disconnexion Syndromes in Animals and Man," and the term is generally attributed to him. Behavioral neurology was particularly concerned with aphasia and similar disorders—related, yes, but quite different in approach and practice from neuropsychiatry.

The INA came up with the following definition, admittedly a compromise: "Neuropsychiatry is a field of scientific medicine that concerns itself with the complex relationship between human behaviour and brain function, and endeavours to understand abnormal behaviour and behavioural disorders on the basis of an interaction of neurobiological and psychological–social factors. It is rooted in clinical neuroscience and provides a bridge between the disciplines of Psychiatry, Neurology and Neuropsychology."[10]

In my original book *Neuropsychiatry*, I had ventured the following: neuropsychiatry is a discipline that references certain disorders that, "on account of their presentation and pathogenesis, do not fall neatly into one category, and require multidisciplinary ideas for their full understanding."[11] Obviously, the definition bears heavily but not exclusively on what at the time was conventional neurology and psychiatry, covering a spectrum of disorders from epilepsy to conversion hysteria.

Cummings in 1994 offered the following: "Neuropsychiatry is a clinical discipline devoted to understanding the neurobiological basis, optimal assessment, natural history, and most efficacious treatment of disorders of the nervous system with behavioural manifestations."[12] He emphasized that neuropsychiatry "did not challenge the viability of psychiatry.... Neuropsychiatry, like neurology, is concerned with disorders of brain function and views behavioural abnormalities from a neurobiological perspective ... based on neuroanatomy and neurophysiology and attempts to understand the mechanisms of behaviour, whereas psychiatry integrates information from psychology, sociology and anthropology to grasp the motivation of behaviour."[13]

Indeed, the waters and the waves between psychiatry and neuropsychiatry are considerable; the latter is not simply "the predominant incarnation of psychiatry."[14] It may well be that there are many psychiatrists who would label themselves as neuropsychiatrists, to single themselves out as having a special expertise, yet they might lack not only the fundamental training for such a moniker but also the historical feel for the special nature of the discipline they are claiming. Perhaps this book will fill in some of the gaps. However, without an intimate acquaintance in particular with neuroanatomy and neuropsychology, simply struggling with the names of various neuroreceptors and transmitters and knowing enough psychopharmacology to adequately prescribe psychotropic agents do not add up to being a neuropsychiatrist.

The book opens with a reminder of the transmission by our early ancestors of nature into culture, of raw meat into the cooked, and the importance of storytelling to *Homo sapiens.* Early chapters look at the Greek, Roman, and Arabic legacy but reflect only poverty of intellectual inquiry with regard to the brain. Interest in its ventricles held pride of place, but for some such as Aristotle the heart was considered to be the prime mover. Chapter 2 explores the so-called Dark Ages, which were not at all dark culturally, but in which religious strictures bound any rigorous exploration of the mind. Anatomical dissection was either forbidden or confined to animals: the soul was God's business.

The word *anatomy* entered the English lexicon in the early sixteenth century. Andreas Vesalius performed anatomical explorations of the human brain, Thomas Willis took the brain out of the skull, and René Descartes kept the mind business away from the body, giving scientists a freedom to explore the brain. Although Descartes's work opened a way to examine reflex function, his separation of mind and body has held generations of intellectuals in thrall ever since.

The story moves forward at a more rapid pace through the time periods addressed in chapters 3 and 4, when the foundations of a neuroscience were apparent. Martin Luther shook the shackles of the Catholic Church by asserting the value of individual freedom from authority, the kernel of the romantic revolution that is a central theme of this book. A neurologically based understanding of the individual's expressive self was hardly entertained by the philosopher-scientists of the so-called Enlightenment era. Then there was a shift from a passive view of the brain's engagement with the world, simply as a receiver of sensations, to an active one. The liberation of the individual came with romanticism, but with an anti-Enlightenment Platonic bias.

Alan Richardson noted, "Most work on the Romantic mind continues to be informed by a disembodied version of associationism, by psychoanalysis, or by epistemological issues that link Romantic literary figures to a philosophical tradition running from German idealism to phenomenology and its deconstruction. The Romantic brain, however, has been left almost wholly out of account."[15] This book is an attempt to address this lacuna, and I hope it will be of interest to any neurothinker who might wonder how the prominent split between neurology and psychiatry came about.

Chapters 5 and 6 introduce some neuroscientists and their ideas, which now may seem only shadows extinguished by the light of progress, yet they embraced a broad understanding of human nature and were interested in a romantic vision. Some were poets (such as Erasmus Darwin) or artists (such as Charles Bell), but they emphasized links between life and nature, seeking harmony between parts and a holistic view of how the brain functioned. It was the nature of unity and the unity of nature that were paramount.

The story continues in chapters 7 and 8, with travels around Europe. Paris became the center of activity in the nineteenth century, and then the focus shifted to Germany, with English contributions being represented by Hughlings Jackson, who may be considered the father of modern neuropsychiatry.

It is in chapters 9 and 10 that the major developments splitting a newly developed discipline of neurology from the psychologically based psychiatry are outlined. There were some, however, who refused to give up on what became today's neuropsychiatry, as discussed in chapters 11 and 12. This involved not only the discovery of the power of a seizure to dramatically alter the course of some mental disorders but also the discovery of the electroencephalogram (EEG). Epilepsy became the meeting point of neurology and psychiatry, bolstered by new discoveries in neuroanatomy and neurochemistry (chapter 13).

In many ways the definitions of neuropsychiatry given above do not conjure up the most important aspect of the rebirth of neuropsychiatry, namely, that the discipline arose out of a clinical need. There were so many patients whose problems were at the interface between the developed, encapsulated, segregated fields of neurology and psychiatry who were just badly served by the conceptual and administrative gulf that arose. The enterprise was catalyzed by the growing understanding of neuroanatomy, neuropsychology, and neurophysiology; the use of newer methods of investigation of the brain, especially the EEG and later brain imaging; and a growing awareness that a clinical understanding of the signs and symptoms of central nervous system (CNS) dysfunction could not be embraced by localizationalist neurological theories. Doctrines of psychoanalysis and behaviorist approaches likewise fell on fallow ground. Neuropsychiatry implies that its practitioners— neuropsychiatrists—are very familiar not only with the above-cited clinical skills and methods of investigation but also with the signs and symptoms of a range of CNS disorders and the psychology of human motivation and desire. What is it that makes us tick, and when does the tic become pathological?

Throughout the book, certain terms have become interchangeable. *Mental illness, madness, insanity,* and the like are often used, relating to the words that were current in the times of the quoted authors. However, they are also used quite loosely, likely to irritate anyone who, harboring a potent Aristotelian tradition, believes that committee-produced sets of diagnostic criteria should be adhered to since they explain the essence of clinical syndromes.

Clinicians generally, but certainly those with an interest in the neurosciences, can be crudely divided into the "classical" and the "romantic." The former tend to concern themselves more with parts than wholes, reduction rather than richness, Aristotle rather than Plato. Their preference is for life's experiences to be reduced to mathematical formulas and rating scale instruments, using wire and box diagrams to explain how the brain works. The subtleties of careful clinical explorations of patients' histories, signs, and symptoms come after laboratory testing and brain imaging. Yet any neuroscience understanding of human experiences, either pathological or within the realms of all of us, must take account of those that might be labeled as mystical or transcendental, or even trans-ascendental. Those interested in neuropsychiatry cannot ignore such phenomena.

The Aristotelian perspective is never far behind, and the tensions between the philosophies of Aristotle and Plato reveal discords that themselves are not easily harmonized. To leave Plato out of things does not seem possible.

NOTES

1. Lishman WA, 1992, 984.
2. Donley JE, 1904, 1–11.
3. Herman A, 2013, 327.
4. Steiner G, 2015, 56.
5. Burckhardt J, 1929, quoted in Carr A, 1964, 27.
6. Nietzsche F, 1873, 1876/1997, 72.
7. See Berrios GE, Markova, IS, 2002, 629, 630.
8. Richter D, 1950; Trimble MR, 1981; Cummings J, 1985.
9. Berrios GE, Markova IS, 2002, 629. On behavioral neurology, see Benson F, 1993.

10. Sachdev P, 2005, 191.
11. Trimble MR, 1981, xiv.
12. Cummings J, Hegarty A, 1994, 209.
13. Cummings J, Hegarty A, 211.
14. Berrios GE, Markova, IS, 2002, 636.
15. Richardson A, 2001, 1.

REFERENCES

Benson F. The history of behavioural neurology. *Behavioural Neurology* 1993;11:1–8.

Berrios GE, Markova IS. The concept of neuropsychiatry: A historical overview. *Journal of Psychosomatic Research* 2002;53:629–638.

Carr EH. *What Is History?* Penguin Books, London; 1964.

Cummings J. *Clinical Neuropsychiatry.* Grune and Stratton, New York; 1985.

Cummings J, Hegarty A. Neurology, psychiatry, and neuropsychiatry. *Neurology* 1994;44: 209–213.

Donley JE. On the early history of cerebral localization. *American Journal of the Medical Sciences* October 1904;1–11.

Herman A. *The Cave and the Light: Plato versus Aristotle, and the Struggle for the Soul of Western Civilisation.* Random House, London; 2013.

Kuhn TS. *The Structure of Scientific Revolutions.* University of Chicago Press, Chicago; 1962.

Lishman WA. What is neuropsychiatry? *Journal of Neurology, Neurosurgery and Neuropsychiatry* 1992;55:983–985.

Nietzsche F. On the uses and disadvantages of the history of life. In: Breazeale D, ed. *Untimely Meditations.* Cambridge University Press, Cambridge; 1873, 1876/1997.

Richardson A. *British Romanticism and the Science of the Mind.* Cambridge University Press, Cambridge; 2001.

Richter D. *Perspectives in Neuropsychiatry.* H. K. Lewis, London; 1950.

Sachdev P. International Neuropsychiatric Association. *Neuropsychiatric Disease and Treatment* 2005;1:191–192.

Steiner G. *The Idea of Europe: An Essay.* Overlook Press, London; 2015.

Trimble MR. *Neuropsychiatry.* John Wiley and Sons, Chichester; 1981.

Woolf V. *Mrs Dalloway.* Penguin, London; 1925/1996.

CHAPTER ONE

Origins of the Romance

The cardinal mystery of neurobiology is not self-love or dreams of immortality but intentionality.

Edward Wilson, 1978, 75

Beginnings

Somewhere between 1 million and 200,000 years ago, the first traces of the development of the human brain, which allowed for the emergence of our creative brain, were left scattered and buried for later generations to discover. *Homo sapiens*, the wise, the thinker and actor, and the creator of stories, evolved. For what did our ancestors do, all that while ago, sitting around the campfires? Unknowingly blessed by a visit from Prometheus, the Titan and trickster, who defied Zeus and was punished most dreadfully for giving emergent mankind fire, they were able to cook the proceeds of their hunting and to commune together. Meat when cooked provided many more calories per gram of protein eaten, enhancing the potential for cerebral energy—nature's energy transmitted to the body and the brain. Cooked meat, nature transformed to culture, is perhaps one of the first examples of a metaphorical transformation that became the cornerstone of human creativity. Metaphor ignited kindled abstract thought.[1]

Of course, the building blocks of our brains today were set in place many years before the era of the hunter-gatherers, perhaps the result of nature sculpting for over a billion years, as the CNS of the vertebrate and then the mammalian brain became the blueprint of the later primate and thus our own brains. For we are primates, descended from ancestors that split away from the apes, perhaps some 7 million years ago. The timing is quite unclear, and it is the case—exciting if you like surprises—that with each year of scientific and anthropological exploration, new findings emerge that place our cultural origins farther and farther back in time.

Campfires inspired early Stone Age hunter-gatherers, who had, as far as we can know, no sophisticated language in a form we would recognize today. These ancient people lived with natural environmental hazards which ensured that only swift actions would lead to survival. Communal activities provided some safety in numbers and aided the success of the hunt.

Early emotional expression was an arousal to prepare for action, and, as I discuss in a later chapter, at some point we became what the evolutionary biologist Michael Corballis referred to as a "lopsided ape," unbalanced in brain and later in mind.[2] However, throughout this ancestral development, brain size increased, from a cranial capacity around 450 cubic centimeters in Australopithecines, to around 600–800 cubic centimeters in *Homo habilis*, to 930 cubic centimeters in *Homo erectus*. The mean brain size of current humans is around 1,400 cubic centimeters, a size estimated to have been reached about 200,000 years ago.[3]

Then things really began to happen: *Homo* became *sapiens*.

What kind of communication these past descendants of ours possessed is quite unknown, but we can hazard a guess about three things. They would have had a sophisticated communication system, as do all primates that are alive today. They had developed ritual behaviors, such as disposal of the dead with burial objects, implying a developing symbolic thought and the dawning of religious experiences. Linked with these would have been the daybreak of self-consciousness: they began telling stories.

On the Shores of the Crimean Peninsula, a Long Time Ago

A young shipwrecked sailor was cast ashore on the lands of the Tauri on the south coast of the Crimea, having been pursued from his homeland by vengeful spirits. His behavior was observed by some herdsmen: "He was jerking his head violently up and down. He groaned aloud as his hands shook and he rushed about in a frenzy of madness. And he shouted like a hunter, 'Pylades, do you see this one? Can you not see how that one, a she-dragon from Hell, wants to kill me and turns her weapons, the fearful vipers of her hair against me? . . . Where can I go to take refuge?'" The observers noticed how the youth misinterpreted the lowing of their cattle and the barking of dogs as similar to the imagined sounds of the furies. He drew his sword and rushed at the heifers "like a lion," and wounded them "until the sea bloomed red with blood." Naturally alarmed and concerned for the safety of their cattle, the herdsmen approached the raving youth, but when they got to him, "the stranger's pulse was stilled and he fell down, his chin dripping with foam . . . the stranger then regained consciousness."[4]

Orestes, brother of Iphigenia, killer of his mother and her lover, was the first recorded case of epileptic psychosis in the written literature. The play *Iphigenia among the Taurians* was one of the tragedies of Euripides (480– 406 BC), performed sometime between 415 and 412 BC. This was the century in which Western culture as we know it today truly began. The ancient art of tragedy evolved into a dramatic spectacle, uniting myth, religion, and music, probing the relationship between gods and men, and establishing a form of art which evolved into medieval church music, secular dances and concerts, opera, and the theater and cinema of today.

Ancient Concepts of the Body and Mind

There were earlier stirrings and developed civilizations, some of which lasted for longer than the time between the age of classical Greece and the present, but we have relatively little information about many of their cultural activities. Further, even though religious artifacts and buildings have been discovered and preserved, from the pyramids of Egypt to the Hammurabi law code of the Babylonians, we know virtually nothing about their medicine and even less about their thoughts on the brain and its diseases.

The Babylonians provided accounts of conditions we now recognize as epilepsy, psychoses, obsessive-compulsive disorder, anxiety, and depression, and, according to Edward Reynolds and James Kinnier Wilson, they were the first to provide us with the clinical foundations of Western neurology and psychiatry.[5] The descriptions of seizures include gelastic attacks (laughing) and interictal emotional disorders and psychoses. The text also includes a note on a patient with no desire for females, paranoid ideation, and anger against the gods and who had developed his own religion. This has echoes, which we will later come to, of personality changes in people with epilepsy. The disorders were attributed largely to supernatural causes, and any indication of the involvement of the brain, or psychological insight, was missing.

There are other lingering ancient associations. The ancient Chinese words for epilepsy, *dian xian*, linked to madness, with *dian* still being used to denote a crazy person. This etymological link traveled to Korea and Japan, as in *gan* and *tenkan*, respectively. These associations with madness also remain today in Malayan, Burmese, Thai, and some other Asian languages.[6] However, in Oriental contemplations about consciousness, the brain seems to have no place, and neither was there a search for the seat of the soul.

Early Chinese beliefs that mankind is made up of the same things as the universe, a microcosm of the macrocosm, continue to hold sway in one form or another in various cultures. Tao, the Way, divided Chaos into opposing forces of Yin and Yang. Five elements came forth: water, fire, wood, metal, and earth. Balance between them was preserved by Toa, which looked after the macrocosm, while that between Yin and Yang determined human fate. Yin and Yang are conveyers of psychic balance which are distributed throughout the body, and as such, separation of body and mind was not conceivable and disharmony between them was associated with disease. The Chinese were aware of the heart's response to stress, an organ that became a "guardian" for the mind's activities, but the culture was one of ancestor worship and veneration, so human dissection was forbidden.

Ancient Indian medicine grounded in Hindu and Buddhist beliefs acknowledged the death of the body, but not that of the soul, transmigration of which, as a part of the Universal Intellect or World Soul, means that it is restored and rejoined to a bodily frame. In some variants, individuals must work out their own way to salvation, *nirvana*, at the end of successive existences having no relation to time or space. Even though ancient Indian texts describe disorders such as epilepsy, depression, and states of possession, reference to human anatomy or physiology is absent, although again the body is somehow viewed as a microscopic image of the universe.[7]

Hippocrates

Iphigenia among the Taurians was not the only tragic play of Euripides to portray some form of madness. His *Heracles*, written around the same time, also gives Heracles the signs and symptoms of a postepileptic psychosis, although others might interpret these events as descriptions of mania. This confusion over the causes of and links between neurological and psychiatric disorders is obviously a problem of retrospective speculation, which dogs any historical exercise, but it heralds long-standing debates that percolate through this book.

Euripides was credited as changing the style of Greek drama, bridging the world of the gods and humans, as in *Heracles*. He portrayed his characters as speaking an everyday language, with a philosophical and psychological introspection, and he was interested in madness.[8] He was a realist and an observer of things around him, and epilepsy then as now was a common disorder. In Greece at that time, it was called the "Sacred Disease."

Hippocrates (460–370 BC) was born in Kos but settled in Athens. He was perhaps the most renowned physician of the age, whose name still attends us through the Hippocratic oath, by which the art of medicine and its ethical values are laid down. His most famous statement about the brain and mental illness is as follows: "and men ought to know that from nothing else but thence [from the brain] comes joys, delights, laughter and sports, and sorrows, griefs, despondency and lamentations. . . . And by the same organ we become mad and delirious, and fears and terrors assail us."[9]

For Hippocrates epilepsy was not sacred and was somehow related to the brain. For Euripides the madness of epilepsy provided a part of the human tragedy for the dramatic tragedy, the meaning behind the play being a mirror to human life. Madness was close to the surface of existence, even in a society bent on proclaiming the importance of philosophical debate and logic. As George Steiner put it, "Tragic drama tells us that the spheres of reason, order and justice are terribly limited."[10]

Euripides was part of the contemporary intellectual ferment, embodying science, philosophy, and medicine, and he must have been aware of the developing Hippocratic approach to diseases and their causation. In his plays and in the Hippocratic writings, the gods are set in the background: diseases were associated with imbalance of the bodily fluids, heredity (not in a way envisaged by today's genetics), the physical environment, and diet.[11]

Hippocrates emphasized that health was related to the correct mixture of bodily humors—black and yellow bile, phlegm, and blood. Such a scheme and its resonances are shown in table 1.1. As we shall see, such humoral constructs echo throughout Western medicine, and ideas of underlying treatments based on balance and harmony keep recurring. Hippocrates introduced the idea of disease having a longitudinal course; static, distinct entities were not his view, and he put emphasis on clinical examination. He inaugurated a tradition so important to later medical practice, namely, writing down his observations and findings, in contrast to the oral tradition used at that time. He summed up his philosophy in one of his famous aphorisms: "Life is short, the Art long, opportunity fleeting, experience false, judgement difficult."[12] His understanding of medicine for that era must have been profound; some say he actually lived to be over 100 years old.

The Sacred Disease not only was the first monograph on epilepsy but also gave a brain-based orientation to the disorder. Too much phlegm caused the

TABLE 1.1
The Bodily Humors and Their Associations

Humor	Organ	Season	Quality
Blood	Heart	Spring	Warm/moist
Yellow bile	Liver	Summer	Warm/dry
Black bile	Spleen	Autumn	Cold/dry
Phlegm	Brain	Winter	Cold/moist

seizure, triggered by environmental changes, and the condition was cured not by magic or incantations, but by diet and medications. Other observations of interest for the time were that a head injury on one side could lead to seizures on the opposite side of the body, a forerunner of later anatomical discoveries.

Another often-quoted epithet from the Hippocratic corpus is, "Most melancholics usually also become epileptics, and epileptics melancholics. One or the other [condition] prevails according to where the disease leans: if towards the body, they become epileptics; if towards reason, melancholics."[13] This fascinating observation is a forerunner to an important biological link between seizures and mood, hardly discussed again until the twentieth century.

The attacks of epilepsy could be brought on by fright, anger, or other psychic precipitants. This intertwined with another very important disorder, the name of which has been handed down to us from the Greeks—hysteria.

Hysteria

Hysteria was observed in the Egyptian medical texts, but the name derives from the Greek *hystera*, meaning uterus. Originally this referred to a concept of the origin of symptoms, found exclusively in women, and caused by a wandering womb. The idea was that the womb, being frustrated by lack of proper use, leaves its anatomical position and travels around the body, causing pressure in anomalous places and hence symptoms. On reflection, such ideas must have related to empirical observations, namely, that of all the bodily organs, the uterus can be seen to move, when prolapsed.

There has been an academic debate about what the Egyptians and subsequently the Greeks actually were referring to when they discussed the wandering womb, and the term *hysteria* was not used in the Hippocratic writings. But the early history of this malady reflects on three important points: the same symptoms that are seen today were described over 2,000 years ago, across at least two different cultures, and the postulated mechanism was gender related. Examples of the reported symptoms included the classical globus hystericus—caused by pressure from the wandering uterus on the throat—paralyses, blindness, and convulsions, the last being central to the evolving history of the subject.[14]

Philosophical Bridges

The theories of the Greek physicians were closely allied to those of philosophers, beginning with the pre-Socratics. They began an intellectual exploration of the existence and nature of things, and they tried to explain the origins of the earth, what things were made of, and what the essence of mankind was. For Thales everything came from water, whereas Heraclitus speculated on the unity of opposites and the idea that everything is in flux. There is a becoming, and a change in all things, but a unity in duality; an example he gave was that the road up and the road down are the same thing. But underlying everything was fire, for when fire left the body, earth and water were useless. For Pythagoras, who believed in the migration of souls, numbers and mathematics were the key.[15]

Heraclitus and Pythagoras had important influences on Plato and hence Aristotle, the latter two philosophers playing such an important role in the intellectual patterns of Western thought, religion, and medicine, as will become apparent. The philosophical traditions of idealism and dualism were founded in this era. Idealism held that deduction was the way to understand the world, that knowledge is available only to the rational mind, and that the senses can only deceive. It was based on Plato's ideas asserting the existence of forms. The forms of the first archetypal level belonged to the timeless, unchanging world of perfection and were accessible only by those who devoted themselves to the practice of philosophy. These forms were mind independent, eternally existent, and nonvisible. The second level consisted of the mutable, imperfect forms of the empirical everyday world. The forms of the third level were represented in mimetic art. Dualism, the idea that body and soul were of differing substances, was one consequence of such theories.[16] The tensions between the first and the third levels reveal the "schism of consciousness" which comes down to the neurosciences from Plato to today. The body became a burden to the soul, in some philosophical schools meaning a war between the two, if the soul were ever to be redeemed. Along with idealism, dualism has been a main argument that has absorbed philosophers, scientists, and laypeople since that time. Dualism in particular lies at the heart of the mind–brain problem.[17]

Despite an interest in what we now call the mind and the contributions of Hippocrates, there is a lack of interest by the Greek physicians in the brain itself, and anatomical descriptions of the brain in health or disease seem entirely lacking. In Homer's *Iliad* battles raged and there were many broken skulls, and the Hippocratic writings refer to trephination in patients with closed head injuries. Yet the brain as an object of interest was hardly discussed. Aristotle did dissect animals, but he opined that the heart, not the brain, was the center of the nervous system.[18] Nevertheless, he was much more earthbound than Plato. He was concerned with things that existed, their properties, their essence, and the relations between them. The formal discipline of logic began with Aristotle, and he revelled in classification.[19]

To Rome

The Greeks colonized far and wide around the Mediterranean and Black Seas, but it was to Italy, notably in Rome, that the power base shifted, leading to an empire that lasted some 450 years. The Romans absorbed much of Greek culture, their empire embracing former Greek colonies, encircling the Mediterranean and beyond, and establishing laws and a social structure that was enduring. But much of their science and culture was derivative, echoing Greek origins, and their best physician who took an interest in the brain was Galen (AD 129–200). Anatomy was his subject. He had studied medicine in Alexandria, an important stopping point on the way from Greek to Roman medicine, since there human dissection was permitted and one Erasistratus (310–250 BC) had described the ventricles. Experienced at treating gladiators, Galen knew about head injuries, and he described in epilepsy what we would now refer to as an aura and the motor march of the later-denominated Jacksonian seizure.

Galen adopted an Aristotelian approach in his work, relying on the senses and observation for knowledge, but unfortunately confined his dissections to animals, including primates, human dissection being forbidden by Roman law. He described the cranial nerves, the elements of what became the sympathetic nervous system, and the ventricles. However, although empirical at heart, his mind was inclined to follow much from the philosophies of Plato and Aristotle. He disagreed with the latter on the centrality of the heart, placing the brain as central to the mind. He adhered to a system of spirits, which influenced bodily function, and while natural spirits reigned over basic bodily needs, vital spirits were generated in the heart, which in the brain become animal spirits, which could be made and stored in the ventricles.

But there was a fundamental error of his anatomy, which remained neurological dogma for centuries.[20] Galen observed a web of vessels surrounding the pineal gland, which was referred to as the *rete mirabile*—a feature not found in the human brain, but which he opined was the site of the conversion of the vital into animal spirits.

The meaning and function of these various spirits are unclear to us, but they conjure up concepts of a life force that—via the nerves, which were regarded as hollow—initiated movement and sensation. The brain was the seat of voluntary power and sensation, and Galen identified the frontal areas as being the seat of the soul. Yet he was curiously agnostic on such questions as the immortality of the soul and how and by whom the universe was created, as these issues were of no use to medicine. Sadly, Galen believed that the cerebral convolutions had no relevance, relegating them out of neurological history until prominence was given to them by phrenology at the beginning of the nineteenth century.[21]

Aretaeus of Cappadocia (AD 50–130), about whom so little is known, separated medical disorders into two broad classes, acute and chronic, and although he briefly touched on hysteria in men, he, along with other Roman physicians, held on to the uterine origins and elaborated on the metaphor, first imputed to Plato, of the wandering womb as an animal. With his insistence on longitudinal study of illness, he importantly defined a form of mental disorder in which manic and depressive phases occurred in the same person. How well the delineation of this disorder has stood the test of time.

Caesars were not immune to illness. The epilepsy of Julius Caesar is discussed by several authors, from Suetonius to Shakespeare, and according to Suetonius, Caligula suffered from the falling sickness. Claudius seems to have had a tic disorder. "[Claudius] stumbled as he walked along, owing to the weakness of his knees, and also because if excited by either play or business, he had several disagreeable traits. These included an uncontrolled laugh, a horrible habit under stress or anger of slobbering at the mouth, and running at the nose, a stammer, and a persistent nervous tic—which grew so bad under emotional stress that his head would toss from side to side." Of particular interest in this context are Suetonius's other comments on Claudius's inept remarks:

> He often showed such heedlessness in word and act that one would suppose that he did not know or care to whom, with whom, when, or where he was speaking. When a debate was going on about the butchers and vintners, he

cried out in the house: "Now, pray, who can live without a snack," and then went on to describe the abundance of the old taverns to whom he himself used to go for wine in earlier days. . . . Every day, and almost every hour and minute, he would make such remarks as these: "What! do you take me for a Telegenius?" "Scold me, but hands off!" and many others of the same kind which would be unbecoming even in private citizens.

Taken together, Suetonius's descriptions of Claudius's tics and repetitive behaviors could represent the first recorded documentation in history of what we now refer to as Gilles de la Tourette syndrome.[22]

Legacies

The ancients should be given more due than is customary. It is not unusual to hear quite derogatory remarks made about their ancient ideas of anatomy or treatment, yet these are made with such an arrogant hindsight and ignorance as to how knowledge about the world is discovered, or indeed what "knowledge" actually may be. Science as we know it was not part of the Greek or Roman intellectual tradition; metaphorical interpretation of illness was common. The Egyptians had used Egypt itself as a metaphor for the human body, giving prominence to the Nile and the necessity of moisture, and the Greeks used metaphors, such as fire or the wandering womb, but the concept of balance between humors and elements dominated. The philosophers placed much emphasis on *logos*, asserting that there are axiomatic truths, which can be deduced by strict logic, giving absolute knowledge.[23] There was a distrust of the world as given through the senses. Proportion and numbers were divine and essential to interpreting the world, including illness and health. Classical Greek beauty, of perfect proportions as represented in sculpture and architecture, was associated with virtue and goodness; physical and psychological health corresponded. Although the philosophy of Aristotle paved the way for the later development of the scientific method, the nagging history of idealism and the philosophical legacies of especially Plato will be with us up to the final chapters of this book.

NOTES

1. The origins of cooking by fire are unclear; charred wood or charcoal does not survive as long as bones. However, fire was used by *Homo erectus* perhaps 2 million years ago. The earliest human hearths date to about 250,000 BC, but the recent discovery of charred bone in

South Africa which may date to 1 million years ago suggests a much earlier use. The arguments for the benefits of fire for human evolution are discussed in Wrangham R, 2009. He argues that in addition to increasing the calorie release from raw nutrients, fire allowed early hominids to keep warm and increased social cohesion around what later becomes the hearth and the bond-fire. The last point has considerable symbolic and metaphorical value in our history. Simply cooking starchy foods increases the net energy gain by 30%.

2. Corballis MC, 1991.

3. These can only be estimates based on size of cranial capacity from skull remains and attempts to reconstruct what the underlying brain may have looked like. *Homo habilis*, endowed with smaller teeth and a larger brain than the earlier living Australopithecines, a handy man with a bipedal gait, used tools to hunt perhaps 2 million years ago. Acheulean stone tools, dating from about 1.7 million years ago, have been linked to our pre-ancestors *Homo ergaster* and *Homo erectus*. They were more sophisticated, leading to the even more intricate creations of the middle and later Stone Age periods. The dating varies, and various terms are used by different authors. According to Barnard A, 2012, the early Stone Age is represented by the Oldowan and Acheulian eras (2.6 million to 300,000 years ago), the middle Stone Age was from around 300,000 to 50,000 years ago, and the late Stone Age is from 50,000 years ago to recent times.

4. *Iphigenia among the Taurians*, in Euripides, 2000, 8–9.

5. Reynolds EH, Kinnier Wilson JV, 2008. Together these authors have been translating the cuneiform script of the Babylonian medical tablets in the British Museum and uncovering new insights into neuropsychiatric disorders in Babylonian times, including descriptions of illnesses that resonate with those seen today.

6. For a detailed analysis of these etymological associations, see Lim KS, et al., 2012. The authors note how these linguistic associations have contributed to and continue to exacerbate stigma against people with epilepsy in these countries.

7. See, e.g., Bhugra D, 1992; Veith I, 1973.

8. The interested reader can also try *The Bacchae*, *Orestes*, and *Hippolytus*. It is said that Euripides was prosecuted for portraying a demigod, Heracles, as mad. In the play, madness is personified as Madness, the Gorgon daughter of the night. Heracles kills the tyrant Lycus, before killing his own children. He hallucinated that he had a chariot that he mounted, claiming he had reached a distant city. He stripped off his clothes and wrestled with a nonexistent opponent, proclaiming himself the winner of the fight. Heracles is seen tossing his head, rolling his eyes, panting wildly, and foaming at the mouth. His laughter is manic. After these excesses, he sleeps, and when he awakes, he is confused and amnesic. See *Heracles* in Euripides, 2003. Descriptions of the madness of the title character in *Orestes* are akin to those described in Euripides's *Iphigenia among the Taurians*.

9. Adams F, 1939, 366. As with Homer, the exact identity of Hippocrates is unclear. There is no doubt that he was known in Athens, and Plato refers to him as a famous physician who came from Kos. Aristotle referred to him as the wise physician. There were many medical texts attributed to Hippocrates, but who wrote what is quite unclear (Tsiompanou E, Marketos SG, 2013, 288).

10. Steiner G, 2008, 8.

11. There is a chronological overlap with at least some of the Hippocratic writings (which are not all attributed to one man in any case) and the time of Euripides, although not so with Aeschylus, who died around the time Hippocrates was born. See Jouanna J, 2012.

12. *Aphorisms*, sec. 1, in Adams F, 1939, 299.

13. Quoted by Temkin O, 1971, 55.

14. See Veith I, 1965; Merskey H, 1995; Micale M, 1995, 18–29. There are arguments especially over the role that the uterus played in Egyptian concepts of "hysteria."

15. Heraclitus (535–475 BC) was famous for his view that "everything is in flux"—it is not possible for a man to enter the same stream twice; Pythagoras (ca. 570–495 BC) was a scientist and a mystic, whose name became the foundation of a quasi-religious following.

16. Plato argues that the soul shares the properties of forms, it is eternal and incorruptible, it exists before birth and continues after death, and its only activity is pure reasoning (*Phaedo*). In other texts Plato refers to three souls: the appetitive, the passional, and the rational (*Republic*). Plato did not want relativism, or the philosophies of Heraclitus, and he was scornful of the arts on account of their potential to arouse feelings that could interfere with *logos*.

17. Steiner G, 2011, 57.

18. These ideas echoed down to Shakespeare's *Merchant of Venice*: "Tell me where is fancie bred / or in the heart, or in the head" (3.2.64). However, even before the Greeks, the Egyptians espoused the heart as the seat of the soul. At the time of death, the heart was weighed against a feather, and if it was heavier, because of guilt, the spirit would not go to heaven.

19. Raphael's famous painting *The School of Athens* (ca. 1510/1511) shows images of several philosophers, but central are Plato, with his finger pointed upward, and Aristotle, pointing to the ground.

20. The historian Stanley Finger noted, "Yet the duration of his grip on anatomy, physiology, and medicine persisted well after his death. Remarkably, Galen's teaching served as the guiding force of science and medicine for over thirteen hundred years" (2000, 51). Likewise, his descriptions of the ventricles remained unchanged for over 1,000 years (Meyer A, 1971, 9). Galen delineated various anatomical areas familiar to us today, such as the corpus callosum, the fornix, and the pineal and pituitary glands.

21. Clarke E, Dewhurst K, 1972, 8.

22. For Julius Caesar's seizures, see Shakespeare, *Julius Caesar*, 1.2. For Claudius, see Suetonius, 2007, v. 30, v. 40. Telegenius is a mythical or historical figure from Roman times who was famous for his stupidity.

23. *Logos* has had several meanings and does not equate with logic. It denotes reason and has acquired meanings such as "discourse" or has been related to mathematical "ratio." In philosophy "it refers to a cosmic reason which gives order and intelligibility to the world." Runes DD, 1965, 183–184.

REFERENCES

Adams F. *The Genuine Works of Hippocrates*. Williams and Wilkins, Baltimore; 1939.

Barnard A. *Genesis of Symbolic Thought*. Cambridge University Press, Cambridge; 2012.

Bhugra D. Psychiatry in ancient Indian texts: A review. *History of Psychiatry* 1992;3:167–186.

Clarke E, Dewhurst K. *An Illustrated History of Brain Function*. Sandford, Oxford; 1972.

Corballis MC. *The Lopsided Ape: Evolution of the Generative Mind*. Oxford University Press, Oxford; 1991.

Euripides. *Bacchae and Other Plays*. Trans. Morwood J. Oxford World Classics, Oxford University Press, Oxford; 2000.

Euripides. *Heracles and Other Plays*. Trans. Waterfield R. Oxford World Classics, Oxford University Press, Oxford; 2003.

Finger S. *Minds behind the Brain*. Oxford University Press, Oxford; 2000.

Jouanna J. *Greek Medicine from Hippocrates to Galen*. Brill, Leiden; 2012.

Lim KS, Li SC, Casanova-Gutierrez J, Tan CT. Name of epilepsy, does it matter? *Neurology Asia* 2012;17:87–91.

Merskey H. *The Analysis of Hysteria*. 2nd ed. Royal College of Psychiatrists, London; 1995.

Meyer A. *Historical Aspects of Cerebral Anatomy*. Oxford University Press, Oxford; 1971.

Micale M. *Approaching Hysteria: Disease and Its Interpretations*. Princeton University Press, Princeton, NJ; 1995.

Reynolds EH, Kinnier Wilson JV. Psychoses of epilepsy in Babylon: The oldest account of the disorder. *Epilepsia* 2008;49:1488–1490.

Runes DD, ed. *Dictionary of Philosophy*. Littlefield Adams, Totowa, NJ; 1965.

Steiner G. Tragedy re-considered. In: Felski R, ed. *Rethinking Tragedy*. Johns Hopkins University Press, Baltimore; 2008.

Steiner G. *The Poetry of Thought: From Hellenism to Celan*. New Direction Books, New York; 2011.

Suetonius. *The Twelve Caesars*. Penguin Classics, London; 2007.

Temkin O. *The Falling Sickness*. 2nd ed. Johns Hopkins University Press, Baltimore; 1971.

Tsiompanou E, Marketos SG. Hippocrates: Timeless still. *Journal of the Royal Society of Medicine* 2013;106:288–292.

Veith I. *Hysteria: The History of a Disease*. University of Chicago Press, Chicago; 1965.

Veith I. Non-Western concepts of psychic function. In: *The History and Philosophy of Knowledge of the Brain and Its Functions*. B. M. Israel, Amsterdam; 1973.

Wilson EO. *On Human Nature*. Harvard University Press, Cambridge, MA; 1978.

Wrangham R. *Catching Fire: How Cooking Made Us Human*. Basic Books, London; 2009.

CHAPTER TWO

The Middle Ages

Darkness and Light

Darkness and light, the founding metaphor of Western philosophy.

Jacques Derrida, 2001, 31

On April 26, 1336, the poet Petrarch (1304–1374) walked up Mont Ventoux in the Province region of France, and on reaching the summit, after admiring the splendid view of all that he could see around him, he read a text from Saint Augustine's Confessions. *It urged him to consider not only the beauty of the outer world of nature but also his inner world. From that moment on he avowed to turn his eye inward, to contemplation.*[1]

Historical ages are always arbitrary, and what is considered the middle depends on the bookends. If the time gap between AD 500 and 1500 seems about right, it covers what Petrarch referred to as the "Dark Ages"—dark not only because of economic decline but also in reference to his view of the decline of intellectual activity which followed the achievements of the enlightened Greeks.

Yet in Western Europe during these times, the foundations of modern science were laid down. Paper and printing presses, spectacles, and the mechanical clock spread widely throughout Europe, aiding dissemination of data and measurement. There was the rise of universities—Bologna in 1158, Paris in 1200, Padua in 1222, and Oxford in the early thirteenth century. This period marked the building of some of the greatest artistic achievements of Western art, such as the Gothic cathedrals beginning in France in Orleans, Rheims, Rouen, Notre Dame, and Saint-Denis. These elegant structures, built for God and to last forever, portrayed His glory.

Philosophical ideas such as the famous razor of Ockham, stating that multiple entities should never be invoked unnecessarily, promoted clarity in thinking and pointed to an empirical approach for investigating nature, but, as we shall see, it was Aristotle who achieved most attention. Before considering these historical developments, another cultural invasion from the east needs to be considered—the rise of Islam.

Arabic Influences

The influence of Islam, founded after Mohammad (571–632), spread from Mecca through Persia and overtook the Byzantine culture from the seventh century, finally putting paid to what was left of the Roman Empire. The Romans left a lack of knowledge of the Greek heritage and hence a loss of so much early science and philosophy. However, via the Byzantine and then Arabic cultures, much was effectively saved and reconstituted.

Although there are well-known Arabic physicians, their original contributions to medicine and neuroscience in particular are very limited. Strict adherence to the teachings of the Koran forbade human dissection, examination of the naked body of a woman, and the artistic representation of living beings, essentially placing a straightjacket on intellectual freedom.

Two of the most famous physicians of the times were Rhazes (aka Abū Bakr Muhammad ibn Zakariyya al-Rāzi, 854–925), a musician and philosopher who was Hippocratic by inclination, and Avicenna (aka Abū Ali al-Hussein ibn Abdullah ibn Sīna, 980–1037), whose book *The Canon* proved a blend of Hippocrates, Galen, and Aristotle. Rhazes's *al-Kitab al-Hāwi* comprised 23 volumes (in modern script), and although controversially critical of several of Galen's ideas, notably the theory of humors, it remained, along with *The Canon*, in use in Europe for several centuries.[2]

If anything, the texts reveal more about an approach to psychiatry than neurology. There is an episode of dream interpretation in the Koran,[3] and the works of both Rhazes and Avicenna reveal psychological insights, even adopting treatment techniques we would now refer to as psychotherapeutic. There are interesting case histories. For example, Rhazes was called to see a caliph suffering allegedly from bad arthritis. He advised the man to take a hot bath. But when the caliph was in the bath, he threatened him with a knife, at which point the frightened man got out of the bath and fled. Avicenna wrote of the case of a woman who was unable to use her arms. A doctor went to examine her, stripped off her veil, and went to lift up her skirt, which led her to resist his actions with her arms. Both of these episodes were given physiological explanations by the respective physicians, but perhaps they were early recorded cases of "hysteria" cured by exposure.

Avicenna noted changes in pulse rate with emotion, and he described a disorder equivalent to depression. Ishac ibn Imran wrote *The Disease of Melancholy* in the ninth century, attributing the condition to black bile.[4] Psychotic disorders were recognized, such as the patient of Avicenna who had the delusion that he was a cow. Avicenna told him that he was coming to him as a butcher but declared him to be too lean to be slaughtered. The man began to eat, gained weight, and lost his delusion.[5]

Epilepsy certainly featured in early Arabic medicine. Rhazes's compendium of medicine contained a chapter on brain diseases, including epilepsy, and Avicenna referred to the equivalent word as meaning a person being possessed by an outside force. Epilepsy was also called the diviner's disease, because of an association between the falling sickness and religious visions or prophesies.

But at the end of it all, humoral theories still dominated: earth, air, fire, and water, with counterparts dry, cold, heat, and moist. Blood, phlegm, and yellow and black bile were the cornerstones of the prevailing physiology. The humors and outside forces seem to have somehow coincided and interacted, in an unsteady conflagration of scientific possibility and fantasy. While some writers, such as Maimonides (aka Mūsa ibn Maymūn, 1135–1204) in Córdoba and Averroës (aka Abū al-Walīd Muhammad ibn Ahmed ibn Rushd, 1126–1198), tried to tread the difficult line between scientific inquiry and religious dogma, supporting the cause of reason, this gained no advantage. The anatomy and any understanding of the brain could not advance beyond Galen.[6]

Early Christianity and the Opening Up of the Mind–Brain Split

Constantine (272–337) adopted Christianity in AD 313 and moved the capital of the Roman Empire to Byzantium (Constantinople), and over time Orthodox Christianity became the firmly established religion in Europe. The ideas of Plato vied with those of Aristotle, and mind and body became dissociated as never before. Bodies were debased, souls needed to be saved, and as a consequence, the church became the arbiter of knowledge. Illness became viewed as a punishment for sin; insanity was the worst of all, because of the loss of reason in the afflicted. New disorders were posited, such as lycanthropy,[7] and ideas of the powers of possession to cause madness grew stronger.

The full Aristotelian corpus became available in Latin in the thirteenth century, and earlier prohibitions against promoting his works eased. The

difficulty was amalgamating Christian theology with non-Christian philosophy, proving the truths of revelation in the light of reason. Thomas Aquinas (1225–1274), a Catholic priest and philosopher, attempted to bring together
the ideas of Aristotle and his theology. His writings on the soul included its
separation from the body, yet it was the principle of life. Indeed, he considered souls as present in all living things, but only the human soul was created by God. As with Aristotle, it was the intellect, a part of the soul, that
characterized and separated man from the rest of nature, but for Aquinas,
the intellect was both passive, a blank slate (*tabula rasa*), and active. He uses the
term *potentia*, sometimes translated as "power," reflecting that entities either
could be affected by others or not, and have the potential to act on things or
not. In his philosophy, the sense organs were passive receptors of sensations,
related to the brain and altered by objects, while "agent" intellect was transformative. The active intellect is the force triggering intellectual activity
in the human mind and causing thoughts to pass from the potential to the
actual.

Following Aristotle, Aquinas designated the heart as the source of
common sense and as the first principle of movement, the organ respon
sible for distributing sensory power to the brain, from where it is distributed to the senses of sight, hearing, and smell. He simply was unable to
believe that the brain itself could be involved with the intellect and
will. He attempted to explain the unity of the soul and the body, but
for him the soul was an unmoved mover analogous to the "non-bodily"
nature of heat, essentially incorporeal. The mind–body split was a feature of his influential philosophy, even if by some accounts imperfectly
argued.[8]

Spaces in the Brain

Galen had described the cerebral ventricles and noted the effects of damaging them in animals. Injury to the fourth ventricle (cerebellar) produced
stupor, while the effects of compression of the first, anterior ventricles were
only slight. The "cell doctrine," as it was called, was proffered in the fourth
and fifth centuries by early church fathers, including Saint Augustine, who
attributed various functions to the ventricular spaces, or cells. These all
related to aspects of the mind, but the first "cell" was often linked to sense
reception or "common sense," with imagination and image formation taking
place more posteriorly. Judgment, thought, and reason were allocated to the
second cell, and memory to the third.[9]

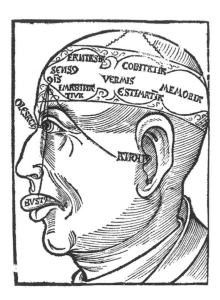

Figure 2.1. Diagram of the ventricular system from 1525. Various psychological attributes for the ventricles are given, as are links to the anterior one between the senses, such as sight, taste, and hearing. Most writings about the brain concentrated on the ventricles from the Greeks through to the sixteenth century. From Clarke E, Dewhurst K, 1972, 36

The first manuscript diagram of the brain dates from AD 1100, with the brain being labeled cold and moist, showing the continuing influence of Aristotle. There were several sketches produced in the Middle Ages, essentially drawings of heads with writing on them noting the ventricles and their supposed functions (see fig. 2.1). Often they were simply shapes and circles, but some had links drawn on them between the eye or heart and the ventricles. This marked the beginning of ideas that would come to dominate neurology, namely, functional localization within the brain. But any real developments in neuroanatomy will have to wait until the next chapter.

Of Devils, Dancing, Witches, and Daemons

Epilepsy has attracted many epithets, and the idiom of falling had more than an attachment to gravity. The "falling evil" carried connotations of demons and *lunaticus*, and in the early Christian era, of ecstasies, trances, and the idea of possession. Visitations from the supernatural, God or ghost, led to convulsions and other strange manifestations, facts not doubted in the Middle Ages. Dante placed the person with epilepsy down in the Inferno:

Like someone who falls down, not knowing why—
Whether a demon has him in a seizure,
Or other blockage binds him physically—
Then stands and looks around as he comes round,
Bewildered utterly by all the anguish
He's undergone, and sighs as he looks round—
That's how this sinner was when he arose.
And oh, the power of God, how stern it is,
Dealing in vengeance such almighty blows![10]

Pagan associations with Selene, the goddess of the moon, linked epilepsy to astrological attachments, with various other periodic disorders, such as "lunacy," being associated with planets such as Mars, Mercury, and Saturn. The influence of the moon altered the moisture within the ventricles of the brain, leading to attacks. Epilepsy also became attached to various saints, especially Saint John and Saint Valentine, and amulets or talismans with the names of the saints were thought to have healing powers, as did uttering the names of the saints in the ear of the sufferer.

There was another influence that held sway for many years in Western medical history, entwined with religion and politics, namely, witchcraft. Pope Gregory in 1200 authorized the killing of witches, and the hunt was on as to how to identify them.[11] *Malleus Maleficarum* (referred to as "The Witch Hammer") was published around 1486 and went through many translations and editions. The authors were two monks, Heinrich Kramer and James Sprenger, and this became a textbook for witch hunting, an influence lasting for the next three centuries. The two monks were dispatched to travel around Europe and seek out possible offenders on behalf of the church. Oddly, it seems that many accused actually confessed their alleged carnal activities with the devil, either under duress or out of some wish to be famous and have local notoriety.

Hysteria and epilepsy both become involved with this astonishing story. Hysteria ceased to be something for physicians to treat and became the province of the church, as one manifestation of heresy. Women and sex were central to these theories. Sexual aberrations, which included too much pleasure from sex, especially in women, and the involvement of men with incubi and succubae were surely the devil's work.[12]

Most of the text describes how the devil did what he was considered to do, as well as how to identify witches, which was largely by interrogation, although torture was encouraged. It is of considerable interest that certain stigmata became associated with identifying a witch, including abnormal sensory areas, referred to as "witches' patches" or marks of the devil's making. Examiners were instructed to prick the skin, often with long needles, to find these patches, and to see whether they bled after the pin was removed. Convulsions were diagnostically problematic, since they seemed to be prominent manifestations of epilepsy, hysteria, and possession. We shall see how this story evolved in later chapters, but it was inevitable that epilepsy came to be seen as yet another disorder induced by the devil. To be fair, the authors of *Malleus Maleficarum* recognized and frequently mentioned that epilepsy also arose from physical predispositions and defects. But they note, for example, that "we have often found that certain people have been visited with epilepsy or the falling sickness by means of eggs which have been buried with dead bodies, especially dead bodies of witches."[13]

Another link in this story involves some early descriptions of movement disorders. *Chorea lascivia*, so-called dancing mania, also linked to Saint Vitus's dance and later chorea major, appeared in descriptions from the thirteenth century in central Europe. Robert Bayfield in 1663 gave the following account: "The Lascivious dance is a malady, arising from a malign humor, with which whosoever is taken, can do nothing but dance till they be dead or cured. Tis strange to hear how long they will dance, and in what manner, over stools, forms, tables, even great bellied women sometimes (and yet never hurt their children) will dance so long that they can stir neither hand nor foot, but seem to be quite dead."[14] The dancers were insensible to external impressions and had visions, often of a religious nature.

It seems unlikely that Dr. Bayfield ever witnessed these events himself, but he describes their origins in Germany, coinciding with the infectious epidemic called the "Black Death," a form of the plague, which was widespread in Europe and estimated to have killed some 100–200 million people.

The best descriptions of the dancing manias have been given by the German physician J. F. C. Hecker (1795–1850) and portrayed visually in the paintings of Pieter Bruegel the Elder (1525–1569).[15] Hecker describes how in 1374 an outbreak of the "demoniacal disease" spread from Aix-la-Chapelle, over all the Netherlands, and then to Belgium and Germany. Religion and guilt were involved, as was malingering: "Gangs of idle vagabonds, who understood how to imitate to the life the gestures and convulsions of those really affected, roved from place to place seeking maintenance and

Figure 2.2. De Gaper, by Pieter Bruegel the Elder. This image is highly suggestive of the orofacial-mandibular dystonia syndrome.

adventures. . . . Among the numerous bands many wandered about, whose consciences were torments with recollection of the crimes which they had committed during the prevalence of the black plague."[16] However, he conflated some of these dancing episodes with epilepsy: "Those affected fell to the ground senseless, panting and labouring for breath. They foamed at the mouth, and suddenly springing up again began their dance amidst strange contortions."[17]

The best-known image of this affliction is that of Bruegel's *Dancing Mania on a Pilgrimage to the Church at Sint-Jans-Molenbeek*. In this painting a group of dancers with obvious dystonic postures is seen on its way to church. In fact, Bruegel portrayed many images of abnormal movements. For example, in his painting *De Gaper* (fig. 2.2) he captures the features of what later became referred to as Bruegel's syndrome, or the blepharospasm-oromandibular dystonia syndrome.

The dancing manias are but one of many examples of epidemic or communicable hysterias, outbreaks of which are still often reported on. This usually leads to discussions of causation, with the demonic being less invoked now, but with the organic-psychogenic enigma dominating.

How Dark Were the Dark Ages?

For those of us living in societies that have crossed the Rubicon of the Renaissance and the Enlightenment, it is difficult to comprehend the profound and absolute influence of the church in the Middle Ages. The Ten Commandments and the Catholic hierarchy held all in thrall, limiting intellectual development and condemning heretics and dissenters to a threat of hell and eternal damnation. As George Mora put it, "Medieval thought superimposed its allegorical world upon the world of experience, subordinating the world of sense experience to the ideal world of religious truth."[18]

The prevailing philosophy embraced a mind–body dichotomy, but Christ was viewed as a healer, churches and shrines were places of healing, and there was the beginning in one form or another of hospitals. With little to go on except observation of the body, feeling the pulse, and looking at urine, with bleeding and purging as popular treatments, at least these offered a humane orientation to care, and this included some respite for those with mental disorders. The first such hospital building in Europe was the forerunner of the Bethlem Royal Hospital in London, founded in 1247. Even if the ideas of Galen and Aristotle were the underlying principles that guided treatment, and beliefs in miracles, demons, witches, and theurgist arts were in opposition to science, there was, as the Middle Ages progressed, an emerging humanism, empiricism, and rationalism.

This was abruptly curtailed by the plague. Epidemics of various disorders, now recognized as typhus, cholera, and forms of the plague, were a constant threat to life in the Middle Ages. But the so-called Black Death (bubonic plague) that erupted in Europe in the mid-fourteenth century, spreading from Asia and Africa, killed one-quarter of the population, contributing to the deaths of many scholars who may have advanced the cause of science further. It also had profound psychological effects, leading to great insecurity. The realization of the suddenness of death, man's inability to combat the forces of visitations from outside agencies, and fear of punishment and hell all necessitated a need for magic and a call for protection from the saints associated with healing.

Light around the Corner

At the end of the Middle Ages, a new kind of literature emerged in Europe, in which emotion and feelings became more readily expressed. This included tales of courtly love, at once Neoplatonic and with aristocratic leanings.

These were expressed in images of knightly behavior and codes of conduct in such legends as those of the Arthurian Round Table. Poems and a secular music of expressive individuality were composed. William Langland (1332–1386) and Geoffrey Chaucer (1343–1400) in England wrote poems about ordinary folk, in demotic English. Langland's ploughman falls asleep and has visions; stanzas of Chaucer's "tragedye" *Troilus and Criseyde*—a story of love and war, without a happy ending—reveal some of Criseyde's intimate thoughts.[19] European texts written in these times which have so profoundly affected Western thought include Dante's (1265–1361) *Divine Comedy*, Cervantes's (1547–1616) *Don Quixote*, and Sir Thomas Malory's (?–1471) Arthurian romance *Le Morte d'Arthur*.

Music was taught as part of the quadrivium, along with arithmetic, geometry, and astronomy.[20] Saint Augustine had discussed the significance of Pythagorean audible and visible harmonies, which represented the *kosmos*, literally "order" but metaphorically opposed to chaos. Advances in math and geometry went hand in hand with astrology and alchemy, but architecture was the predominant art. Along with mathematics, these were the link between God and the world; for example, Christian and Platonic ideas were bound up in the design of the cathedrals. Metaphors abound between the human body and the house of God, both resonating as sacred places. Gothic cathedrals symbolized the body of the cosmos, with pointed arches, narrow pillars, flying buttresses, interior decorations of outstanding beauty, and ever-developing patterns of rib vaulting; they became centers of learning, alongside universities. Large stained-glass windows, narrating the past for the present, let in the light; the verticality of the arcade columns fused with the fanlike vaulting, taking the eye onward and upward to Him. Light began pouring through the windows down to the cathedral walls and floors, form revealing function, a metaphor that rebounds in neuroscience. They were the achievement of what later became referred to as *Gesamtkunstwerk*, complete art works, in which all the senses were involved. The architectural design held within it, as the house of God, light, music, color, smell, and harmony, a metaphorical vision of the brain as a cathedral of perfection and complexity. The very concept of these structures, many still with us after nearly 1,000 years, portrayed a reverence and an elegance scarcely achieved since. They embodied Western civilization.

The introspective, restless Petrarch suffered from what he called *acedia*, a psychological disorder with somatic symptoms, akin to melancholy. For him the Mont Ventoux experience led him to a reevaluation of his life and faith, but this also represents an early example of a literature of introspection, examining the tussles between the feelings and desires of our inner life and our social and religious obligations.[21] The meaning of words was enhanced in the Middle Ages, for words themselves meant power, but with new printing technology they could be widely disseminated to a growing literate class in a relatively short period of time. In the beginning was the word, but the word was still with God.[22]

NOTES

1. Petrarch gave up his vocation as a priest. The sight of a woman named Laura in the church of Sainte-Claire at Avignon stirred his passion, and she is celebrated in his poetry. He developed a form of the sonnet referred to as the Petrarchan sonnet.

2. A detailed account of the Arab legacy to Western science and philosophy is given in Al-Khalili J, 2011. *The Canon* became a standard textbook in many European medical schools and was used until the end of the sixteenth century, giving it equivalent importance to the works of Galen.

3. Surah 12 of the Koran is devoted to a discussion of Joseph's dream, the same Joseph of the Christian Bible. In the dream Joseph saw the sun, stars, and the moon prostrate themselves to him. He was an important prophet of Muslim history. Dreams were very important to Mohammad, as his reception of the sacred scriptures occurred in dreamlike conditions. Dream interpreters were highly regarded, and in some societies they still are.

4. Howells JG, 1975, 560.

5. Case histories recounted in Alexander FG, Selesnick ST, 1966, 62; and Inglis B, 1965, 43, 44.

6. What the humors actually represented is unclear and must have shifted over time.

7. "Wolf-madness is a disease, in which men run barking and howling about graves and fields in the night, lying hid for the most part all day, and will not be perswaded [sic] but that they are Wolves, or some such beasts." Bayfield R, 1963, 168.

8. Kenny A, 1993, 130; Pasnau R, 2002.

9. Those ventricles we identify as the lateral ventricles were the first cell; the second cell was today's third ventricle; and the third cell, our fourth.

10. Dante Alighieri, 2010, *Inferno*, canto 24, lines 112–117. The original Italian verse is as follows: "E qual è quel che cade, e non sa como, / per forza di demon ch'a terra il tira, / o d'altra oppilazion che lega l'omo, / quando si leva, che 'ntorno si mira / tutto smarrito de la grande angoscia / ch'elli ha sofferta, e guardando sospira: / tal era 'l peccator levato poscia. / Oh potenza di Dio, quant' è severa, / che cotai colpi per vendetta croscia!"

11. Witchcraft was first made a statutory crime in England in 1541.

12. Women fare very badly. This all stems back to Eve, coming from the bent rib of a man (hence not straight). The monks considered women evil by nature and a desirable calamity. Witches even had the power to steal a man's penis. Kramer and Sprenger's book itself was of pornographic interest, especially for its sadistic references, its encouragement of torture, and descriptions related to the examination and punishment of the condemned.

13. Kramer H, Sprenger J, 1948, 137.

14. Bayfield R, 1963, 170.

15. Hecker JFC, 1832. See Merskey H, 1995, 414–421.

16. See Merskey H, 417, 419.

17. See Merskey H, 415.

18. Mora G, 2008, 204.

19. Chaucer coined the term *tragedye*. Here is an excerpt from *Troilus and Criseyde*: "Before we part my purpose is to tell / Of Troilus, son of the King of Troy, / And how his love-adventure rose and fell / From grief to joy, and, after out of joy, / In double sorrow; help me to employ / My pen, Tisiphone, and to endite / These woeful lines, that weep even as I write." Chaucer G, 1995.

20. The quadrivium was contrasted with the trivium. The latter embraced grammar, rhetoric, and logic and, along with the quadrivium, formed the cornerstone of Middle Age education. However, the origins go back to Plato and Pythagoras.

21. In Petrarch's *My Secret Book*, written between 1347 and 1353, issues of free will, the temptations of earthly pleasures, and inner conflicts between the sensual and the spiritual are explored. It was not made public until after his death. Petrarch's anguish over Laura, his beloved, is exposed through many of his sonnets. She was unattainable because she was married, and she apparently rejected him. His unrequited love caused him unendurable desires. Petrarch's Laura stands in contrast to Dante's Beatrice, who was his guide (after Virgil) to heaven.

22. John 1:1.

REFERENCES

Alexander FG, Selesnick ST. *The History of Psychiatry*. Harper and Row, New York; 1966.

Al-Khalili J. *The House of Wisdom: How Arabic Science Saved Ancient Knowledge and Gave Us the Renaissance*. Penguin, New York; 2011.

Bayfield R. A treatise de morborum capitis essentiis & prognosticis. In: Hunter R, Macalpine I, ed. *Three Hundred Years of Psychiatry*. Oxford University Press, Oxford; 1963.

Chaucer G. *Troilus and Criseyde*. Trans. Coghill N. Viking, New York; 1995.

Clarke E, Dewhurst K. *An Illustrated History of Brain Function*. Sandford, Oxford; 1972.

Dante Alighieri. *The Divine Comedy: Inferno*. Trans. Nichols JG. Alma Classics, London; 2010.

Derrida J. *Writings and Difference*. Routledge, Oxford; 2001.

Hecker JFC. *The Dancing Mania, an Epidemic of the Middle Ages: From the Sources by Physicians and Non-physicians*. Berlin: Enslin; 1832.

Howells JG, ed. *World History of Psychiatry*. Ballière Tindall, London; 1975.

Inglis B. *A History of Medicine*. Weidenfeld and Nicholson, London; 1965.

Kenny A, 1993. *Aquinas on Mind*. Routledge, London; 1993.

Kramer H, Sprenger J. *Malleus Maleficarum*. Trans. Summers M. Pushkin Press, London; 1948.

Merskey H. *The Analysis of Hysteria: Understanding Conversion and Dissociation*. 2nd ed. Gaskell, London; 1995.

Mora G. Unusual mental states and their interpretation during the Middle Ages. In: Wallace ER, Gach J, eds. *History of Psychiatry and Medical Psychology*. Springer, New York; 2008.

Pasnau R. *Thomas Aquinas on Human Nature*. Cambridge University Press, Cambridge; 2002.

CHAPTER THREE

The Renaissance

Let no one enter here who is ignorant of mathematics

Inscribed above the entrance to Plato's Academy

On July 2, 1505, an intelligent and industrious law student, at the village of Stotternheim, on his way to Erfurt, Germany, encountered a thunderstorm. With nowhere to shelter, a bolt of lightning struck next to him and threw him to the ground. This frightened him so much that he avowed that if he were saved, he would become a monk. Already struggling with his inner self over his mundane life, and fearing damnation, on July 17 he knocked on the doors of the Black Monastery in Erfurt and asked for admittance. After a one-year probation, Martin Luther (1483–1546) was ordained as a priest on April 3, 1507.

Perhaps an early case of posttraumatic stress disorder, Luther's conversion had huge implications, not only for the history of the Western world but also for science and philosophy.[1] Together with the artist Lucas Cranach (1472–1553) at Wittenberg, Luther shaped the future of religion and art in Germany.[2] On October 31, 1517, Luther posted his 95 theses on the door of the Castle Church of Wittenberg, and the Western world was never the same again. He attacked the indulgences of the Catholic Church and the infallibility of the Pope (identified as the Antichrist) and affirmed the necessity for faith in God.[3] His translation of the Bible into vernacular German and the dissemination of copies of it made by the newly invented printing presses led to his views becoming widely known. Not only was this the beginning of the Reformation, but it also helped drive the German Renaissance and scientific exploration.

Solid versus Cavities: The Opening Up of the Brain

Scholasticism began to give way to humanism, as well as nostalgia to discover the lost treasures of the past, especially from Greek and Roman times.[4] The intellectual step to secularism and intellectual independence was represented by Petrarch, the scholar-poet and careful collector of books and manuscripts, who, while no scientist, considered medicine to be fettered by the Arabs. Over the next century or so, the heavens opened up with the discoveries of Nicolaus Copernicus (1473–1543), Galileo Galilei (1564–1642), Tycho Brahe (1546–1641), and Johannes Kepler (1571–1630). The earth was no longer the center of the universe; the moon was not spherical, not perfect, but rough and full of cavities and prominences. There were many more stars in the heavens than dreamed of in any previous philosophy.

The Reformation asserted the right of individuals over papal authority in matters of religion, and revival of Plato's philosophies shook the Aristotelian hegemony. Luther had sympathy for medicine, and his wife, the nun Katherine von Bora (1525–1552), set up a hospital to look after the sick, where one of his sons would serve as a physician.[5] Once the infallibility of the Pope had been challenged, other authorities were put on guard. This was the beginning of a new era in which the motto would be, as Alexander Pope (1688–1744) would later put it, "Know then thyself, presume not God to scan; / The proper study of Mankind is Man."[6] The human body became an object of study, as dissection, forbidden in pagan Rome and by the Muslims, provided a stimulus to neurological thinking. Galen began to take a backseat as ideas of the humors as guides to health and disease became circumscribed. The word *anatomy* as referring to the study of the body entered the English language in 1528, *cranium* in 1543, and *emotion* in 1660. In this period of time, study of the human brain opened up a new world of both reality and imagination, William Harvey (1578–1657) showed how the heart pumped blood around the body, and René Descartes (1596–1650) laid down his philosophy.

Leonardo da Vinci (1452–1519) is well known as both a painter and an inventor, but he was also an anatomist. His admission that he had undertaken 10 human dissections led to him losing papal favor and to him leaving Rome in 1515. He injected wax into the ventricles of animals to get a realistic cast of their shape, and there is a well-known illustration of a sagittal view of the head showing not only the three ventricles but also the connecting links between them and the ears and the eyes. However, his artistic eye still attributed Galenic properties to the ventricles (see fig. 3.1).

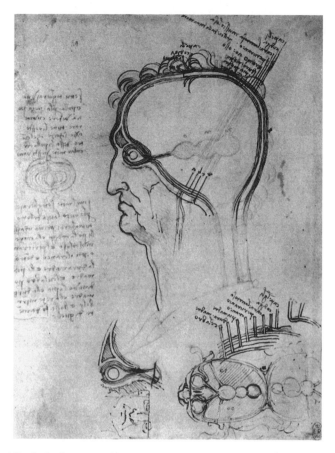

Figure 3.1. The links between the eye and the ventricles, by Leonardo da Vinci. The ventricles are faintly labeled O, M, and N. The first is directly linked to the eyes via a representation of the optic nerve. The links to the auditory channels are also shown. From Clarke E, Dewhurst K, 1972, 33

The first significant breakthrough came with Andreas Vesalius (1514–1564), who published *De Humani Corporis Fabrica* (On the structure of the human body) in 1543, the same year that Copernicus published his *De revolutionibus orbium coelestium* (Concerning the revolutions of heavenly bodies).[7] He was professor of surgery at the University of Padua and asserted that without dissection no physician could produce anything of value, as anatomy was the "basis and foundation of medical art."[8] Often using corpses of hanged criminals or from cemeteries, Vesalius broke so many years of tradition and dogma that he was regarded by some to be as dangerous as Luther. His teacher Jacobus Sylvius (1478–1555), a defender of Galen, referred to him as "an insane innovator whose poisonous breath infected the whole of Europe."[9]

Vesalius identified over 200 errors in Galenic anatomy. The *rete mirabile*, still in evidence in the drawings of Leonardo, was at first accepted by Vesalius but was then rejected (Vesalius openly accepting his earlier error). He produced the best illustrations of cerebral convolutions up to his time.

Indeed, it has to be of some astonishment that the parts of the brain which would seem most visibly obvious to us today—the parenchyma, and the undulating display of the sulci and gyri—seem to have been virtually ignored by earlier physicians, philosophers, and artists. The gyri were likened to the intestines (by Erasistratus of Alexandria [304–250 BC], who probably did human cadaveric dissections) and seem to have been considered somewhat functionless. Distinctions between gray and white matter did not become illustrated until 1586, albeit in a highly stylized sketch by the Roman Archangelo Piccolomini (1526–1586), who used the terms *cerebrum* and *medulla*. Vesalius recognized no obvious cortical patterning and portrayed the gyri like clouds.

Vesalius effected a fundamental change to the teaching of anatomy. Instead of sitting in a high chair reading Galen's text while others (barbers) did the dissection, he worked around the corpse with his students. This was hands-on anatomy. His book contained the first illustrations of the base of the brain, showing some of the cranial nerves and the various basal ganglia. However, he remained perplexed as to how the different parts of the brain functioned, and he had no anatomical ideas about the reigning soul.

Vital Spirits

Paracelsus, otherwise called Theophrastus Bombastus von Hohenheim (1493–1541), was an itinerant Swiss self-taught medical man and the author of *On the Diseases That Deprive Man of His Reason*, one of the first books dedicated entirely to mental illness. From all accounts he was a very arrogant man, who was not much liked by his contemporaries. He prided himself in reading no books for 10 years and gave value only to his own writings. However, his motto was that the proper teacher of medicine is nature, echoing Hippocrates. He abjured anatomy as it was being developed, especially by the Italians, his own interest being in the analysis or anatomy of essence—what was it that man was composed of? So bombastic was he that he was also compared with Luther, and he publically burned some ancient texts, including those of Galen.

He introduced the idea of a "natural spirit," alchemically linked to three substances, sulfur (soul), salt (body), and mercury (spirit), which made up the body, the latter of which was connected with the soul. This was the beginning of what became referred to as the iatrochemistry movement, the idea that diseases were less related to the internal balance of substances, but could arise from external, natural (but not supernatural) things, and hence could be caused by but also cured by chemical substances. This led to therapies based on the use of heavy metals and the origin of what became our modern pharmacopoeia of pills and potions. It was also linked to what was referred to as the doctrine of signatures, the idea that herbs resembling body parts could be used for treatment of diseases of those parts. There were religious overtones here, the idea being that if God has given diseases to mankind, He would also give a sign to identify a means of cure. Paracelsus in his travels used plant extracts in treatment, and herbalism is still widely used in certain cultures. With reference to the head, the walnut attracted attention, possessing a signature of the brain.

Believing that personality was closely linked to mental disorder, Paracelsus advocated a form of psychological intervention, even using the word *unconscious* for the first time in a discussion of *chorea lascivia*. Among diseases that deprive man of his reason, he listed hysteria, but he reverted to a womb-laden etiology: if this organ touched the heart, it would lead to a convulsion similar to epilepsy and all its symptoms.

In Praise of Folly

Folly, born in the earthly paradise of Plutus, was brought up by Drunkenness and Ignorance and aided by Flattery, Oblivion, Laziness, Wantonness, and Pleasure. Folly is full of self-love and argues that life would be dull without her since her divine powers gladden the hearts of men. The Renaissance satire *In Praise of Folly* was written (in Latin) by Erasmus of Rotterdam (1466/1469–1536) and published in 1511. The book lampoons all pedants, religious superstitions, corrupt religious practices, and followers of Aristotle, although it ends with Folly, who, being wise, takes herself seriously, and with praise for Christian ideals.[10]

The era was at the cusp of so many social and religious insecurities, a veritable age of anxiety, with displays of human weakness and follies in literary and artistic representation. Folly was portrayed by Hieronymus Bosch (1450–1516) in his *Ship of Fools*, in which a group of hapless, witless souls are at sea with a destination unknown; madness, by Bruegel the Elder in *Dulle*

Figure 3.2. Dürer's *Melencolia I*. For interpretation of some of the symbols, see text.

Griet, in which Mad Meg and her grotesque companions, surrounded by Bosch-like monsters, are about to storm hell. There are early representations of a neurosurgical operation (craniotomy) by both Bruegel and Bosch, showing the removal of a stone from the head, based on a belief that madness was caused by a stone in the body, but most often by one in the head (the stone of madness).

This was also the age of the rediscovery of melancholy. The most famous image is probably Albrecht Dürer's (1471–1528) engraving *Melencolia I* of 1514, the year of his mother's death (see fig. 3.2). With the subject lost in thought and surrounded by objects representing the sciences, magic, measurement, and architecture, such as an hourglass, scales, and keys, and living creatures that are all immobile, it is an image of self-conscious reflection and despondency. Melancholy herself has turned away from the light, and a flying bat carries the sign of her condition.

Dürer gave us something else: his full-face self-portrait of 1500. Painted when he was 28, it has a monogram A.D. and an inscription that reads (in translation), "I, Albrecht Dürer of Nuremberg, portrayed myself . . ."[11] This theme is returned to shortly.

The Elizabethan Malady[12]

Petrarch's *acedia*, as well as the continuous use of terms equivalent to *melancholia* from ancient times, got its imprimatur from the publication in 1621 of Robert Burton's (1577–1640) *The Anatomy of Melancholy*. While sounding like a medical textbook, this remarkable work, citing over 1,000 authors, half of them medical, became the most frequently reprinted psychiatric text ever published. A self-sufferer of *gravidum cor* and *foetum caput*, he wrote the text as a form of therapy. He titled himself Democritus Junior, after the cheerful philosopher Democritus, teacher of Hippocrates.

Melancholy was viewed across a wide spectrum, from a disposition to a habit or disease. All possible causes were explored, from the demonic, to the humoral, to the familial and effects of upbringing. He noted the various arguments for and against the heart, brain, midriff, or other bodily parts being affected but decided that "the *Brain* must needs primarily be mis-affected, as the seat of *reason*: and then the *heart*, as the seat of *affection*." He had a subdivision of "Head melancholy," but other parts of the body were influenced in sympathy with the brain. The whole was likened to the workings of a clock: "For our body is like a clock; if one wheel be amiss, all the rest are disordered; the whole fabric suffers."[13]

Other ideas were percolating around this time, linked to observations on melancholia and clinging to the old humoral theories. One related to the melancholic temperament, linked to creative activity, religious thought, and philosophical ruminations. Another sufferer, Marsilio Ficino (1433–1499), who influenced Dürer's works, described scholarly melancholy. This was closely linked to black bile and the influence of the planet Saturn, under whose influence he considered he was born. Melancholy enabled a genius to delve more deeply into his intellect, and Dürer even equated spontaneous creativity with divine creation, a spark of hope appearing as the rays of bright light in the upper left side of his image. This echoed Platonic ideas of the poet being divinely inspired, but the creative temperament came with a risk of the development of melancholy, since the scholarly way of life itself promoted more black bile.[14]

Melancholy in the Literature

Lovers and madmen have such seething brains,

Such shaping fantasies, that apprehend

More than cool reason ever comprehends.

The lunatic, the lover and the poet,

Are of imagination all compact.

This well-known quote from Shakespeare's *A Midsummer Night's Dream* sheds light on these ideas.[15] The first great age of English literature belonged to the sixteenth and early seventeenth centuries.[16] Although Henry VIII was much maligned for his treatment of his six wives and his political friends and enemies, his reign and that of his daughter by Anne Boleyn, Elizabeth I, saw a flowering of English arts and popular culture, culminating in the age of the great dramatists, William Shakespeare (1564–1616) being an icon among several. The separation from Rome, the translation of the Bible into English, and the inclination of Protestantism to deepen and intensify individuality altered not only the English language and the arts but also the sciences. Secular drama became popular, and "madness" was portrayed in many plays of the Elizabethan era. These portrayals included states of mind that would correspond to mental illness as seen today, but ideas of madness became part of lay parlance. The terms *Bedlam* and *bedlamer*, for example, were spin-offs from the first psychiatric hospital in England, the Bethlem hospital, and the word *lunatic* passed into common speech.[17] Demons, the influence of the moon, shock, involutional change with aging, and associations between physical deformity and character defects are all found in the plays. The rudely stamped, deformed, unfinished Richard III competes with Timon of Athens as representative misanthropic personality disorders.[18] *Othello* is an exploration of jealousy. The play *King Lear* depends on the madness of the king, which now might be interpreted as a case of dementia. In *Hamlet*, Ophelia's loss of reason and suicide are associated with the death of her father and Hamlet's rejection of her, while the state of mind of Hamlet himself has been the cause of much debate and speculation, especially the reasons for his actions or inactions—melancholy or play acting?[19]

Hamlet jests as he examines the skull of poor Yorick, the empty cranium that once held his gambols, songs, and merriment. Hallucinations are also found, as in the "dagger scene" of *Macbeth*, in which Macbeth asks of the dagger if it is not the false creation of a "heat-oppressed brain."[20] In *As You like It*,

Shakespeare reminds us that all the world is a stage, as the melancholy Jaques documents the seven ages of man, ending in "sans teeth, sans eyes, sans taste sans everything."[21] Shakespeare also mentioned epilepsy, referred to as the falling-sickness in *Julius Caesar* and noted in *Othello*, and there are many references in his plays to the effects of depression, alcohol, and drugs on the mind.

Shakespeare made dramatic use of alterations of the mental state, in comedy and drama, which became part of popular entertainment.[22] For some critics, such as Harold Bloom, Shakespeare invented the Western personality, his great exemplars being the exuberant Falstaff (that "trunk of humours") and Hamlet ("Hamlet's tragedy is at last the tragedy of personality").[23] Hamlet was a man at the cusp of intellectual ferment and was a university student in Luther's Wittenberg, along with his fellow student Horatio. Hamlet sees a ghost. While ghosts were compatible with Catholic theology and a belief in Purgatory, for Protestants and for the cerebral Hamlet, a more skeptical approach needed to be taken—Hamlet asks, "be thou a spirit of Health or goblin damned?" His rational sense asks whether what he has seen may be the devil, since "There are more things in heaven and earth, Horatio / Than are dreamt of in your philosophy."[24] Yet he brings to us with his introspective conundrums the oldest philosophical question, "To be or not to be?" Just what is the essence of existence? Hamlet also says that "conscience doth make cowards of us all."[25] The word *consciousness* does not occur in Shakespeare's works, having first entered the English language in 1678. *Conscience* etymologically indicates a sharing of knowledge, with others or within oneself (*con + scientia*), and refers to a need either to be free from sin or to hide one's sins from others. The word *conscious* therefore became associated with self-knowledge about whether a charge against one was true or false, consciousness then reflecting knowing what one was accountable for; *self-consciousness* appeared later, in 1690. However, we can see in Renaissance times how the English language was formulating words and ideas about introspection and the self, about the mind being conscious of the self as an inner witness.

The New Testament: From Deduction to Induction

"Elementary, my dear Watson," the phrase never actually uttered by Sherlock Holmes, reflects on Holmes's method of arriving at the conclusion of a mystery—eliminate the impossible, and whatever remains, however improbable, must be the truth. The claim is, of course, false, and his reasoning is so often from effects to causes (a mark on a watch case implying that the watch had been at a pawnbrokers, leading to a conclusion, with other evidence, of the owner's personality). However, he considered it a major mistake to theorize before one had data, and he did not want guessing to interfere with the faculty of logic. Induction takes us from particular instances to the general; deduction, from the general to the particular.[26] Induction requires perceptual experience; seeing is believing, as the saying goes.

The Renaissance brought new ways to explore the world and the heavens beyond human eyes. Francis Bacon (1561–1626), a powerful politician in the court of Elizabeth I, pursued the advancement of science by the method we know today; he considered that scientific knowledge could give mankind power over nature.[27] *Novum Organum Scientiarum* (New instrument of science), published in 1605, and his *Essays* established him as an Aristotelian and naturalist, who believed that instead of studying the Book of Revelation, man should study the Book of Nature, forging a cleft between metaphysics and physics. Observation, recording of data, and seeking regularities that will reveal the laws of nature led him to seek causal rather than teleological explanations. The deductive method, requiring great play of the imagination, allowed for many productive ideas to emerge involving the mind, but the problem was the failure to apply methods of verification, avoiding experimentation, and remaining bound by theological dogma. Bacon's principles minimized error, at least in theory, hence his renown as the father of modern science.[28]

Bacon was a friend of William Harvey, whose contributions to understanding circulation are well known. Overthrowing Galen's ideas and using inductive principles, Harvey came to view the heart not as an organ of suction (of air from the lungs) but as an organ of propulsion (of blood). With analogies to the circular movements of the heavenly bodies (the microcosm mimicking the macrocosm), he was interested in movement, the principals of motion and rest in all things: pulse and respiration continuing indefinitely, but other deeds, such as those performed "in accordance with Nature's demands" or by the emotions, being discontinuous. Harvey had an interest in neurology and recognized that the state of the mind influenced the heart. He thought it noteworthy that damage to the right side of the brain should cause paralysis on the left side of the body, and vice versa. He considered the brain to be the *sensorium commune* to which all sensations were referred, and his own anatomical observations provided a good description of the ventricles—he thought that epilepsy originated there. But he posed the question in relation to the mental faculties, "Is the substance of the brain or the ventricle the chief part?" The ventricles were, after all, full of water, and he did not wish the soul to "mingle with excretions."[29]

Harvey also had a practice in obstetrics and gynecology, and so he was familiar with hysteria. He held on to the uterine theory, which was still popular at the time, as shown by a remarkable book published in 1603 called *A Briefe Discourse of a Disease Called the Suffocation of the Mother*. The author was Edward Jorden (1569–1632), and the title has obvious uterine allusions, stimulated by the case of Mary Glover, who suffered from convulsions.

King James I of England and VI of Scotland in 1597 published *Daemonologie, in Forme of a Dialogue*, an attempt to reemphasize the dangers of witchcraft.[30] This had the effect of increasing enthusiasm for witch hunting, and more and more witches were being identified, especially by anesthetic "witch" patches, and their cases coming before the courts.

Jorden's book came out a month after that of the king, providing a counterbalance. Thus, medical evidence was now often called for in the trials, and Jorden was an expert, even having been called as a witness by King James himself. Jorden considered the passions of the body to have natural causes, and while the uterus was the seat of the pathology, "sympathy" explained distant effects. He recognized perturbations of the mind as relevant to the development of symptoms and recommended treatment by friends and attendants, rather than harsher measures.

Elizabeth Jackson, "an olde Charewoman," was accused of bewitching Mary Glover. The latter had paralyses of the left side of the body, loss of speech, periodic blindness, movements of the belly, and episodes of prolonged seizures. Under examination, Dr. Jorden firmly stated that the attacks were "naturall . . . *Passio Hysterica*." After hearing all the evidence from both sides of the argument, the judge commented that "the Land is full of Witches: they abounde in all places; I have hanged five or sixe and twenty of them myself." He commented on "divers strange marks, at which (as som of them have confessed) the Devill suck their bloud." He brusquely rejected the evidence of Dr. Jorden with the retort, "I care not for your Judgement: geve me a natural reason, & a natural remedy, or a rash for your Physicke." Jorden could not give a cause or a cure, and the jury found Elizabeth Jackson guilty, but Mary Glover was passed on for exorcism.[31]

With Jorden's account we had not only the first English book on hysteria but another example of medical opinion breaking into supernatural theories of causation. Here was the beginning of ideas of conversion of symptoms from one part of the body to another, with sympathetic reactions between one organ and another and hints at a psychotherapeutic treatment. Jorden opined that the brain was involved, the animal faculty becoming disturbed, accounting for sensory and motor symptoms—a beginning of the shift of the origin of such symptoms away from the uterus to the brain. These ideas, as we will see, become well developed with time. It is of further interest that this trial was probably the first record of a testimony of a psychiatric expert witness in English, and there were two sides of the case presented to the court, as is the practice today.

Jorden was not alone in his beliefs. The Dutch physician Johann Weyer (1515–1588), who had a major interest in mental disorders and in diseases of women, is referred to by the historian Gregory Zilboorg as the true founder of modern psychiatry. Weyer published *De Praestigiis Daemonum et Incantationibus ac Venificiis* (On the deceptions of the demons and on spells and poisons) in 1563. This was an attack on "The Witch Hammer." He examined a number of the accused himself and reported his findings, which were early examples of the psychiatric method of examination. He discussed differences between the stigmata and symptoms of medical illness and malingering, which he uncovered in certain patients. Notorious was the case of Barbara Kremers, a 10-year-old who, it was claimed, had not eaten or drunk, nor passed urine or feces, for a year and had a six-month-long attack of mutism. Many people visited her and left her money, and the city council, proud of their famous citizen, had given her a certificate testifying to the miracle. But her parents allowed her and her 12-year-old sister to stay at Weyer's house, and the ruse was soon discovered: her sister had been supplying her with food and water. Weyer was able to relieve her of her malady, much to the discomfort of her parents and the council.[32]

The links between devils, deception, and convulsions are noted in the play *The Devil Is an Ass* by Ben Jonson (1572–1637). In part this is a play about fraud, mistrust, and financial scams; in it he suggests that by feigning diabolical possession, a deal could be invalidated:

> MERECRAFT. It is the easiest thing, sir, to be done.
> As plain as fizzling; roll but wi' your eyes,
> And foam at th' mouth. A little castle-soap
> Will do't, to rub your lips.[33]

Decapitation and Descartes

At first Harvey's ideas were considered crazy, but he had many defenders, including the philosopher whose concepts have cast such a wide penumbra over neurology and psychiatry, namely, René Descartes (1596–1650). Although famous for his "cogito ergo sum" and his selection of the pineal gland as the seat of the soul, his broader scientific and philosophical scope is often ignored. In 1633, Galileo was arrested, and copies of his book *Dialogue concerning the Chief World Systems*, promoting the views of Copernicus, were burned in Rome. Descartes was much disturbed by this, as he also agreed with the views of Copernicus. Since Galileo was held under "grave suspicion of heresy," Descartes was fearful that his own writings might fall afoul of the church. Born in France, he was a devout Catholic, but eventually, after many travels, he settled in Holland, where there was much less limitation on individual freedom of thought. His *Discourse on Method* was published in 1637, *Meditations* in 1641, and *Treatise on Man* in 1664.

Descartes was after a new method of exploring the nature of knowledge but came up with a solution radically different from that of Bacon. By doubting everything, he came to the logical conclusion that it was not possible to doubt his own existence; as a mathematician he was seeking mathematical-like certainty for the nonmathematical.[34] Visual impressions could be deceiving, logical fallacies abounded, and how could he be certain that his daily experiences were not akin to dreams or that a demon was not deceiving him and making him have his experiences? What Descartes could not doubt was that he had conscious experiences, and therefore he could not doubt that he existed: "But what then am I? A thing which thinks. What is a thing which thinks? It is a thing which doubts."[35]

He considered the essence of matter to be extension in space but asserted that thought was unrelated to matter; it was not extended and required no place to exist. Humans, he opined, are composed of two substances: *res cogitans* (thinking mind) and *res extensa* (the body). What makes us human could not be derived from the body, or more importantly, from the brain. The mind and the body simply did not directly connect, and here was a clear distinction between humans and animals (i.e., the latter did not think—as automata). But he had a problem, one that has spiked neurology, psychiatry, and philosophy ever since, which became referred to as Cartesian dualism. Descartes thought it necessary to know how the soul was joined to the body, which he explored in his *The Passions of the Soule in Three Books* (1650).

Descartes discussed the relative merits of the heart as opposed to the brain as the locus of contact, and he designated the pineal gland as the location for the soul. Situated centrally, above the third ventricle, this gland had special interest for the Alexandrian School around 300 BC, in their views regulating the flow of "thought" from the ventricles. The unpaired nature of the pineal impressed him, and it was from this location he concluded that the soul "diffuseth her beames into all the rest of the body by intercourse of the spirits, nerves, yea and the very blood." He continued, "The little strings of our nerves are so distributed into all parts of it [the body] that upon occasion of several motions excited therein by sensible objects, they variously open the pores of the braine, which causeth the animall spirits contained in the cavities thereof, to enter divers ways into the muscles." This led to muscle movements. The pineal hung "between the cavities which contained these spirits, that it may be moved by them . . . [and] moved several ways by the soul too. . . . As also on the other side, the machine of the body is so composed, that this kernel [the pineal] being only divers wayes moved by the soul . . . it drives the Spirits that environ it towards the pores of the brain, which convey them by the nerves into the muscles, by which means it causeth them to move the members."[36]

This conception is portrayed by Descartes in his *Treatise on Man* and is illustrated in figure 3.3. At a stroke Descartes had liberated the body for physical science. By clearly separating the soul from the extended matter, he paved the way for the future scientific exploration of the physical world, free from theological prohibition. God may have created everything, but the mechanical laws of nature could be explored, and indeed were done so by Descartes. He studied anatomy intensely, viewing the body as a machine. The illustration in figure 3.3, in which reflex arcs are shown, prefigures the wiring diagrams of brain connections, which are still so popular today. His drawings of the brain clearly show gyri and sulci, and he provided detailed descriptions of the eye.

Another philosopher of this time who exhibited neurological resonances is Benedict Spinoza (1632–1677). The neuroscientist Antonio Damasio, who much criticized Descartes in his book *Descartes' Error*,[37] introduced neurologists to Spinoza in his *Looking for Spinoza*. Damasio's interest was that Spinoza was familiar with developments in science, and while he agreed with Descartes about the importance of applying mathematics to understanding reality, Spinoza was firmly against the Cartesian division of mind and matter, considering both to be bound to a single substance. Damasio,

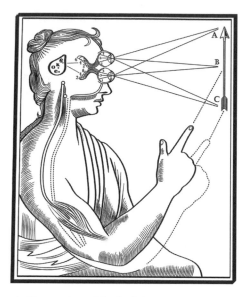

Figure 3.3. Descartes's illustration of the link between sensory perception and muscle movement and the role of the pineal. This is the first illustration of reflex action related to the brain, and it influenced future thinking about the reflex nature of the stimulus-response reaction.

whose own work we will visit in a later chapter, was impressed by Spinoza's words "The object of the idea constituting the human Mind is the Body. . . . The object of our mind is the body as it exists, and nothing else." In this philosophy, the mind was united with the body, and Damasio's interpretation of Spinoza's doctrine was that he acknowledged that "mind and body are parallel and mutually correlated processes . . . two faces of the same thing."[38]

Descartes died of pneumonia and was buried in Sweden in 1650. In 1667 his body was exhumed and taken to France. The man who so clearly separated the mind from the body was himself to be sundered, as ironically his skull was separated from his body and appears to have traveled around Europe, changing hands between several apparent owners.[39]

New Horizons—New Diseases

An older belief that epilepsy was associated with genius was resurrected, and several famous great men in Western history were thought to have suffered from epilepsy, a theme that returned again in the twentieth century. However, many people with epilepsy simply found themselves as poor beggars, but simulated epilepsy for financial gain was recognized by the moniker the "Counterfet Cranke," named after a man shown to be an impostor.

There was another complicating factor. In 1492 Christopher Columbus (1451–1506) sailed westward to find the sea route to Asia. He ended up initially in the Bahamas and then explored the Caribbean. Subsequent voyages took him farther south into Central America. His crew took with them many diseases unknown to the natives they would encounter, and many millions of the latter perished from such illnesses as measles, typhus, typhoid, and smallpox. The expeditions brought back to Europe many riches, including gold, spices, and tobacco, but also the spirochete linked to epidemics of syphilis. The name *syphilis* was first used in 1530 by the Italian physician and poet Girolamo Fracastoro (ca. 1478–1553), in a poem about a shepherd named Syphilus who was inflicted with the disease by the god Apollo as retribution for defiance that Syphilus and his followers had shown him. When an outbreak occurred in Naples in 1495, it was referred to as the *Morbus Gallicus*, the French disease. Needless to say, syphilis was called the "Italian disease" in France, the "Spanish disease" by the Dutch, and the "British disease" by the Tahitians.[40] It is likely that the symptoms and signs of syphilis in those times were different from later descriptions, and it may well have been more virulent; some have suggested that Columbus himself may have been one of the first sufferers.

Convulsions were one sign of the disease; another condition linked to seizures was scurvy, which was common in sailors as a result of the lack of vitamin C in their diet.[41] Different patterns of seizures were described, from brief episodes associated with some confusion, to unilateral seizures, to those with obvious associated psychiatric manifestations. The definition of epilepsy thus broadened, in the writings of one physician "to any affection of the body where the victims are disordered in their minds, while the members (of the body), be it all, or some, or only one, are moved against their will." One classification separated "perfect" from "imperfect epilepsy," and there was increasing interest in childhood forms of the disorder.[42]

So things were on the move. There was more and more emphasis on clinical observation, careful documentation of individual cases, deliberations on causes of disorders, and attempts to classify different forms of similar clinical pictures. Even the idea of localization of functions in the brain was in the popular imagination. Queen Elizabeth I was faced with deciding the punishment of the Duke of Norfolk, who was involved in a conspiracy against her. He was the premier English nobleman at the time. Twice she altered her mind about consigning him to execution, telling her trusted Lord Burghley (1520–1598) that the "hinderpart of her brain did not trust the forward sides of the same." Passion and logic were in conflict for the queen, but the duke was not beheaded.[43]

The psychology of the times contained many contradictions. Words like *melancholy* had varied connotations, and there was reliance on the ideas of the old-time masters. There was a lack of imperative for exactness, and old habits of thinking were hard to shift. The predominant philosophy of the times was Platonic, promoted especially in Florence by Cosimo de' Medici. He provided Ficino, who had translated many of Plato's works, with the wherewithal to found a Platonic Academy. The cult of courtly love was revived, echoing back to Dante and Petrarch, inspiring paintings such as those of Sandro Botticelli (ca. 1445–1510), whose *Birth of Venus* depicts the goddess being blown ashore by the gods and emerging from the sea on a shell. A veritable *Venus pudica*, attention is drawn to her genitals by her flowing hair and contrapposto, but she is modestly covered by a diaphanous cloak, illustrating the transformation from the pagan to the Christian, from the profane to the sacred. Beauty and the pursuit of perfection were high on the agenda.

The body had become the object of the anatomist and the artist alike, as more careful renderings of the structure of the brain were published. Raphael (1483–1520) used live models and perfected the use of perspective. Along with other artists (I have already noted the confident self-portraits of Dürer), he portrayed full-face expressions, giving some personality to his sitters. Michelangelo's proud David offered us a giant slayer of magnificently perfect proportions, the inner life portrayed as one of freedom and energy poised to strike. For such a Platonist, the perfect form was already extant in the block of marble, just waiting to be released by his tools. The rise of the self-conscious individual as an object of study had begun. Luther, when he appeared under an edict from Charles V at Worms before a court hearing and was asked to recant his views and disown his writings, refused by stating, "Here I stand; I can do no other."[44]

As the Elizabethan era came to an end, in England the arts, like diseases, flourished. The legend of Dr. Faustus came to the stage with Christopher Marlowe's (1564–1593) play *The Tragical History of Dr. Faustus*. This morality play explores the dangers of seeking hidden knowledge, of selling one's soul to the devil, and of seduction by an incubus. Marlowe's Dr. Faustus had, like Hamlet, been educated at Wittenberg and wished to search all corners of the newfound world. He implores, "Sweet Helen, make me immortal with a kiss. . . . Instead of Troy shall Wittenberg be sacked."[45] But as the clock strikes 12:00, 24 years after he signed his pact with Mephistopheles for practicing more than heavenly power permits, the damned Faust calls out, "Come not, Lucifer! I'll burn my books"; as his house burns down, he is borne to hell to burn.[46]

NOTES

1. The author has seen cases of posttraumatic stress disorder which have led to religious conversion, either in one not previously religious, or resulting in an intensification of previously held convictions. Luther's father and mother were strict disciplinarians, demanding absolute obedience, often beating their son until he bled. Luther remained terrified of death throughout his life.

2. Lucas Cranach, a close friend of Luther, was a follower of Albrecht Dürer and court painter to the electoral Saxon dynasty in Wittenberg. His art embraced "Old and New Testament snake law." Cranach's images and Luther's sermons "conveyed the gospel message with immediacy, transparency and power." Ozment S, 2011, 3, 134.

3. Indulgences were the granting of donations in lieu of penance for sins and to avoid the fires of purgatory. Luther wrote *The Ninety-Five Theses on the Power and Efficacy of Indulgences* in 1517. He was charged with being a heretic and threatened with excommunication.

4. Scholasticism emerged via the growth of the theological educational institutions and the universities. The philosophy was associated with a critical logical method of defending the Christian beliefs, largely based on Aristotelian principles. Thomas Aquinas, William of Ockham, and Peter Abelard (1079–1142) were central proponents of this thinking.

5. Luther married Katherine von Bora on June 13, 1525, an act of defiance to the church (a monk, disgraced at that, marrying a nun) and the beginning of the secularization of legal marriage. Cranach was his best man.

6. Pope A, 2006, 281, epistle 2, lines 1–2.

7. The illustrations in Vesalius's book were done by Jan van Calcar (1499–1546), a student of Titian (1488/1490–1576).

8. Vesalius was 28 when he published his book. Surely, this dictum must remain sound today. It is of some astonishment and with some sadness that the author finds that when he goes to various medical schools, hands-on dissection of the human frame has virtually disappeared. Use of wax or other models, illustrated texts, and, for example, only brain scans to understand brain anatomy would seem like stepping back to the sixteenth century.

9. See Moon RO, 1909, 119.

10. Erasmus wrote the text in a week while staying in England with his friend Thomas More (1478–1535), later to be Lord Chancellor in the reign of Henry VIII. Erasmus considered this to be a minor piece of work, yet today it is the book he is most known for.

11. Dürer completed several self-portraits, but only in three-quarter views before this one. Except for images of Christ, full facial views were not usual in art before this time.

12. This is the title of Lawrence Babb's excellent book on melancholy in Elizabethan times.

13. Burton R, 1837, 1:167.

14. Needless to say, Luther could not accept such a formulation and considered that melancholy was the devil's work.

15. *A Midsummer Night's Dream*, 5.1.4–8.

16. This statement does not ignore that Geoffrey Chaucer (1340–1400) is acclaimed as the first English poet. However, the era discussed here is often referred to as an English Renaissance.

17. The original Bethlem Hospital was founded in 1247 in Bishopsgate, London. In 1676 it was transferred to an area in London now occupied by Liverpool Street Station, and in 1815 to the site of the Imperial War Museum. Today it is in West Wickham, south of London. The terms related to inmates who were released from the hospital and roamed the countryside, for example, "Bedlam beggars with roaring voices" (*King Lear*, 2.2.185).

18. Timon's gravestone reads," "Here lie I, Timon; who alive all living men did hate; / Pass by, and curse thy fill; but pass and stay not here thy gait." *Timon of Athens*, 5.4.82–83. He was a man of much wealth who had many apparent friends, until becoming penniless. Disillusioned, he develops hatred of his friends and Athens.

19. For a detailed overview of mental illness on the stage, see Oyebode F, 2012.

20. *Macbeth*, 2.1.39.

21. *As You Like It*, 2.7.166; according to Babb L, 1951, this is one of the most frequently quoted passages in English literature (93).

22. Matthews PM, McQuain J, 2003, have written an adventurous attempt to bring the Bard into contemporary neurology.

23. Bloom H, 1998, 431. "Humours" refers to the humoral theory. *Henry IV*, pt. 2, act 4.

24. *Hamlet*, 1.5.166–167.

25. *Hamlet*, 3.1.83; see also *Richard III*, 1.4.36–37: conscience "makes a man a coward."

26. Sherlock Holmes in *The Sign of the Four*; see Doyle AC, 1890, 111. Holmes incidentally refers to his method as deduction.

27. The word *scientist*, however, was not used until 1833 by the scientist-philosopher William Whewell (1794–1866). Bacon referred to his inductive method as the New Testament.

28. Well, at least possibly one of them. Experimenting led to his death. He wanted to find out whether refrigerating meat would preserve it, so he stopped his carriage, purchased a fowl, and stuffed it with snow. He caught a cold, which led to bronchitis, and he died a month later.

29. Brain R, 1964, 16, 18, 20.

30. It was republished in 1603 when he acceded to the English throne. He introduced a new witchcraft act in 1604; this replaced that of Elizabeth I of 1563, which had recognized witchcraft as a crime of the greatest magnitude. Later in life he became more skeptical on the matter.

31. See Hunter R, Macalpine I, 1963, 47, 68–75. Perhaps on account of Jorden's defense of her, Elizabeth Jackson was not hanged but imprisoned, but she also had to stand on the pillory to confess her trespass.

32. Zilboorg G, 1941, 207–235. There were many who disagreed with the likes of Jorden and Weyer, but the battle against witch hunting was eventually won. It was officially stopped by the Witchcraft Acts in 1736.

33. Jonson B, 2012, 5.3.1–4.

34. Descartes had considerable mathematical skills: he invented the graph with Cartesian coordinates.

35. *Meditations*, Descartes R, 1997, 29. The famous "I think" is to be found in the *Discourse on Method*, pt. 4. It was originally published in French as "je pense, donc je suis."

36. Hunter R, Macalpine I, 1963, 134.

37. Damasio A, 1994. Descartes's error was to fail to understand that "Nature appears to have built the apparatus of rationality not just on top of the apparatus of biological regulation, but also *from* it and *with* it" (128).

38. Damasio, A, 2003, 214, 217.

39. A number of collections claim to possess Descartes's skull. The specimen in the Musée de l'Homme is engraved with the names of several former owners.

40. For an excellent account of the history and alleged famous sufferers, see Hayden D, 2003.

41. James Lind (1716–1794) demonstrated that scurvy could be treated with citrus fruit, as described in his book *A Treatise of the Scurvy* (Lind J, 1753).

42. Temkin O, 1971, 192.

43. Ackroyd P, 2012, 366.

44. Cahill T, 2013, 182.

45. "Was this the face that launched a thousand ships, / And burnt the topless towers of Ilium? / Sweet Helen, make me immortal with a kiss. / Her lips suck forth my soul: see where it flies. / Come, Helen, come, give me my soul again. / Here will I dwell for heaven is in those lips, / And all is dross that is not Helena." *Dr. Faustus*, 5.1.93–102 (1616 text), Marlowe C, 1996.

46. *Dr. Faustus*, 5.3.191, Marlowe C.

REFERENCES

Ackroyd P. *The History of England*, vol. 2, *The Tudors*. Macmillan, London; 2012.

Babb L. *The Elizabethan Malady: A Study of Melancholia in English Literature from 1580 to 1642*. Michigan State College Press, East Lansing; 1951.

Bloom H. *Shakespeare: The Invention of the Human*. Riverhead Books, New York; 1998.

Brain R. *Doctors Past and Present*. Pitman Medical, London; 1964.

Burton R (Democritus Junior). *The Anatomy of Melancholy*. Longman, Rees, London; 1837.

Cahill, T. *Heretics and Heroes: How Renaissance Artists and Reformation Priests Created Our Modern World*. Doubleday, New York; 2013.

Clarke E, Dewhurst K. *An Illustrated History of Brain Function*. Sandford, Oxford; 1972.

Damasio AR. *Descartes' Error: Emotion, Reason, and the Human Brain*. Penguin Putnam, New York; 1994.

Damasio AR. *Looking for Spinoza*. Harcourt, New York; 2003.

Descartes R. *Key Philosophical Writings*. Trans. Haldane ES, Ross GRT. Wordsworth Classics of World Literature, Wordsworth Editions, London; 1997.

Doyle AC. *The Sign of the Four*. Lippincott, London; 1890.

Hayden D. *Pox: Genius, Madness, and the Mysteries of Syphilis*. Basic Books, New York; 2003.

Hunter R, Macalpine I. *Three Hundred Years of Psychiatry, 1535–1860*. Oxford University, London; 1963.

Jonson B. *The Devil Is an Ass*. In: *The Cambridge Edition of the Works of Ben Jonson*. Cambridge University Press, Cambridge; 2012.

Lind J. *A Treatise of the Scurvy. In Three Parts. Containing an Inquiry into the Nature, Causes and Cure, of That Disease. Together with a Critical and Chronological View of What Has Been Published on the Subject*. Sands, Murray, and Cochran for A Kincaid and A Donaldson, Edinburgh; 1753.

Marlowe C. *Dr Faustus B-text*. Drama Classics. Nick Hern Books, London; 1996.

Matthews PM, McQuain J. *The Bard on the Brain*. Dana Press, New York; 2003.

Moon RO. *The Relation of Medicine to Philosophy*. Longmans, London; 1909.

Oyebode F. *Madness at the Theatre*. Royal College of Psychiatrists Publications, London; 2012.

Ozment S. *The Serpent and the Lamb: Cranach, Luther, and the Making of the Reformation*. Yale University Press, New Haven, CT; 2011.

Pope A. *An Essay on Man.* In *Alexander Pope: The Major Works.* Oxford World Classics. Oxford University Press, Oxford; 2006.

Temkin O. *The Falling Sickness.* 2nd ed. Johns Hopkins University Press, Baltimore; 1971.

Zilboorg G. *A History of Medical Psychology.* W. W. Norton, New York; 1941.

CHAPTER FOUR

The Enlightenment

Miranda: O Wonder!
How many goodly creatures are there here!
How beauteous mankind is! Oh brave new world . . .

William Shakespeare, *The Tempest*, 5.1

On March 23, 1832, the German poet and author Johann Eckermann (1792–1854) walked to Frauenplan in Weimar to pay his last respects to the man whom he had had convivial conversations with for almost a decade. The ailing Johann Wolfgang von Goethe (1749–1832), on the evening before he died, had discussed some of his theories of color with his daughter-in-law. His scientific theories and experiments were published in Theory of Colours *(1820), in which he was very critical of the views of Isaac Newton (1642–1727). Perhaps as William Blake (1757–1827) also felt, he thought that Newton had unwoven the rainbow by splitting light with a prism. For Goethe light was undivided and homogenous; he was interested in unity, in color harmony and aesthetics. He was less interested in color as a physical phenomenon, in the way that light entered our eyes, and more interested in how light is processed by the brain, as well as the interplay between color and mood. For Goethe, the primordial phenomena were light and dark, the latter being not just a passive absence of light but an active participant in what we see and feel.*

During the morning of March 22, his strength gradually left him, and his last words were reported to be "Mehr Licht" (more light). Eckermann went to his bedchamber, and Goethe's servant drew back the white sheet covering his naked body. Eckermann said, "A perfect man lay in great beauty before me." He placed his hand on Goethe's heart and burst into tears.[1]

Over some two and a half millennia, we have seen how the geographical axis of ideas about science in general but the brain in particular have shifted—from the East, to Greece, to Rome, and then to Spain and France.

The shift now continues in a northerly direction, as we come to the so-called Enlightenment era.[2] While being kindled in France, it was with Thomas Willis (1621–1675) in England that the brain became the subject of exploration almost unfettered by the church. However, first we need to retrace some steps. With the discoveries of Harvey and the material philosophies of Bacon, the way forward for two main schools of medicine arose: the iatrochemical and the iatromechanical.

The Fluid and the Solid—the Wet and the Dry

We have noted that Damasio championed Spinoza from the past galaxy of philosophers to help underpin his modern-day understandings of the mind–brain problem. Sir Charles Sherrington (1857–1952), whom we will meet later, opted for another paragon, namely, the physician-philosopher Jean Fernel (1495–1558). Fernel gave us the word *physiologia*, which told of the causes of the body's actions, distinct from anatomy, which gave a location where the causes took place. He was taking an essentially Aristotelian approach and discussed bodily activity in terms of animal spirits: even though nature was God's principal, man was part of nature. Fernel recognized that a number of the body's actions take place without the action of the will, which was an idea introducing a physiology of the reflex. He also described for the first time the spinal canal.

Spirit is a word that we can see has echoed through the discussions of the brain since chapter 1, and it will continue to appear throughout the book. For Galen and the early theorists, spirit was one interaction between the body and the soul; it provided a mechanism for movement. It was bound up with concepts by analogy, such as innate moisture or heat, or other very basic life properties. Descartes used the term *animal spirits*, but in a material sense, akin to air or wind rather than an ethereal vapor.[3] But the ground was now prepared for a contrast between two schools of thought: those that looked more to physiology, and those that preferred anatomy, or the iatromechanists. The latter were encouraged by Descartes's ideas, conceiving the body as a "Cartesian machine," as well as his localizing significant brain activity outside the ventricles to the brain's substance, even if it was the pineal gland, which he opined was capable of being buffeted from side to side by either the soul or the animal spirits.[4]

Franciscus Sylvius (1614–1672), a Dutch physician and anatomist of aqueduct and fissure fame, and supposed inventor of jenever (gin), was a supporter of the ideas of Harvey and Descartes but took the iatrochemical line. He followed some of the concepts of Paracelsus, and he explained the chemical phenomena of the living body as the result of interactions between acids and alkalis, since salt was made up of them and effervescence and precipitations were chemical reactions. The carotid and cervical arteries carried blood from the heart to the brain and cerebellum, which provided material for the animal spirits. This penetrated into the cortex and underlying white matter, being purified along the way. The nerves carried the animal spirit to all parts of the body, that which was not used up being taken into the lymphatic system before returning to the blood. These spirits gave the affectation of the mind to the body and the affectations of the body to the mind.[5] The concept of the spirits was taking on physiological meanings with which we can begin to identify.

The Circle of Willis

The chief iatromechanists were to be found in Italy, and the school found little favor in England.[6] This was surprising, considering the influence of Bacon and Harvey, yet Thomas Willis, the man who gave us the word *neurology (Neurologie)*, was an enthusiastic iatrochemist. He determinedly studied anatomy, relying on his own observations and, in theory at least, not adopting the opinions of others. However, in order to understand his achievements, it is important to note not only the historical backdrop to his life and times but also his collaborators.

Born only 25 years after Descartes, Willis seems to have come from a different world, scientifically and politically. One of the most shocking events in English history, the beheading of King Charles I, happened in 1649. It was shocking because at that time the divine right of kings held that the king was subject to no earthly authority, was ruled directly by God's will, and could only be judged by Him. The country was in the grip of the English Civil Wars (1642–1651), which led to so many deaths and much destruction of property and religious artifacts. In this, the supporters of the monarchy were pitted against the Parliamentarians and Oliver Cromwell.

Willis was born near Oxford and was a royalist. His medical education was brief, being interrupted by the war; he got his degree in 1646. Oxford became a royalist garrison, and soldiers brought diseases with them as they marched across the countryside. Willis described the effects of "the fever," probably typhus, with neurological complications, including convulsions, which carried away several members of his own family. He was in the army for about two years and possibly met Harvey, who was in Oxford in his capacity as a physician to Charles I. Exactly what impact the experience of the war had on Willis is unclear, but he often used anecdotes and metaphors, including military ones, for his iatrochemical views, such as a hemorrhage being stopped by the touch of a stick taken from an ash tree, or explaining convulsions and muscular action as akin to explosions of gunpowder.

As the royalist fortunes declined, he left Oxford, but he returned to Christchurch in 1646, with his license to practice medicine. He successfully maintained a stance against the Puritan reforms, and he came into contact with many who later obtained high positions after the restoration of the monarchy in 1660. Willis avidly studied chemistry. While he did not discard humoral theory and was accused by some as being Galenic, he viewed himself as experimenting in the Baconian tradition with Harveian physiology. He believed that all matter could be broken down into five chemical elements (spirits, sulfur, salt, water, and earth), and he used metaphors such as fermentation, agitation, and effervescence. Analogies to explain health and disease were with baking, wine making, combustion, and the like. The work of the physician was like that of the vintner: "But it seems to me that the Brain with the Scull over it, and the appending Nerves, represent the little Head or Glassie Alembic, with a Spunge laid upon it, as we used to do for the highly rectifying of the Spirit of Wine: for truly the Blood when Rarified by Heat, is carried from the Chimny of the Heart, to the Head, even as the Spirit of Wine boyling in the Curcurbit, and being resolved into Vapour, is elevated into the Alembic."[7]

Willis fell in with an elite intellectual circle. Christopher Wren (1632–1723) came to Oxford after his family had been stripped of land and possessions, his father having been dean of Windsor. He, along with Willis, the clergyman, mathematician, and astronomer John Wilkins (1614–1672), and other "virtuosi," as they called themselves, formed the Oxford Experimental Philosophy Club. Willis was referred to as "our chymist."[8] Later collaborators included the polymath Robert Hooke (1635–1703), the philosopher-chemist-alchemist Robert Boyle (1627–1691), Richard Lower (1631–1691), and the physician-philosopher John Locke (1632–1704).

With his great skill at drawing, Wren did many of the anatomical plates for Willis's publications. He carried out dissections, including vivisection, and initiated a method of injecting substances into the blood as a way of outlining blood vessels. He delighted in what he saw using the microscope. Later he would become one of the world's most famous architects. Along the way he was also, at one time, professor of astronomy.

Lower did the model anatomical dissections and is celebrated as the first person to transfuse blood. It was initially decided to try this on a mad person from the Bethlem Royal Hospital. They chose "a man that is a little franetic, that hath been a kind of minister . . . that is poor and a debauched man, that the College have hired for 20s. to have some sort of the blood of a sheep let into his body." The man was reported to be healthy at the time, but there were debates as to the possible harmful effects on him. The transfusion was carried out in November 1667, apparently without adverse effects, and was repeated on several occasions on the same person. Lower went on to discuss his results, concluding that blood transfusion could not be recommended for many cases, best reserved for "arthritic patients and lunatics, whose bodies are strong and viscera firm, the composition of whose brains is not yet spoilt."[9]

After the Restoration, Willis was a favored man, and he took up the chair of Oxford professor of natural philosophy. He resolved to "unlock the secret places of man's mind. . . . I addicted myself to the opening of heads."[10]

Willis and Lower had done many autopsies, and they used a different method to examine the brain. They took it out of the skull; it became a visual object, an independent organ with separate parts, which could be preserved by methods developed by Boyle. The brain could be cut into slices, viewed under a microscope, and its blood vessels outlined. Willis described the circle of vessels at the base of the brain, which, as they discovered by experimenting with dogs and selective ligation of arteries, allowed blood to get around the brain when part of the blood supply was compromised. Willis was ecstatic, and he thanked the creator: "Certainly there can be nothing more artificial thought upon, and that can better argue the Providence of the great creator, than this fit or convenient disposition of blood in the brain." It really seemed to be an alembic that allowed the spirits to be extracted from the blood to the brain.[11]

His *Cerebri Anatome* (Cerebral anatomy, 1664) is perhaps the most classic neurological text of all time. Willis described his new way of dissecting the brain and its coverings (the meninges), penetrating the interior with his inquiring knife. He made remarks on comparative anatomy, noting differences between the human brain and those of fish, fowl, calves, or sheep. In the book illustrations of animal brains are given alongside those of humans.

Willis was concerned with how the spirits are produced and transformed. He opined that they "are only procreated in the Brain and Cerebel: (which it is easy to prove by the symptoms which happen in the Apoplexy and Palsie, as shall be afterwards clearly shewn) and from this double fountain of the animal spirits they flow out into all the rest of the parts, and irradiate, by a constant influence, the whole nervous stock."[12] The *Choroeides*, hung slack and loosely giving nourishing juices to the spirits, and the white matter tracts allowed animal spirits to travel from one hemisphere to the other. Heat from the blood was preserved by the *Choroeides*, acting as a "Stove or Hot-house," which dilated the little spaces that open up, allowing onward passage of the spirits.[13] The "Turnings and Convolutions" allow a "more plentiful reception of the spirituous aliment, and also for the more commodious dispensation of the animal spirits."[14]

He envisaged the substance of the brain as the repository of imagination, memory, fantasy, and appetite, as well as the seat of the senses and motions. As to the ventricles, they were not the ancients' seat of the animal spirits, but were related to "that vile office of a Jakes or sink . . . in the dead to be filled with water . . . open for excretion . . . a mere sink of excrementitious Humor."[15]

Willis poetically describes the system with the following similes:

Indeed the animal Spirits flowing within the nerves with a living Spring, like Rivers from a perpetual Fountain, do not stagnate or stand still: but sliding forth with a continual course, are ever supplied and kept full with a new influence from the Fountain. In the meantime, the Spirits in the rest of the nervous kind especially those abounding in the Membranes and musculous stock, are like Ponds and Lakes of Water lately diffused from the channels of Rivers, whose waters standing still are not much moved of their own accord; but by being agitated by things cast into them, or by the blasts of winds conceive divers sorts of fluctuations.[16]

As a comparative anatomist, Willis corrected Vesalius and contradicted many of the views of Aristotle and Descartes. He was highly critical of the latter's view on the pineal, a gland he referred to as "the *Anus* or Arse-hole."[17]

He illustrated the stria terminalis and the fornix, suggesting that the latter was linked to the imagination. These are the first descriptions of what later become components of the limbic system.

Although what we refer to as the basal ganglia were outlined in the anatomical plates of Vesalius, it was Willis who gave the first detailed description of the *striate body.* He suggested that they had a role in movement, via animal spirits traversing this area, and noted abnormalities at postmortem in some who had died of long-term motor weakness. He saw many patients with abnormal movement disorders: "some, who had all the muscles and tendons through their whole body afflicted with contractions and leapings without intermission . . . others whose thighs arms and other members, were perpetually forced into various bendings and distortions. And also others I have seen, who of necessity were compelled to leap and run up and down, and to beat the ground with their feet, and hands, and if they did it not, they fell into cruel convulsions." One girl, with very violent motions, had a "bemoaning and very noisful sobbing"—perhaps yet another early description of Gilles de la Tourette syndrome?[18]

Willis was one of the first physicians to link pathology in death to signs and symptoms in life. He had extensive clinical experience and was mainly interested in disorders of the brain which included mental illness. According to Ida Macalpine and Richard Hunter, he gave the most extensive account of mental disorders which had appeared to his time, and he recognized that delusions and hallucinations associated with madness were caused by faults in the brain, thus first clearly aligning abnormal mental experiences with brain disease.

Willis became one of the richest and most sought-after physicians in the land, and he attended the royal family of Charles I. He took a part share in an inn in Oxford, which he turned into a hospital, applying iatrochemical methods of treatment. He gave the first modern descriptions of migraines, including noting a change of appetite prior to an attack, and suggested that the disorder was due to alteration of the blood vessels in the head. He was the first to note the sweet taste of diabetic urine. He examined the brains of those referred to as stupid or foolish, noting that this sometimes ran in families. In one case he described "a certain Youth that was foolish from his birth," whose brain was illustrated in *Cerebri Anatome* and shown as smaller than usual and misshapen.[19]

Willis was very interested in convulsions. He saw many patients with "fits," some of whom died, others recovering, and some who developed stupidity. In his book on hysteria (*Affectionum quae Dicuntur Hystericae et Hypochondriachae*, 1670), he attacked the long-standing uterine theories and placed the cause in "the brain and Nervous Stock." With empirical observation he noted that the uterus was in its correct place in a woman who died of "mother fits," and he recorded attacks in "maids before their ripe age also in old women after their flowers have left them." Perhaps of more importance was his recognition that males could be affected by the same symptoms and signs, even if women were more likely to suffer.[20]

Willis recognized that epilepsy could be symptomatic and idiopathic; he described the sensation of auras (a breeze) and emphasized the cerebral origin of the condition, as with hysteria. The animal spirits would explode, and while in epilepsy this occurred in the middle of the brain, in hysteria this was at the beginning of the nerves in the head.

His interest in mental states gave a new meaning to the word *"psycheology"* as the study of "Nature and essence . . . parts, powers, and affections of the Corporeal Soul." He introduced the concept of "nerves" into our lexicon, later leading to the characterization of nervous disorders.[21]

Willis described melancholy as a distemper of the brain and the heart, noting that in some patients the course could be toward stupidity or even madness. The afflicted could lie sick with imaginary diseases, talk idly, and be sad and fearful. Treatment required withdrawal from troublesome passions and attempting activities such as jesting, singing, dancing, hunting, and fishing; traveling could also help. He had no time for humoral theories of causation, invoking instead alteration of the spirits.

By *madness* he was referring to a condition in which fantasies and imaginations were "busied with a storm of impetuous thoughts" and "their Notions or conceptions are either incongruous or represented to them under a false or erroneous image": "to their Delirium is most often joined Audaciousness and Fury." In such states the brain and animal spirits were likened to "an open burning or flame"—true mania.[22]

Robert Martensen implies a hidden agenda to Willis's conception of the brain, linked to his royalist outlook and his social position. In 1660, the year of the restoration of the monarchy, Willis received his medical doctorate, with a recommendation from the archbishop of Canterbury, along with his appointment as Sedlian professor of natural philosophy at Christ Church, Oxford. Willis considered that there was a hierarchy embedded in the brain's organization. The cerebellum ("Cerebel") was under the control of the cerebral hemispheres; the rational soul was supreme. He used both military and political metaphors, diseases of the soul being akin to civil wars. Willis's model of the cerebral body was "constructed as it was to emphasise the supremacy of order and reason over passionate enthusiasm." Such hierarchies interlinked with brain anatomy will appear again in the theories of anatomically interested neuroscientists.[23]

Although his anatomical studies showed similarities between the cerebral anatomies of animals and humans, Willis believed in a rational soul, which was placed in the brain by God and survived death. This resided in the corpus callosum and was distinguished from the sensitive soul since it dealt with higher things "related to God, Angels, Itself, and eternity."[24]

Sydenham's Career

Almost contemporary with Willis was his school friend Thomas Sydenham (1624–1689). Called by some the modern Hippocrates, he was not impressed by the formulations of the iatromechanists, the iatrochemists, or the royalists, and he adopted Hippocratic ideals of careful observation and follow-up of patients. He believed that diseases should be classified according to the method of the botanists, introduced by Carl Linnaeus (1707–1778). Diseases had their own identities, and their signs, symptoms, and course could be noted. All diseases could be reduced to species, being naturally occurring classes. He clearly distinguished Saint Vitus's dance (Sydenham's chorea) from the dancing manias of the Middle Ages. Regarding causation, he was interested in proximate causes, for which evidence was certain, whereas remote causes were simply incomprehensible.

Hysteria was one of the most common disorders in his extensive practice, accounting for one-sixth of his patients; he stated that no chronic disease was so frequent. Hysteria was the great mimic, resembling most disorders that afflicted mankind. He recognized convulsions like those of epilepsy, but for him the diagnosis of hysteria was clinched by the associated disturbances of the mind, which included incurable despair and depression. Interestingly, anticipating later theories, Sydenham discussed the personality of those with hysteria. The features included emotional inconstancy, extravagant feelings for others which change without cause, constantly changing their minds, and starting things without finishing them. He recognized the links to life's misfortunes and grief, and he considered hypochondriacal complaints in men as equivalent to female hysteria. He still used the concept of animal spirits in causation, but they rushed down the various organs of the body, creating symptoms in whichever part they ended up. This equated to "a disturbance of nervous energy."[25]

Locke's Lock

A close friend of Sydenham was John Locke. He had been a student at Willis's lectures and made extensive notes of their content. It is in part through him that we have much of our information about Willis's ideas. Locke was a physician, but he is best known as one of the most important philosophers of the Enlightenment. Along with Sydenham, he helped sculpture empiricism, the philosophy that knowledge comes only through the senses, quite opposed to the theories of rationalism, which considered that certain knowledge came only from reason. Locke's debt to Willis in terms of his attention to matters cerebral is important, but unlike Willis, Locke thought that it was not possible to understand the workings of the mind, and he was much less bound to theology.

Locke's *An Essay Concerning Human Understanding* was published in 1690, and in it he developed his view that the mind was a *tabula rasa* at birth, a white paper void of "ideas," a theory he developed while attending Willis's lectures, and which acknowledges Descartes. All knowledge is founded on experience. The contents of our consciousness are derived from sense impressions, as simple impressions combine to make more complex ones. Locke's concept of the mind was that it required a passive vehicle in perception, but it possessed active properties, as "ideas" are combined together, forming complex ideas. The mind directly intuits the mind-independent empirical world, complex ideas matching up unerringly with the world. This was a philosophy of an association of ideas, ideas keeping company as if they were one, but guided by reason, the judge and guide of everything: ideas are "what the mind perceives in itself." Knowledge is mediated through ideas. Our minds are a link between us and external reality, and words are outward markers for these ideas. However, as words are arbitrary, words themselves in Locke's philosophy lose their power; they are labels, not essences.[26]

More on Nervous Disease

Hysteria and hypochondrias were quite in vogue, referred to as the English malady of George Cheyne (1671–1743). Writing from personal experience, Cheyne gave his book the full title *The English Malady; or, A Treatise of Nervous Diseases of All Kinds, as Spleen, Vapours, Lowness of Spirits, Hypochondriacal and Hysterical Distempers* (1734). He attributed these conditions in part to the English weather, the nature of English soil, and the richness and heaviness of English food.[27] He considered that these disorders afflicted one-third of the English population. They were associated with inactivity and sedentary occupations, especially for those living in populous and unhealthy towns. They afflicted neither the fools nor the stupid, but rather were the burden of the more intelligent mind. In considering the main title, he suggested that our Continental neighbors threw such brickbats upon us, and he could not disagree. But he referred to these conditions as due to a "distemper'd brain" and thought them as much a bodily affliction as smallpox or fever. Friend of Newton and physician to the poet Alex-ander Pope, he confessed how difficult it was to actually tell his patients that they had such a diagnosis, as it suggested some kind of disgrace. He feared reprisal from the patient or their family for making such diagnoses—*plus ça change, plus c'est la même chose.*

In Scotland, mainly circulating around Edinburgh, the so-called Scot-tish Enlightenment emerged, which included medical men such as William Cullen (1710–1790). His predecessor as professor of medicine at Edinburgh was Robert Whytt (1714–1766). Whytt is most remembered in neuroscience for his observations on spinal cord reflexes, which he referred to as involun-tary motions, and the discovery of the pupillary light reflex (Whytt's reflex). His book *Observations on the Nature, Causes and Cure of Those Disorders Which Have Been Commonly Called Nervous, Hypochondriac or Hysteric, to which Are Prefixed Some Remarks on the Sympathy of the Nerves* (1765) empha-sized the word *nervous*, which led Cullen to introduce the term *neurosis* and the adjective *neurotic*. These remained respectable psychiatric rubrics until the latter part of the twentieth century, but even today they have their uses.

During these times, a more consolidated view of "nervous disorders" was thus evolving, even if speculations as to the causes of them were quite varied. Cheyne tried to integrate iatrochemistry with iatromechanical hypotheses. Whytt considered "sympathy" between different organs to be important: "However, those disorders may, peculiarly, deserve the name of *nervous* . . . on account of an unusual delicacy, or unnatural state of the nerves . . . in consequence of this, various sensations, motions, and changes are produced in the body."[28] He acknowledged that "hysteric" convulsions were related to "violent affections of the mind," and, along with Sydenham, he thought that "a sudden and great flux of pale urine . . . (was) the patho-gnomic sign of hypochondriac and hysterical disease."[29] Uterine epithets were dwindling, and reference to the "vapours" and involvement of the brain in these afflictions became much more the focus of attention.

Classification was coming to the fore. Cullen was one of the great medi-cal systematizers of the eighteenth century. His classification of diseases separated the general from the local, one of the former being the "neuroses"—alterations of sense and motion, without pyrexia. There were four subdi-visions, namely, *Comata* (reduced voluntary movement, including apo-plexy, paralysis, and unconsciousness), *Adynamiae* (reduced involuntary movement), *Spasmi* (abnormal movement of muscles, including epilepsy and hysteria), and *Vesaniae* (altered judgment without coma or pyrexia).

These orders were then further subdivided into *genera* and *species*. Of course, like the compilers of today's classifications of mental disorders, he had little idea of causes, with only speculation to go on, but the attempt to confine diseases, including mental disorders, to rigid Platonic ideals had begun. As Macalpine and Hunter reflected, over time, Cullen's neurosis dissolved into many distinct neurological and systemic conditions, but the term remained intact until 100 years later, when the prefix *psycho-* was added.

But Cullen added more. In the *Vesaniae*, different parts of the brain were in unequal states of excitement: "as our reasoning or intellectual operations always require the orderly and exact recollection of memory or associated ideas; so, if any part of the brain is not excited, or not excitable, that recollection cannot properly take place, while at the same time other parts of the brain, more excited and excitable, may give false perceptions associations and judgements."[30] He also used the concept of irritability and essentially gave us a physiological approach to these disorders, with the brain as central and different brain areas being differently affected in different states.

The Enlightened Mind

The Royal Society was founded in November 1660, at Gresham College, where Wren was the Gresham professor of astronomy. It was an attempt to repeat the Oxford Experimental Philosophy Club, and among the first fellows were Wren, Wilkins, Willis, Hooke, and Boyle. The purpose was to found "a College for the Promoting of Physico-Mathematicall Experimentall Learning." The members were supposed to dissociate themselves from alchemy and devote themselves to the investigation of natural phenomena in the light of reason, unfolding the glories of God's work. Based on Baconian principles, they believed that collective information and data were better than those derived from one individual.

Not only did the Enlightenment period lead to new facts being discovered, but old facts were looked upon with a new light. Did Descartes really shatter all prospects of building a bridge between the brain and the mind? After all, for him the soul inhabited the body, and there was a conjunction at a cerebral location. His philosophy may be denoted as metaphysical dualism, a conundrum that goes back to Plato and his ideas of the soul. Recall that Plato divided the sensible, material world from the nonsensible transcendent realm of forms—perfect ideas that are somehow unchanging and reveal the true essence of things, unlike the world we apprehend through our senses, which is one of mere image. But Plato in his scheme of things had tried to reconcile human feelings with the intellect, and in doing so he came up with a soul divided into three parts.[31] Such trinities recur throughout Western history and even in neuroscience (see later), but Descartes's perspective was different. His was a dualism that separated mind from body. The latter was divisible, but the mind utterly indivisible—there was only one soul. Whatever his intentions, this split has bedevilled neurology, neuroscience, and philosophy ever since, and in the twentieth century one legacy of Descartes's dualism has been to encourage the divisions between neurology and psychiatry.

However, the empirical writings of Newton, Locke, and others had a considerable influence in France, exemplified by publication of the *Encyclopedia (Encyclopédie)*, whose chief editor was Denis Diderot (1713–1784). This eventually ran to 35 volumes. It professed to contain all human knowledge and challenged the intellectual authority of the church and the Bible. Along with the works of other skeptical Enlightenment thinkers, such as Voltaire (1694–1778), Diderot's project influenced ideas that led up to the French Revolution of the late eighteenth century.

The philosophy and psychology of the physician David Hartley (1705–1757) were important stepping-stones linking empiricism to what was known about the anatomy and physiology of the brain. Although he was fundamentally Christian and at one time in his earlier life considered becoming ordained, his book *Observations on Man* (1749) discussed an issue that had been the subject of much debate for many years, namely, whether or not the nerves were hollow. Willis, among others, had contended that they were; this was necessary to allow for the conduction of animal spirits from the brain. Hartley noted that attempts to inject nerves with fluids had failed, and his preferred model was that the nerves were filled with small particles or molecules. Vibrations of light sent waves along the nerves to the brain, which left "vibratiuncles," lasting traces that became the basis of the association of ideas, memory, and consciousness.[32] Habit developed when, through association, complex combinations developed, linked to pleasure or pain. Further, complex actions became more automatic with repetition, so that the mind could attend to other matters. Since mankind was divinely created for spiritual enlightenment, in development, pleasure would outweigh pain, the intellectual would outweigh the sensual, thus providing the path to a higher station of life, or, as he put it, "Association . . . has a Tendency to reduce the State of those who have eaten of the Tree of the Knowledge of Good and Evil, back again to a paradisiacal one."[33]

In Hartley's scheme, emotion was embedded into the individual's development of associations, a bold attempt to put forward views of the embodiment of thoughts and feelings, albeit within the penumbra of a Christian theology.[34] His ideas, when shaven of their religious embroidery, formed the basis of associationism in psychology and can be followed through to the twentieth-century schools of behaviorism. Along with Locke, he proposed that our self-conceptions, as well as the ideas we have about the self of others, arise in our early development and have a materialistic basis. Psychology by now had become a discipline of science, not theology.

Royal Madness

One consequence was that ideas about mental illnesses were moving away from a bodily affliction (iatrochemistry) and toward psychology. Publication of detailed case histories relating to states of mind proliferated in the late eighteenth century.[35] Private madhouses opened up, allowing for observation of patients by attendant physicians, and so-called moral treatment of mental illness, with attention to the mind and with much less emphasis on physical and chemical remedies, became popular. The mental illness of George III (1738–1820) captured the public imagination.

The Madness of King George, the play written by Alan Bennett and made into a film directed by Nicholas Hytner (1994), revisited a remarkable reign in English history. Grandfather to Queen Victoria, George III ascended to the English throne in 1760 and had his first bout of madness when he was 50 years old. The symptoms were intermittent, lasting weeks to months, with recovery between attacks. There were probably only four serious episodes, and his later long-term illness has been attributed to dementia.

There has been a considerable debate as to the cause of his recurring mental illness. Hunter and Macalpine championed a diagnosis of porphyria, while others considered him to have a manic-depressive disorder. The whole story brought to life Dr. Francis Willis (1718–1807), unrelated to Thomas, and raised all sorts of political and constitutional questions. Willis, as opposed to some of the other attendants, gave the king a positive prognosis. His prognosis proved correct, at least in the short term, but he had to use a straight-jacket (properly called a waistcoat) at times to control the king. Although he was aggrieved that he was never awarded the pension of £1,500 a year and baronetcy he had been promised, Willis became a national hero, and his private practice increased in size such that he had to open a second private madhouse where he practiced in Lincolnshire.[36]

Hunter and Macalpine diagnosed the illness as porphyria based on the symptoms, especially abdominal and cephalic pains, painful weakness of the arms, muscle wasting, and signs of autonomic system abnormalities. But the crux of the story was the color of the King's urine, which had been recorded on six occasions as being discolored ("dark . . . bilious . . . bloody"), and on one occasion was observed to leave a bluish stain on the vessel it had been standing in.[37]

The illness of the king highlighted the treatments of insanity in the public eye, the whole calamity lifting some of the veils off the "mad business" and the private "madhouses." The opening of several private asylums had led to exposure of cases of abuse in the diagnosis of insanity and compulsory detention of the afflicted—a useful way of disposing of troublesome relatives— opening debate about treatment of the afflicted. To isolate and blister the king was one thing, but to bind him in a straightjacket was quite shocking. Willis made the point that his treatment of the king was the same as of any other patient, even though he realized that the queen and others were most offended. For "as death makes no distinction in his visits between the poor man's hut and the prince's palace, so insanity is equally impartial in her dealings with her subjects."[38]

There have been endless arguments as to whether or not such physical treatments were inherently cruel. There is a constant tendency throughout medicine for later generations to scorn the treatments of the past (with a constant innuendo of "how could they have been so stupid . . ."; treatment of madness has been subjected to an overexposure in this regard), but Willis had no other desire than to see the king cured, and he had to use methods that were available at his time—he did his best.

The private madhouses developed alongside workhouses, but these were only a part of the developing interest in mental illness. Bethlem Royal Hospital had been at a site in Moorfields, London, since 1676, in a building designed by Hooke.[39] St. Luke's Hospital in London, exclusively for the mentally ill, was founded in 1751, and several other similarly dedicated institutions opened nationally. Guy's Hospital in London opened a ward for lunatics in 1797.

All this led to the need for hospital managers, as well as a growing medical interest in insanity as a professional discipline. Hospital medical directors, the early alienists or asylum doctors, observed, recorded, managed, and later researched as best they could the causes of mental illness.[40]

One outcome was a revolution in the management of patients, the main protagonists in this regard being Samuel Tuke (1784–1857) in England and

Philippe Pinel (1745–1826) in France. The York Retreat was opened by Tuke in 1796, and his *Description of the Retreat, an Institution near York for Insane Persons of the Society of Friends* (1813) alarmed the governors of the nearby York Asylum. Tuke's rule at the retreat was that "neither chains nor corporal punishment are tolerated, on any pretext in this establishment, patients were encouraged to become a rational self again"; judicious kindness was the centerpiece of what was referred to as "moral treatment."[41] According to Hunter and Macalpine, Tuke's ideas "fired the imagination of reformers the world over."[42]

Tuke reported that people with epilepsy were admitted to the retreat, but these he conceded were incurable, and he noted that such patients were not admitted to some other hospitals for mental illness, such as St. Luke's. Bethlem Royal Hospital initially discharged patients who had not improved after 12 months, but in the 1720s two additional wings were built for those who remained incurable, one for men and the other for women.

Pinel, who was at the Bicêtre and the Salpêtrière in Paris, also changed treatment practices.[43] Much romanticizing has been done about his removing the chains from patients at the Bicêtre and setting them free, which some have argued was less fact and more fable. It is suggested that it was his mentor and supervisor at the Bicêtre, Jean-Baptiste Pussin (1745–1811), who actually eliminated the chains, replacing them with straightjackets. Pinel, apparently following on from this, did remove the chains from women at Salpêtrière in 1800, an act celebrated in a famous painting by Tony Robert-Fleury (1837–1912).

Pinel was an acute observer of mental illness. His *Traité medico-philosophique sur l'aliénation mentale, ou la manie* was published in 1801 and translated into English as *A Treatise on Insanity* in 1806. He distinguished periodic or intermittent disorders (melancholia and mania) from dementia and idiotism. He did not believe that insanity was incurable, or that it was always associated with organic brain lesions. However, he undertook comparisons of skull sizes in different presentations and considered possible physiological substrates of illness.

He placed people with epilepsy in a separate group, not because they necessarily required special treatment, but because the sight of seizures might upset other patients. "Few objects are found to inspire so much horror and repugnance amongst maniacs in general, than the sight of epileptic fits. They either retire from the scene greatly terrified, or go up to the patient in a violent passion, and if not prevented assault him with furious and fatal blows. . . . The mere sight of epileptic convulsions has excited similar convulsive affections in spectators of acute sensibility . . . mania accompanied with epilepsy is almost always an incurable malady."[44]

Pinel accepted that restraint was sometimes necessary, but he wanted humane treatments, part of the reforms that became associated with the French Revolution. The Revolution was a defining cultural event, which placed Pinel at the forefront of Paris medicine, but also inspired literature on "revolutionary" madness. He considered the profound effects of the bloody events and personal desolation on the mental state, political agitation adding to the numbers of those developing paranoia, delusions, and even simulated insanity. Some commentators went further, implicating the insane in promoting the actions of the agitators. Since, beginning in 1790, people in France who were to be compulsorily detained for mental illness had to have their state assessed by doctors, psychiatry gained an important political foothold.

Pinel was the "father" of a French school of psychiatry, which was empiricist in orientation and spread its influence well into the nineteenth century, as Paris became a leading center of world medicine.

Another French Enlightenment doctor who commented on epilepsy was Samuel August Tissot (1728–1797), who published his *Traité de l'épilepsie* (Treatise on epilepsy) in 1770. He developed the theme that epilepsy was caused by masturbation, and in another book, boldly entitled *L'onanisme* (On masturbation, 1782), he wrote on the severe consequences of the habit. Such ideas had already been around for 150 years, leading to suggestions that clitoridectomy or orchidectomy could be treatments for severe cases. Tissot also considered classification. At this time there were two broad classes of epilepsy discussed, idiopathic and sympathetic. The former originated in the brain, the latter in sympathy from some other part of the body. Tissot used the term *essential* epilepsy for those cases where no definite cause was discovered, as well as the distinguishing terms *grand accès* and *petits accès*, although the term *absence* would not be introduced until Louis Florentin Calmeil (1798–1895) used it in 1814.

Others who pursued classification included Vincenzo Chiarugi (1759–1820) in Florence, another supporter of treatment that employed restraints only when necessary. He carried out brain dissections and, although adhering to the four temperaments, gave a tripartite classification of psychopathology, citing melancholia, mania, and amentia, but with many subcategories.

Such attempts to classify disorders and the growing interest in examining brains of deceased patients were part of the Enlightenment enterprise. But the newer humanistic methods of managing patients were radical. One commentator summed up these developments as follows: "From the time when Pinel obtained the permission of the Couthon to try the humane experiment of releasing from fetters some of the insane citizens chained to the dungeon walls of the Bicêtre to the date when Conolly announced, that in the vast Asylum over which he presided, mechanical restraint in the treatment of the insane had been entirely abandoned, and superseded by moral influence, a new school of special medicine has been gradually forming." Several centers in Europe followed suit, abandoning the more extreme forms of restraint, such as using chains and shackles, and emphasizing the humanitarian approach to mental disorders.[45]

An Enlightened Age?

Anthony Pagden, in his book *The Enlightenment and Why It Still Matters*, affirmed that the Enlightenment was an "exclusively European phenomenon, shared only with Europe's overseas settler populations, it could never have arisen except in a broadly Christian world."[46] The Enlightenment was of profound importance for the development of Western civilization, yet the philosopher Immanuel Kant (1724–1804) noted, "When we ask, do we live at present in an enlightened age? The answer is: No, but in an age of enlightenment."[47] The "enlightenment" itself had not arrived. Philosophy was separating from medicine, and medicine from theology, but many wavered. Willis was unable to reveal how his two souls interacted. Early in life he probably intended to become an Anglican clergyman, and he was later buried in Westminster Abbey. He believed that God gave man his rational soul. The great Isaac Newton (1643–1727) conducted many alchemical experiments and believed that he was working for the glory of God. He was convinced that the Bible was a sacred text, and he read it assiduously every day; his discoveries in religion and mathematics were complimentary. Locke considered that Newton's mind was deranged.[48] Galileo, who rejected Aristotle, proved Copernicus correct, and with the telescope he saw things others had never envisaged, but he seems to have believed in Dante's descriptions of hell in *Inferno*. He set about estimating the height of Lucifer, which he gave as equivalent to 2,000 yards; hell, he calculated, was 1/12 the size of the earth.[49] Locke believed that man was created by God, in His image, endowed with reason. Many seventeenth-century philosophers and scientists were concerned with both first and second causes. The first still had to remain with theology, but the second was the activity of science. John Donne (1572–1631) lamented,

> The element of fire is quite put out;
> The sun is lost, and th'earth, and no man's wit
> Can well direct him where to look for it.
>
> . . . .
>
> 'Tis all in pieces, all coherence gone . . .[50]

The French Revolution was yet to come.

NOTES

1. Eckermann JP, 1998, 426. Johann Peter Eckermann (1792–1854) sought out Goethe in Weimar in 1823, and for almost a decade he became his amanuensis. He was like Samuel Johnson's (1709–1784) James Boswell (1740–1795), except that the heritage of Johnson would be much diminished if it were not for Boswell, whereas by the time Eckermann began to collaborate with Goethe, the latter was aged 74 and already famous. There remains a discussion as to what his last words actually were referring to—perhaps an instruction to his servant to open the curtains, or perhaps a comment on science and philosophy.

2. The Enlightenment broadly refers to a period of the seventeenth and eighteenth centuries, but for neuropsychiatry the era is best encapsulated by the period from around 1650 to the end of the eighteenth century. The Royal Society was founded in 1660.

3. *Passions of the Soul*, Descartes R, 1997, 1–10. See Bennet MR, Hacker PMS, 2003, 27–30. See Sherrington C, 1946, for his discussion of Fernel.

4. It was obvious to his critics that animals other than humans possess a pineal gland, yet apparently did not have souls, and, after all, it was a rather small part of the human anatomy to carry such a huge burden. No one could identify the pores through which the spirits acted. Martensen RL, 2004, suggests that Descartes perhaps was not even familiar with human brains (60).

5. King LS, 1970, 101–102.

6. For example, Giovanni Alfonso Borelli (1608–1679) applied mathematical and physical principles to the understanding of muscle function. The motto was essentially that the whole machine is an assemblage of smaller parts.

7. Dewhurst K, 1981, 26. "Curcurbit"—flute or the lower part of an alembic. "Alembic"—used for distilling chemicals.

8. Zimmer C, 2004, 95.

9. The history is noted by Hunter R, Macalpine I, 1963, 184–186. See also Zimmer C, 209–211.

10. Zimmer C, 174.

11. Zimmer C, 177.

12. Willis T 1681/1971, 55. The 1681 edition was published in 1664; the English translation of Samuel Pordage (1681) is the most quoted.

13. Willis T, 55.

14. Willis T, 59.

15. Willis T, 65, 66–67, 71.

16. Willis T, 106–107.

17. Willis T, 20.

18. Zimmer C, 2004, 105–106.

19. Willis T, 1681/1971, 27–28.

20. Macalpine I, Hunter R, 1969, 190. Hypochondriasis was the male equivalent of hysteria.

21. Macalpine I, Hunter R, 187–189.

22. See Hermsen LM, 2011, 19.

23. Martensen RL, 2004, 118.

24. Zimmer C, 2004, 222. It should be noted that by "corpus callosum" Willis was referring to the whole of the white matter, not the more well-defined structure abridging the two hemispheres for which the anatomical term is used today.

25. Payne JF, 1900, 144.

26. Most of this discussion is to be found in Locke J, 1690, bk. 2, chaps. 2, 8, and 11. The quotation is chap. 8, para. 8. There was and still remains, at least from a neuroscience perspective, what meanings can be attached to words like *ideas, images, objects of thought, representation*, and a host of other metaphors that are commonly used by philosophers and neuroscientists.

27. Cheyne gives the last chapter over to a full history of his own ailments, as well as treatments. At one point he reached a weight of 32 stone (448 pounds). The first edition was in 1733.

28. Whytt R, 1768/1984, qqq (between pp. 486 and 489).

29. Whytt, R, 602, 595.

30. Hunter R, Macalpine I, 1963, 473–475.

31. The arguments are laid out in *Phaedrus*; see Plato, 2004, 385–401. The way forward to the truth was by being a philosopher. His metaphor was a chariot with two horses and a charioteer. One horse was good, the other bad, one being white, upright, and handsome (on the right—courageous, virtuous), and the other being black, crooked, and lumbering (on the left—the appetitive). The soul in his philosophy was nourished by the divine and preexisted the body, but imbalance between the steeds caused the charioteer (seeker of wisdom) to have to reign in the equine imbalance to achieve harmony. This emancipated the virtuous elements of the soul, yearning to leave its bodily prison and find its heavenly abode.

32. As Chris Smith explains, "Whenever vibration A is set up, it radiates through the medullary substance, gradually dying away into vibratiuncles. But if, before it totally expires, it meets another active vibrating region, region B, then it will, provided it is not too attenuated, modify that vibration, and itself be modified. When this has occurred sufficiently often, the substance of the medulla in region B will become adapted, modified to respond to A. Hence whenever A reaches it, it will resonate—feebly: hence once again vibratiuncle b is induced, or idea b." Smith CUM, 1989, 54.

33. Porter R, 2003, 353.

34. The full title of Hartley's book is *Observations on Man, His Frame, His Duty, and His Expectations*. A brilliant discussion of Hartley's works and the importance of his writings can be found in Porter R, 347–373.

35. Porter R, 305–309.

36. In later episodes the king was treated by Willis's sons.

37. There are a collection of disorders called the porphyrias, inherited disorders that involve altered metabolism of porphyrin. In the condition, the formation and excretion of porphyrins is increased, causing the CNS signs and the discoloration of the urine. Macalpine I,

Hunter R, 1969, 172–175. The authors point out that the illness and its possible causes, with hereditary implications, were hardly discussed after the king's death, and the journals of the Willises which are in the British Museum did not become available until 1959. He ended his life quite insane, living in seclusion in Windsor Castle, by then having dementia, which Hunter and Macalpine suggest was an unrelated disorder, but which clinical picture led to confusion over reaching the diagnosis.

38. Macalpine I, Hunter R, 1969, 281.

39. The building was so splendid with its north-facing appearance across the open fields of Moorfields. The design looked to French architecture. It became a tourist attraction, and notoriously visitors could pay a small fee to go inside and look at patients. This was stopped in 1770.

40. For a discussion see, e.g., Scull AT, 1979. "Alienist" is from the French *aliéné*.

41. Tuke S, 1813, 141.

42. Hunter R, Macalpine I, 1963, 687.

43. The Bicêtre was for males and the Salpêtrière for females. The latter had both curable and incurable patients, which boosted Pinel's enthusiasm for research on mental illness.

44. Pinel P, 1806, 204.

45. Editorial, 1853, vol. 1, 1. Georges Auguste Couthon (1755–1794), a French politician and lawyer, served as a deputy on the Legislative Assembly during the French Revolution. He was responsible for drafting a law that led to a vast increase in the number of counterrevolutionaries being guillotined. John Conolly (1794–1866) was a physician at the Middlesex County Lunatic Asylum in Hanwell. He advocated that all qualified doctors should be as familiar with disorders of the mind as with other disorders, and his book *The Treatment of the Insane without Mechanical Restraints* (1856) "marked the successful conclusion of a movement which commenced with Pinel and which created a new epoch in the lives of the insane and a new approach to insanity not only in the British Isles but throughout the civilized world." Hunter R, Macalpine I, 1963, 1031.

A note on Richard Hunter (1923–1981). Richard was one of my colleagues, holding a joint appointment at the National Hospitals, Queen Square, and at the Friern Hospital, Middlesex, originally the Colney Hatch Asylum. The latter was in a large Victorian building, regarded as England's largest and most modern asylum when it opened in 1851. Richard was an inveterate seeker after organic lesions in psychiatric patients but an upholder of the principles advocated by Conolly, the Hanwell asylum being close by. He was quite against the use of physical methods of treatment, even psychotropic drugs, and believed literally in asylum. He carefully documented patients' signs and symptoms and accumulated a large number of brains from patients who had died. His book *Psychiatry for the Poor* (Dawsons of Pall Mall, 1974), written with his mother Ida Macalpine, who was a psychotherapist, described in detail the hospital, its history, and the diseases and autopsies he investigated. With his mother he wrote many books on the history of psychiatry, their most important legacy being *Three Hundred Years of Psychiatry, 1535–1860*. He was a most keen collector of old books and had a vast library from which he could source their work.

46. Pagden A, 2013, 83.

47. Kant I, 2001, 140.

48. See, e.g., White M, 1997; Keynes M, 1980. Of Newton, White wrote, "Religious zeal was one of his prime motivators, and, like almost all other thinkers and uneducated folk of his day, he lived and died believing wholeheartedly in divine guidance. . . . He never swayed from his assertion that God was responsible for maintaining planetary motion through the device of gravity" (White M, 1997, 149). He was considered a solitary figure, with occult ideas, who had episodes of depression.

49. See White M, 2007, 55–56. Lucifer was 43 times larger than a statue of a giant in St. Peter's Square, Rome, and this statue was 43 times taller than Dante.

50. John Donne, *The First Anniversary*, lines 206–209, 213.

REFERENCES

Bennet MR, Hacker PMS. *Philosophical Foundations of Neuroscience*. Blackwell, Oxford; 2003.

Cheyne G. *The English Malady; or, A Treatise of Nervous Diseases of All Kinds, as Spleen, Vapours, Lowness of Spirits, Hypochondriacal and Hysterical Distempers*. London, 1734.

Descartes R. *Key Philosophical Writings*. Trans. Haldane ES, Ross GRT. Wordsworth Classics of World Literature, Wordsworth Editions, London; 1997.

Dewhurst K. *Willis Oxford Casebook (1650–1652)*. Sandford, Oxford; 1981.

Eckermann JP. *Conversations with Goethe*. Trans. Oxenford J. Da Capo Press, New York; 1998.

Editorial, *Asylum Journal*, 1853.

Hermsen LM. *Manic Minds: Mania's Mad History and Its Neurofuture*. Rutgers University Press, New Brunswick, NJ; 2011.

Hunter R, Macalpine I. *Three Hundred Years of Psychiatry, 1535–1860*. Oxford University Press, London; 1963.

Kant I. *Basic Writings of Kant*. Ed. Wood AW. Modern Library Classics, New York; 2001.

Keynes M. Sir Isaac Newton and his madness of 1692–93. *Lancet* 1980;1(8167):529–530.

King LS. *The Road to Medical Enlightenment: 1650–1695*. MacDonald, London; 1970.

Locke J. *An Essay Concerning Humane Understanding*. Basset, London; 1690.

Macalpine I, Hunter R. *George III and the Mad Business*. Pimlico, London; 1969.

Martensen RL. *The Brain Takes Shape: An Early History*. Oxford University Press, Oxford; 2004.

Pagden A. *The Enlightenment and Why It Still Matters*. Oxford University Press, Oxford; 2013.

Payne, JF. *Thomas Sydenham*. T. Fisher Unwin, London; 1900.

Pinel P. *A Treatise on Insanity*. Trans. Davis DD. W. Todd, Sheffield; 1806.

Plato. *Essential Thinkers*. CRW, London; 2004.

Porter R. *Flesh in the Age of Reason*. Penguin Group, London; 2003.

Scull AT. *Museums of Madness*. Allen Lane, London; 1979.

Sherrington C. *The Endeavour of Jean Fernel*. Cambridge University Press, Cambridge; 1946.

Smith CUM. Neurology and mental atomism: Some continuities and discontinuities. In: Rose C, Smith F, eds. *Neuroscience across the Centuries*. Gordon, London; 1989.

Tuke S. *Description of the Retreat, an Institution near York for Insane Persons of the Society of Friends*. Alexander, Bristol; 1813.

White M. *Isaac Newton: The Last Sorcerer*. Addison-Wesley, Reading, MA; 1997.

White M. *Galileo, Antichrist*. Weidenfeld and Nicholson, London; 2007.

Whytt R. *The Works of Robert Whytt*. The Classics of Neurology and Neurosurgery Library, Birmingham, AL; 1768/1984.

Willis T. *The Anatomy of the Brain*. USV Pharmaceutical Corp., New York; 1681/1971.

Zimmer C. *Soul Made Flesh*. William Heinemann, London; 2004.

CHAPTER FIVE

Romanticism

Early Developments

The mind is its own place, and in itself / Can make a Heaven of Hell, a Hell of Heaven.

John Milton, 1977, *Paradise Lost*, book 1, lines 254–255

The following was written by the poet/philosopher Samuel Taylor Coleridge (1772–1834) in a letter to his neighbor Thomas Poole on November 15, 1796: "Charles Lloyd has been very ill; and his distemper (which may with equal propriety be named either somnambulism, or frightful reverie or epilepsy from accumulated feelings) is alarming. He falls all at once into a kind of night mair, and all realities round him mingle with, and form part of the strange dream. All his voluntary powers are suspended, but he perceives everything and hears everything, and whatever he perceives and hears he perverts into the substance of his delirious vision. He has had two principal fits, and the last has left a feebleness behind and occasional flightiness." The description is nearly perfect for a state of postictal psychosis.

Charles Lloyd (1775–1839), scion of the banking family, around 1793 had the first of a succession of seizures that were to besiege his life and blight his talent. He became fascinated by Coleridge when the latter visited Birmingham in 1796. On account of Charles's illness, his father offered Coleridge a stipend to tutor his son, who had already shown poetic talent, publishing a volume of work in 1795. Part of the arrangement was that Lloyd should live with Coleridge, and from that domiciliary arrangement we have descriptions not only of his seizures but also of his episodes of delirium. Coleridge considered him "a man of great genius" who, along with himself and Charles Lamb (1775–1834), could be one of "three Bards walking up Parnassus." Yet his fate was one of recurrent episodes of postictal psychosis, spending his last days in an asylum in Paris.[1]

Where, when, and why the romantic era came to be is a matter of much speculation, and many people have their favorite contenders. The Enlightenment period may be said to have continued until around 1800, when forces in the arts and sciences provided a seismic shift, in that minds, motivations, and memories of individuals became preoccupations of study. It was the era of personal identity, of heroes and heroines. The very word *romantic* has embedded in it the idea of the novel—*roman* (French). There was a shift away from empirical philosophies toward idealism, with not only ideas seeking holistic as opposed to particulate perspectives on life and society but also a reemphasis on man's attempts to overcome forces of oppression. It was the era of revolutions, notably in America and France, and it engendered the cult of celebrity. The undermining of rationalism came especially from German-speaking countries. We have seen how Luther, who considered reason to be a whore, began the process of liberation from the Catholic oppressions of his time, perhaps the beginning of a mind-set that later flourished into romanticism. The use of printing presses and publishing in demotic languages opened up information to a much wider spectrum of the population, especially the growing middle classes, with a new kind of wealth created by the Industrial Revolution. And by now, the brain had literally come out of the skull, becoming an object of dissection and depiction as never before.

However, before returning to the brain, there is a feature of these times which is often overlooked—the blend between poetry and science. Not only were many of the scientists of the romantic era poets, but some of the poets contributed to our developing understanding of the brain. Milton's superb epic *Paradise Lost*, spanning eons of time and the whole universe, has Satan as the first character to appear. William Blake (1757–1827) concluded that Milton himself was of the devil's party. The poem raises awareness of the striving of Satan's consciousness and issues of free will.[2] Like Oedipus, the blind Milton wanted to tell of things invisible to mortal sight—insight, since the sighted only see surfaces.

Neural Romanticism

In his book *British Romanticism and the Science of the Mind*, Alan Richardson portrays the romantic associations with neuroscience which were a feature of this era. Let us start with Coleridge. Born in 1772, he was deeply religious. His father was a vicar, and Coleridge himself became a convinced Unitarian. He was against war and the slave trade and espoused fundamen-

tal Christian values, but in revolutionary times he had a revolutionary outlook. Suffering from neuralgia, he began to use opium and may have had a
drug-induced episode of delirium, in which he visited Kubla Khan, the land
of milk and honey. When Charles Lloyd was living with him, he observed
firsthand his guest's abnormal mental states brought about by seizures. In
addition to these experiences, he was confronted by the tragic matricide
by the sister of his friend and fellow poet Charles Lamb in a fit of insanity.
Coleridge was the "Mariner" exploring another world of the mysterious
oceans of the mind, where "a thousand thousand slimy things" were found—
the creative forces not only in nature but also within the human mind. "My
mind feels as if it ached to behold and to know something *great*—something
one & indivisible—and it is only in the faith of this that rocks or water-
falls, mountains or caverns give me the sense of sublimity or majesty! With
music 'loud and long, / I would build that dome in the air, / That sunny dome!
Those caves of ice!' "[3]

Dr. Thomas Beddoes (1760–1808) was much concerned with the treatment of tuberculosis and set up the Pneumatic Institute at Hotwells, Bristol,
with a view to using inhaled gases to cure various diseases. Humphry Davy
(1778–1829), a scientist and a poet, was the first superintendent, and he was
experimenting with nitrous oxide, a gas synthesized by the chemist Joseph
Priestley (1733–1804), who also discovered "dephlogisticated air," namely,
oxygen, as well as soda water. The four fundamental elements of earth, air,
fire, and water were deposed and decomposed, presenting new avenues for
understanding the laws of life itself. Davy foresaw that the action of fluids
on solids could explain mental activity, essentially neurochemistry, leading
to an entirely new way of understanding and treating mental disorders.

Davy's own inhalations of nitrous oxide gave him very pleasurable sensations, sometimes leading him to laugh and dance and raising his spirits.
He began experimenting on some of his friends, including the poets Robert
Southey (1774–1883), later to become poet laureate, and Coleridge. The latter
was briefly in Bristol and, with Davy, joined in with inhalation sessions. He
described the delightful effects and was intrigued by the implications of
mind-altering gases. These experiences, as well as his touches of and with
madness noted above, led him to think about and explore further the human
mind and creativity.[4]

Coleridge had been entranced by the plays of Friedrich Schiller (1759–
1805) and the philosophy of Kant. In 1798 he traveled to Germany with
William Wordsworth (1770–1850),[5] learned German, and, by speaking and

translating between German and English, gained much insight into the variable images given by words, their fluidity, and their variable meanings. While he was away, his newborn son Berkeley died of epileptic seizures. His other son was also named after a philosopher, Hartley. Initially enthusiastic about associationism, Coleridge came to realize that he could not square its ideas with his interest in the mind's creative processes. He rather debunked Hartley's hypothetical vibrations and vibratiuncles, noting that they would mean that "our whole life would be divided between the despotism of outward impressions and that of senseless and passive memory."[6] He trashed Newton as a pure materialist, a lazy observer of the external world, always passive.

Coleridge's *Biographia Literaria*, published in 1817, is part autobiography, part poetical criticism, and part philosophy. He was deeply influenced by the philosophy he had read in Germany and could not accept that the mind was a *tabula rasa*; it was for him an active principal with powers to shape the world of individual experience. He invented the word *esemplastic*, meaning "to shape into one": between the active and the passive lay the imagination, that which Coleridge considered gave us a Milton, a Shakespeare, or a Wordsworth.[7] Attempting to unite the subjective and objective, he affirmed the *sum* (I am), but not of the Cartesian dictum. His formula was *sum quia sum* (I am because I affirm myself to be . . . I know myself only through myself . . . self-consciousness).[8] He wrote that "a sort of stomach sensation" was attached to all his thoughts.[9] He introduced the English reader to the terms *unconscious* and *psychosomatic* and to the notion that "thought is *motion*."

A further interesting link between the brain and poetry is to be found in the life and compositions of John Keats (1795–1821). He trained in medicine at Guy's Hospital and was well aware of the effects of medicaments such as opium on the senses. Richardson describes him as the most "visceral" of the English poets and points out that before the nineteenth century the brain was hardly used as an image in poetic verse; if any body organ was eulogized, it was the heart. But "Keats provides some of the most striking and suggestive examples of the newly lyricized image of the brain." Fearful of death, and worried that "When I have fears that I may cease to be / Before my pen has glean'd my teeming brain," he refers often in his poetry to the brain, since "dim-conceived glories of the brain / Bring round the heart an undescribable feud."[10] As the memory, like ancient monuments, fades with the passing of time, the brain—with age and its potential failings—becomes closely bound with mortality. As Percy Bysshe Shelley (1792–1822) lamented, " 'Ozymandias, King of Kings; / Look on my Works, ye Mighty, and despair!' / Nothing beside remains."[11]

These ideas that stimulated the poets' imaginations were reflected in the ideas of some neuroscientists at this time. Richardson discusses the origins of what he calls "neural Romanticism," and the key protagonists are Erasmus Darwin (1731–1802), Franz Gall (1758–1828), Pierre Cabanis (1757–1808), and Charles Bell (1774–1842).

Early Embodiment

Erasmus Darwin had set the ball rolling, at least in England, with his *Zoonomania*. This remarkable book, which was published in two volumes in 1794 and 1796, covered what seems like nearly everything that was known at that time in medicine, including anatomy and physiology, natural science, and links to current philosophies. He made observations on the way images and colors impinged on vision, discussing afterimages, which suggested the brain's active involvement in perception. Further, he noted that "after the amputation, of a foot or a finger, it has frequently happened, that an injury being offered to the stump of the amputated limb, whether from cold air, too great pressure, or other accidents, the patient has complained, of a sensation of pain in the foot or finger, that was cut off." This was further proof for him that "all our ideas are excited in the brain, and not in the organs of sense."[12] He too was critical of associationism and the theories of Locke; for Darwin, ideas were "animal motions of the organs of sense."[13] The mind was embodied, a view that, as we shall see later, has reemerged in neuroscience and philosophy. He was drawn to the idea that in order to survive, an organism had to adjust to its environment, and even if his explanation involved what are now referred to as Lamarckian principles, namely, the inheritance of acquired characteristics, he placed evolution and its importance firmly before his scientific contemporaries. As with several other scholars of his time, Darwin was a poet. The first of his major works, *The Botanic Garden*, was a science poem composed in Augustan couplets.

The following comes from *The Temple of Nature* (1803):

Next the long nerves unite their silver train,
And young SENSATION permeates the brain;
Through each new sense the keen emotions dart,
Flush the young cheek, and swell the throbbing heart.
From pain and pleasure quick VOLITIONS rise,
Lift the strong arm, or point the inquiring eyes;
With Reason's light bewilder'd Man direct,
And right and wrong with balance nice detect.

Last in thick swarms ASSOCIATIONS spring,
Thoughts join to thoughts, to motions motions cling;
Whence in long trains of catenation flow
Imagined joy, and voluntary woe.[14]

This is an assertion that our volition can be spontaneous and unconscious.

Charles Bell's *Idea of a New Anatomy of the Brain* (1811) was an early summary of his ideas of the functions of the brain and spinal cord, based on his own careful dissections. He examined the comparative brain anatomies of man and other animals, and he studied the brains of patients with, for example, dementia. His intention was to unravel the puzzling maze of the anatomical and conceptual confusions that he considered abounded at the time. He is well known for his separation of the peripheral nerves into the posterior sensory and anterior motor divisions, and for the observation that the nerves were not like a string band, but were bundles of single nerves, "distinct in office." Like Darwin, he made observations on alteration of sensory input, for example, by pressing the eyeball or touching parts of the tongue with a pin, and also on phantom limbs. He stated that the cerebrum and cerebellum had different functions, noting that "the operations of the mind are seated in the great mass of the cerebrum." Further, the "cineritious and superficial parts of the brain are the seat of the intellectual functions. . . . All ideas originate in the brain." He "established a connection by which the ideas excited have a permanent correspondence with the qualities of bodies which surround us."[15] The body was not just a convenient residence for the mind, but its citadel.

Wordsworth, in his great autobiographical poem *The Prelude*, referred to the "primordial feelings" that form "our first sympathies" of childhood. He recalled how, as a child, he "held unconscious intercourse" with nature, which was "a pure organic pleasure." This allowed "To those first-born affinities which fit / Our new existence to existing things."[16] The subject/object division is elided:

while my voice proclaims
How exquisitely the individual Mind
(And the progressive powers perhaps no less
Of the whole species) to the external World
Is fitted:—and how exquisitely, too.[17]

Surely, this is where embodiment begins.

Bell examined facial expressions and developed his ideas further in *The Anatomy and Physiology of Expression, as Connected with the Fine Arts* (1806; later expanded and revised), in which he was able, using lithography and the development of new coloring techniques, to depict the effect of emotions on the facial musculature, including studies of animals, work that later was used by Charles Darwin (1809–1882) in his studies of the emotions. Bell was interested to understand how it was that an artist could accurately draw or paint images that affected the viewer. He was an active participant at the Battle of Waterloo, treating the injured, and urged artists to render the differences between a dead body that may have been seen on an anatomical table and the dead of battle: "It may be sometimes necessary to give the rigidity of death to the figure, but more frequently either the convulsive tension of expiring life, or the relaxation of death; as Homer describes his heroes, rolling in death, with limbs relaxed and nerveless."[18] He encouraged painters to move away from the statuaries of antiquity, devoid of "the representation of that minuteness or sharpness of feature, and of those convulsions and distortions which are strictly natural; and indeed it is scarcely consistent with the character of a statue to represent the transitory emotions of violent passion."[19] The painter should avoid too close attention to the "academy figure," and a depth of anatomical knowledge, especially of the muscles and their play when thrown into action, was essential. He saw human expressions and gestures as an unconscious language that influenced our opinions and educated the mind, whose powers were developed through the influence of the body via "its connection with an operation of the features which precedes its own conscious activity"[20]—surely early hints of what later becomes Damasio's somatic marker hypothesis.

Something Else Afoot

Another careful observer of expressions was James Parkinson (1755–1824), a physician to Holly House in Hoxton, a private madhouse. Although shaking palsy was first observed by Galen, and tremors were well described by physicians well before Parkinson, he described the disorder that bears his name in some detail in his *Essay on the Shaking Palsy* (1817). There was "in the present instance, the agitation produced by the peculiar species of tremor, which here occurs, is chosen to furnish the epithet by which this species of Palsy, may be distinguished." The small booklet must be one of the most important observational contributions to a disorder, which seems to have remained largely unnoticed until Jean-Martin Charcot (1825–1893) gave it the title *la maladie de Parkinson* in 1861. Parkinson's observations were made on just eight cases, even if one was seen in the street, and another "at a distance."[21] Aside from excellent descriptions of the tremors and gait abnormalities, Parkinson recorded sleep disturbances, constipation, salivation, and drooling, but he opined that "the senses and intellects (are) uninjured."[22]

Parkinson was a political radical with sympathies for the French Revolution and an amateur geologist with fossils bearing his name. In his role as physician to Holly House, he was involved in agitating for reform of certification of the insane, insisting that this should require medical certification of their lunacy, distinguishing cases of madness from such conditions as epilepsy.

The Coming Together by Falling Apart

Meanwhile on the Continent, especially in France, similar ideas were afoot. Cabanis emphasized the preeminence of the brain as being active and, as famously quoted, an organ that digests information, as the stomach digests food. But it was Gall and his pupil and collaborator the German physician Johann Spurzheim (1776–1832) who have historically made more waves.[23] Gall is most famous for his development of "phrenology," which became a trivialized and much misused concept in the public eye, but which profoundly altered neuroscience. He was interested in differing personalities and intellectual abilities among his colleagues (with a common style of upbringing), and he developed thoroughly materialistic views of the brain's psychological laws—for him psychology was a brain-based discipline. He adduced that his fellow students with prominent eyes had excellent memories, and he proceeded to study relationships between function and anatomical structures in people with various deficiencies or special gifts, from imbeciles and criminals to musicians and poets. He also examined brains of animals, looking at the evolutionary development of brain shapes and sizes, with differing naturalistic skills. He was impressed by the increasing complexity of the cerebral convolutions over time, and in his own dissections, instead of looking at horizontal and vertical brain slices, he dissected along the line of the white matter, showing that it issued from gray matter, and establishing once and for all the decussation of the pyramids.

Gall elaborated five principles: (1) the brain is the organ of the mind, (2) the mind can be analyzed into independent faculties, (3) these are innate and have their seat in the cortex of the brain, (4) the size of each cerebral organ is an indication of its functional capacity, and (5) there is a correspondence between the contour of the skull and the cortex of brains such that the size of the organs and their potential role in the psychological makeup can be determined by inspection.

The cortex was no longer a disorganized mass, but composed of a subset of organs, which could be revealed by the shape of the head. In Gall's original scheme there were 27 organs, which Spurzheim increased to 35. Gall stated that 19 elements of his typology were shared with animals, while others such as wisdom, satire, wit, poetry, kindness, and the like were distinctly human. These individual organs could be subject to damage or disease. As Hunter and Macalpine pointed out, this "provided the first psychological framework within which mad-doctors struggling unguided with their patients could understand insane behaviour, and so gave a powerful fillip to the psychological approach. For the first time it became meaningful to get to know patients as persons, if only to be able to interpret bumps on their heads."[24]

As Richardson makes clear, Gall was perhaps the most romantic of the neuroscientists of this era, seeking to understand a unity of structure and function within diversity, not only from an evolutionary standpoint but also within the individual brain. Influenced by themes from *Naturphilosophie*, he was after the unconscious forces and principles that united the faculties of the human brain.

A related area to phrenology which developed in the early nineteenth century was the physiognomy of the poet-philosopher Johann Kaspar Lavater (1741–1801), that is, the study of the face and its link to personality. This was taken up by artists such as Edgar Degas (1834–1917), who depicted the faces of two young murderers with slanting foreheads, a profile even noted in his quite famous statue of the *Little Fourteen Year Old Dancer* (1880), which some critics simply abhorred for its ugliness and bestial resonances. When George Combe (1788–1858), who approved of Gall's views, traveled in America, he visited the artist Rembrandt Peale (1778–1860), who had painted a portrait of George Washington. Combe commented that "Washington's head as here delineated, is obviously large; the anterior lobe of the brain is large in all directions; the organ of Benevolence is seen to rise, but there the moral organs disappear under his hair."[25]

By the end of his life, Gall had amassed a huge collection of skulls and casts of distinguished people, to which was added, by his own bequest, his own skull; the collection was eventually given to a Paris museum. Perhaps not to be out done, Spurzheim, who died while lecturing in America, left his skull to Harvard Medical School.

Cerebral Advances

The observation on the students with prominent eyes drew attention to the frontal parts of the head. Actually, the relevance of the frontal lobes to human faculties had already been suggested by the Swedish religious mystic Emanuel Swedenborg (1688–1772). An early interest in mathematics gave way in midlife to interests in the spiritual world, but he became paranoid, with religious hallucinations and delusions, ultimately declaring that Jesus Christ had made his second coming through him; he founded the Church of the New Jerusalem. Swedenborg led an eccentric solitary existence, writing extensively, his eight-volume *Arcana Coelestia* (1749–1756) being more than 2 million words long. On account of some of his behavioral features, including an episode when he emerged from self-imposed incarceration with foam around his mouth, he has been considered to have suffered from epilepsy.[26]

He also studied medicine, and he was especially interested in exploring the nervous system to gain a better understanding of the soul. He was influenced by the anatomical studies of the Danish investigator and later bishop Niels Stensen (aka Steno; 1638–1686). The latter's *Lecture on the Anatomy of the Brain* (1669) castigated current anatomists, telling them to revise their methods, and he set out his preferred way to examine the brain. He considered the speculations of Willis and Descartes about the various parts of the brain to be useless, as careful dissection and observation came before, not after, theorizing.

In spite of this advice, Swedenborg seems to have done little experimenting himself, but through extensive traveling and reading, he collated knowledge about the brain. He can be credited with the first anticipation of the idea of a neuron, small elements (*cerebellula*, or "little brains") connected by thin fibers which traveled in white matter and extended all the way down from the brain to the spinal cord. He theorized that different brain areas had different functions, as this allowed the brain to function without confusion; without some organization, there would be chaos. However, his most significant contributions related to the cerebral cortex.

After the contributions of Willis, interest in the functions of the cerebral convolutions had diminished. Investigations by Albrecht von Haller (1708–1777), who was particularly interested in "irritability" as a property of muscle, had shown that the cortex itself was insensitive, for example, to pinching with forceps or to chemical irritation such as with silver nitrate. However, he was not beyond suggesting that epilepsy originated in the brain and that irritability of the medullary part was somehow involved.

Swedenborg divided the cerebrum into three lobes, the highest, middle, and lowest, and noted that after damage to the frontal areas, imagination, thought, and memory suffered and the will was weakened. With this hierarchical CNS organization he identified the area we refer to as the motor cortex as the seat of voluntary movements, while involuntary ones he suggested related to areas such as the basal ganglia, cerebellum, and spinal cord. His conception that the cortex of the brain was the prime substance for not only sensation and movement but also the will and understanding was a radical shift in thinking. He can even be attributed with an inkling of the idea of somatotopic localization, since with regard to motor areas he stated that the foot area was localized dorsally, the trunk intermediary, and the face ventrally. This anticipated the findings of Gustav Theodor Fritsch (1838–1927) and Eduard Hitzig (1839–1907) by 100 years. However, his attention to the effects of damage to the anterior cortex anticipated a much later and current interest in the frontal lobes. It is of interest that this remarkable man with such prescient ideas for neurology was, and still is, largely ignored for his contributions, with his religious writings continuing to have prominence.[27]

Others were on the trail. The surgeon Sir William Lawrence (1783–1867), who was on the staff at Bethlem Royal Hospital, as well as at St. Bartholomew's Hospital, argued, in a book published in 1819, that function followed form, that there was no subtle or invisible matter animating living bodies, since "*physiologically* speaking, the grand prerogative of man" was merely the expression of "the functions of the brain." He suggested that "a Newton or a Shakespeare excels other mortals only by a more ample development of the anterior cerebral lobes, by having an extra inch of brain in the right place." He considered mental disorders to be the consequence of a diseased brain.[28]

As Richardson pointed out, those seeking a "vital" force that distinguished the living from the nonliving were opposed to the view that structure and organization were the most important elements of life. Dualism also was under attack. Lawrence was chided to refrain from publishing his materialistic views; his book was considered blasphemous, and he withdrew it. He was dismissed from his lectureship at the Royal College of Surgeons, and "Lawrence's public defeat meant that the tradition of 'materialist psychology' running from Darwin to Lawrence now had to move 'underground' . . . or to be shorn of its more radical implications and reconciled with religious sentiment and the new bourgeois ethos."[29]

Phrenology in England had a major champion in George Combe. He was most impressed with Spurzheim, having watched his anatomical dissections, and became convinced of the truth of phrenology. He founded the Phrenological Society and *The Phrenological Journal and Miscellany*, which was published first in the United Kingdom and later in America and contained detailed case studies. Combe's *On the Constitution of Man and Its Relationship to External Objects* (1829) was a best seller, and in 1846 he was asked to go to Buckingham Palace to examine four of the royal children, examinations that were subsequently repeated.[30]

The popularity of phrenology had many interesting resonances. The skulls of famous people became objects of scrutiny, and some of the less famous left their skulls for Dr. Gall's cabinet. That of the composer Joseph Hayden (1732–1809) was stolen, cleaned, and examined eight days after his funeral by two people who knew him and were interested in phrenology. They declared that he had a fully developed bump for music. When the remains of Franz Schubert (1797–1828) and Ludwig van Beethoven (1770–1827) were exhumed for reburial in the new Vienna cemetery, Schubert's skull was referred to as feminine in contrast to the masculine thickness of that of Beethoven.

But what about Napoleon? He objected to a German teaching anatomy in France. The Académie Français was commissioned to test Gall's theories, and Jean Pierre Flourens (1794–1867), professor of medicine at the Collège de France, undertook the task. It was in 1848 that one Phineas Gage, working on a railway line in Vermont, tamping gunpowder down into a hole, had the tamping iron blast out of the hole and go into his head. It entered at the left orbit, going out through the top of the skull. He survived the incident and has since become one of neuropsychiatry's most famous cases. His workmates and friends thought he was a changed man, "no longer Gage," yet he retained his intellectual faculties. Flourens did not consider that the injury had affected him mentally. This was in part because Flourens, following many experiments ablating various parts of the brains of animals, considered that the cortex of the brain possessed equipotentiality; the brain acted holistically, and each part had an action in common with others. It was not surprising that the Collège rejected Gall's ideas.

After the Revolution, patients with delusions of grandeur, claiming to be Napoleon, were not uncommon, and Napoleon himself consulted Pinel on the question of the rise in numbers of the insane in France. Cesare Lombroso (1835–1909), who we will meet again later, considered that Napoleon was a case of "psychic epilepsy," with megalomaniacal delusions and a lack of moral sense. A cast of Napoleon's skull was made shortly after his death, which showed a small forehead, but the cast was only of the right side.[31]

The contamination of phrenology by mountebanks and charlatans, along with its link to other fringe disciplines such as magnetism and hypnotism (phreno-hypnotism), led to its temporary decline academically. Nevertheless, there was a kernel within the theories which led to later revivals of phrenology in an altered form, namely, those of a well-defined cerebral localization. For all their faults, Gall and Spurzheim almost struck gold when it came to the localization of language. Gall saw a patient who had a unilateral fencing injury to the left frontal lobe and suffered a right-sided paralysis and impairment of speech, but he missed the crucial link to laterality. Nevertheless, he postulated a speech center (the only "center" to continue its existence to the present in terms of localization); it lay on the posterior half of the supraorbital plate; thus, if the organ was large, it would push the eyes forward. Combe agreed with this localization and recalled that at a brain dissection he attended of a patient who had expressive speech difficulties, there was a lesion in the corpus callosum, just near the "organ of language," but the right hemisphere was normal. The observations of Gall and his immediate followers did not lead them to an explanation of the riddle that although there were two hemispheres, damage to one side of the brain could lead to loss of function. This was yet to come.

The Islands of Reil

The German physician Johann Christian Reil (1759–1813), almost forgotten by modern psychiatry, was professor of medicine at the University of Halle for 23 years. His neuroanatomy held debts to Galen and Willis, yet his affirmation that all diseases were nervous diseases placed the brain as central to much more than just neurological diseases. The brain determined temperament. There was feedback to the brain from bodily organs, such that feelings are colored by the state of the brain. The mind and the body interacted. He introduced the term *Gemeingefühl* (human feelings), and his text *Rhapsodien uber die Anwendung der psychischen Kurmethode auf Geisteszeruttungen* (Rhapsodies about applying the methods of psychological treatment to disorganized spirits, 1803) is considered the most important document of romantic psychiatry. He followed Flourens in rejecting ideas of localization of functions in the brain, yet his neuroanatomy introduced to us such structures as the cortical islands of Reil (the insula), the nuclei we now refer to as the locus coeruleus, and the arcuate fasciculus. He studied the cerebellum, thought that the cerebral peduncles and basal ganglia were of special importance, and placed the center of cerebral activity with the latter. It was here that *Lebensgeist* (life spirit) spread its feeling to every organ and had an ability to react. But *Rhapsodien* were concerned with the cure for insanity by psychological methods. The brain possessed a special energy (*Getriebe*), and if its balance (*der dynamishen Temperature*) was disturbed, mental disorders resulted.

He rhapsodized on the glories of the human brain, drew on comparisons from astronomy, the microcosm reflecting the macrocosm, and was the first to adopt a clear psychotherapeutic approach to management, including attempts at hospital reform. It was through psychological methods that a cure could be affected, and he included occupational therapy, musical therapy, and sex therapy for sexual delusions as part of the regimes. Reil gave us the term *Psychiatrie* (psychiatry), advocating a separate branch of medicine, *Psychische Medizin* (psychiatric medicine), for the study and treatment of psychiatric illness.

In spite of his speculations about the location of the *sensorium commune* (the vital center), Reil was only floundering, as were other contemporaries: Luigi Rolando (1773–1831) in Italy selecting the medulla oblongata, and Alexander Walker (1779–1852) in England the cerebellum.

Dreams

It is well known that Coleridge was rudely disturbed by the man from Porlock while he was writing down "Kubla Khan," inspired from an opium-drenched dream. When he returned from the interruption, he had forgotten the whole and only fragments remained. The poem is subtitled "A Vision in a Dream."[32]

The romantics were not the first to have a special interest in dreams and states of mind which hover between the conscious and the unconscious, either sleep-related or drug-induced reveries. The Babylonians had dream texts, Hippocrates considered that they could be of diagnostic help, and for Galen they even had therapeutic value. Adam's dream of the creation of Eve is described in the Bible: "the Lord God caused a deep sleep to fall upon Adam, and he slept; and He took one of his ribs, and closed up the flesh instead thereof; And the rib, which the Lord God had taken from man, made he a woman, and brought her unto the man." But Milton made much more of this, as he does of the entire creation story. In *Paradise Lost* Adam does more than fall into a deep sleep; he has visions of God:

Who stooping opened up my left side, and took

From thence a rib, with cordial spirits warm,

And life-blood streaming fresh; wide was the wound,

But suddenly with flesh filled up and healed.

The rib he formed and fashioned with his hands;

Under his forming hands a creature grew,

Man-like, but different sex, so lovely fair

That what seemed fair in all the world seemed now

Mean . . .[33]

Eve also had dreams. In the most discussed of dreams, she heard Adam's voice, which led her to the tree of interdicted knowledge. There she was tempted to taste the fruit divine, "and be henceforth among the Gods / Thyself a goddess, not to earth confined."[34]

Prophesy and dream interpretation are part of a continuing history; these existed long before Sigmund Freud's (1856–1939) explorations. Revelations in dreams may have altered world history. Constantine (272–337), the first Christian Roman emperor, on the night of 312, before a battle at which his men were outnumbered, had a dream about an object shaped like a cross with a loop on top. A voice spoke to him, saying that he would conquer with this sign. He commanded his soldiers to paint the labarum on their shields. He won the battle, and he was told that the image was a Christian image and he must have heard the voice of God. He delivered religious toleration to the empire, allowed expansion of Christian worship, declared himself a Christian, and was baptized.[35]

For some, the dream was a *vacatio animae*, a Platonic concept revived by the Neoplatonist Ficino to represent moments of loss of the self and hence of reason. But for him these moments liberated the soul, leading to states of poetic or divine inspiration. In the Renaissance many artists drew dream-like images, although, with singular exceptions, these artists rarely painted their own dreams. They tried to portray dream states of others, often erotic and many quite bizarre, such as the nightmare images of Hieronymus Bosch.

Darwin, Gall, and others held an interest in dreams, consciousness being suspended while bodily sensations and memories exerted their own play of images. The idea that the brain was active in sleep, rather than simply asleep, was quite new. For Gall some of the organs of the brain remained active, while others were on hold; for Darwin the action of the will was suspended, a theme taken up by Coleridge. Richardson refers to "Kubla Khan" as "a case history, it is a brain scientist's dream."[36]

The Electric Spark

It was at the Villa Diodati, close to Lake Geneva, in June 1816 that Frankenstein was born and his monster created. Mary Godwin (later Shelley; 1797–1851) was staying at the villa with Shelley, Byron, Claire Clairmont (1798–1879), and the doctor John Polidori (1795–1821). One evening, sitting around a fire, with tempestuous weather outside, Byron suggested that they each write their own ghost story. The project did not seem to have progressed far, until the night of June 22, when Mary had a hypnagogic hallucination, about a man seemingly brought to life by another with a machine, the "horrid" thing, his creation, terrifying him. The next morning she started to write the tale that made her famous, perhaps the first science fiction novel. In the plot, Victor Frankenstein, grieved by the death of his mother, sees lightning strike and destroy an oak tree and wants to harness such powers of nature to re-create life. He goes to Ingolstadt University in search of nature's secrets, and he creates life from disembodied remnants of corpses but is terrified by what he has created. The story is one of death and destruction. The monster, seeking revenge on Frankenstein for abandoning him in a cruel world, pursues him, but his creator dies of exhaustion. The monster, on finding his maker dead, avows to go off to the northernmost place on the globe, where he intends to kill himself with fire.[37]

To give a "spark" of life to the corpse, Frankenstein had used galvanism. Recall that a preoccupation for neuroscientists was how messages were transmitted around the nervous system. Concepts of spirits, traversing nerves (hollow), vied with ideas such as Hartley's "vibratiuncles," corpuscular vibrations and transmission by contact, these being linked to earlier ideas such as Newton's and Willis's blasts and agitations. Steno objected that no amount of dissection had revealed the nature of animal spirits. In fact, their transmission by fluid presented some formidable problems, not the least being that when nerves were tied there was no obvious buildup of "spirits" behind the ligation. Further, the speed of our actions and reactions was just too quick to be explained by such theories. In any case, the nerves were not hollow, a fact confirmed with the newly conceived microscope.[38]

The idea of electricity had been known for centuries; electric fish, for example, were used by Galen to treat headaches, among other conditions.

The word *electricity* was first used in English in 1646. The power of lightning and its electrical nature had been demonstrated by Benjamin Franklin (1706–1790). Creating electric charges by friction and storing them with a Leyden jar allowed electricity to be harvested for therapeutic potential, and by the middle of the eighteenth century, doctors and laypeople were using weak electric shocks for multiple medical disorders, and electrical friction machines for social entertainment were being developed.

The theory of Haller that muscles (but not nerves) possessed an intrinsic irritability influenced Luigi Galvani (1737–1798), who made frogs' legs twitch with electric sparks. He demonstrated that this action could be brought about by stimulating an exposed nerve, although stimulating the brain did not have the same effect. His nephew, Giovanni Aldini (1762–1834), went further: not only did he collect guillotined heads and show that with electricity he could elicit facial movements, but in public demonstrations in London in 1803 he stimulated movements of the face and limbs of an executed criminal.[39]

Although Galvani revealed that electricity could stimulate muscles, it was not clear whether there was any intrinsic electrical activity in the animal body, an idea that Alessandro Volta (1745–1827), the inventor of the battery, disputed. The arguments were long and heated, but Galvani won the day, the term *galvanism* was introduced in English in 1797, and the spirits were on their way out.[40]

From Cells To Reflexes

Hooke developed the compound microscope and gave us the term *cell*. Jan Evangelista Purkyně (1787–1869) drew the interior structure of animal cells, showing nuclei, and described the structure of cells in the cerebellum, including those that were later named after him. Hermann von Helmholtz (1821–1894) demonstrated that at least some nerve fibers came from nerve cells, and Johannes Müller (1801–1858) opined that connections existed between cells. Bell had described the double root of the spinal cord nerves and the effect of cutting the posterior fasciculus. At around the same time (1822), François Magendie (1783–1855) declared that the anterior roots were specifically concerned with movement, and the posterior ones were allocated to sensitivity (hence the Bell–Magendie hypothesis, which has to do with the separate functions of the anterior and posterior divisions).

Marshall Hall (1790–1857) in 1832 hypothesized that the spinal cord contained reflex arcs "cementing the link" between sensory and motor function. The idea of reflex activity was not new, and as noted, Descartes led the way, and Whytt had asked important questions about the actions of bodily organs and muscles without the intervention of the will. Hall took Haller's ideas a step forward, namely, arguing that "sensitivity" was a property of nerves and "irritability" of muscle. He wrote of "involuntary motions," an early hint at the whole cavernous underworld of unconscious mental activity. There followed many experiments with decapitated or decerebrated animals, revealing that the spinal cord itself had activity, in the absence of the brain—but did it also possess a soul? The debates persisted for the next half century. However, the principle of spinal reflex action was secure and had a profound influence on later developments in psychology.

Melancholia Revisited

In 1771, a young man arrives in Wahlheim, Germany, in part to escape an unhappy romance. He falls in love with the already-betrothed Lotte, leaves town to take up a minor government post, is dismissed, and unhappily returns to Wahlheim. Lotte is now married, and in despair he shoots himself. Such were the *Sufferings of Young Werther* (1774), Goethe's *Sturm und Drang* novella, set out as letters from Werther to his close friend, with additions from the editor who had collected them. The tale prefigured romanticism and caused its own storm. Young Werther spurred with his moniker not only a fashion for perfume, jewelry, clothes, and the like but also a spate of suicides—the so-called Werther effect. Werther's actions were a response to an unfeeling world, one that was bourgeois, was restless, and destroyed the true feelings of nature. He was buried in an unmarked grave, yet he was in fashion.[41]

Melancholia was an *olla podrida* of mental states; we have seen them variously hinted at under terms such as *acedia, the vapors, vesania, the spleen*, and the like. But there was a radical change of emphasis in France and Germany, bringing to the fore some of the great pioneers of what was a developing somatic orientation to mental disorders, whom we meet in the next chapter.

Cullen had separate categories for melancholia and mania, but he distinguished both from hypochondriasis. The Dutch physician Herman Boerhaave (1668–1738) had described three different categories of melancholia, but he believed that they were linked through severity and related to melancholy "juice." Richard Mead (1673–1754), described by Hunter and Macalpine as "one of the great physicians who paid considerable attention

of mental illness," and another who applied himself to the limitation of restraint of mental patients, considered that mania and melancholia were the
same illness, differing only in degree. He noted that melancholia involved
being sad and dejected, a key symptom cluster that henceforth began to
define depression. Mead carried forward a nosological trend, which has
continued up to the present time.[42]

The case of Werther was not the only literary example to publicize disorders of mood. A school referred to as the Graveyard Poets combined gothic
and romantic sentiments. Such poems as Thomas Gray's (1716–1771) "Elegy
Written in a Country Churchyard" (1750), Coleridge's "Dejection: An Ode,"
or Keats's "Ode on Melancholy"—which "falls / Sudden from heaven like a
weeping cloud"—emphasized these sentiments. Samuel Johnson (1709–1784)
was a sufferer, lamenting that he had received melancholy from his father,
and defining melancholy in his dictionary as "a kind of madness, in which
the mind is always fixed on one object."[43] The word *depression* was becoming
attached to a lowering of spirits, even if it was not in his dictionary.

Related to this was the growing medical and literary interest in similar
topics. Suicide, while tolerated by the Greeks and Romans, was harshly dealt
with by ecclesiastical laws up to the eighteenth century. The act became associated with mental illness, but there was the alternative representation
as tragedy, or even heroic. The famed framed death of the poet Thomas
Chatterton, whose beautiful arsenic-tainted body was portrayed by the
pre-Raphaelite artist Henry Wallis (1830–1916), inspired a vast following in
England, not unlike the case of Werther.[44] Personality became a source of
medical interest. Jane Austen's (1775–1817) *Persuasion* is a case in point.
In this novel, Louisa Musgrave's character is changed after a head injury, a description that, according to Richardson, was without precedent
in the literature. This drew readers' attention to an association of states
of mind and personality with the brain. It was feared that she had suffered from a concussion. Post-concussion syndrome had made its first
public appearance.[45]

There was little progression on the pathogenesis of the affective disorders
over this time, but somatic and social factors were ever present, either as
causes or as modifiers. Even so, some clarification of the disorder we now
recognize as depression was emerging. Thus, ever since the earliest texts, a
disorder akin to what we today call depression has been described (see
chap. 1), even if mania was better identified with a more extreme form of
madness. Gradually, an association between the two had been recognized in

Greek and Roman texts, being chronic disorders not associated with fever. Humoral explanations were reiterated, and black bile was the main culprit. How close were the associations between melancholia and mania remained unresolved.

The Converted

Mead also introduced an idea that we have already met in *forme fruste*, namely, that insanity and major diseases of the body are incompatible (such as epilepsy and madness). This phenomenon was referred to by another physician and advocate of nonrestraint, John Ferriar (1761–1815), as "conversion of diseases." He was intrigued to understand how symptoms of one disease could shift in their location. The rather obscure concept of "sympathy" had earlier been used. Ferriar described the situation thus: "A disease is said to be converted, when new symptoms arise in its progress, which require a different designation, and which either put a period to the original disorder, or combining with it, alter the physician's views respecting the prognostics, or the method of cure." He then refers to "hysterical conversions," which can deceive even the most experienced diagnosticians. He notes (as Erasmus Darwin had observed) that "some derangements of the mind cannot be removed without exciting an artificial delirium."[46]

These ideas, based on careful clinical observations, clearly presaged not only the ideas of conversion disorders of later psychoanalysis but also the intriguing possibility that a cure for mental illness might involve inflicting a physical disorder, or that some physical illnesses remit when psychopathology intervenes. It is perhaps of interest that Ferriar himself suffered a mental condition, well known to the present author and many of his colleagues, which he referred to as bibliomania. His poem on this begins thus: "What wild desires, what restless torments seize / the hapless man, who feels the book-disease."[47]

Going West

In America, the first psychiatric hospital was established at Williamsburg, Virginia, in 1775. Benjamin Rush (1745–1813) was the country's first psychiatrist. He had trained in part in Edinburgh and been influenced by Cullen. His was one of the signatures on the Declaration of Independence, and after the Revolution, in which he served as a surgeon in George Washington's military forces, he joined the staff at the Pennsylvania Hospital, which had admitted psychiatric patients since 1752. His *Medical Inquiries and Observa-*

tions upon the Diseases of the Mind (1812) was the first American textbook of psychiatry. He described soldiers who were affected by the war, including cases of hysteria, as well as two cases of hereditary madness in twins.

Rush believed that many diseases, including psychiatric disorders, were caused by alterations of blood flow, especially in the brain. Debility, due to either insufficient or excessive stimulation, required treatment that excited the blood vessels. He was an ardent proponent of physical methods of treatment—and so keen on bleeding that he was referred to as the lancet-loving physician of Philadelphia.[48] He introduced a physical method of treatment based on the above ideas and his understanding of one reportedly successful English method of treating mania by "swinging."

Adopting Mead's ideas on the incompatibility between physical ailments and mental illness, Joseph Cox (1763–1818), a physician who spent his professional life trying to help patients with mania, wanted to provide them with a new disease and thus new symptoms, hence hoping to bring an end to the bout of insanity. Patients were swung, either oscillating or in a rotatory fashion, until they became vertiginous, and the swinging could even go on until they became unconscious. These Herculean remedies, as Cox referred to them, were adapted by Rush, with his "tranquiliser" and the spinning chair.[49] The former was a heavy armchair in which patients were strapped, and then a boxlike apparatus was placed over their heads to avoid visual distraction. The spinning chair allowed patients to be strapped in and spun with the head turned sideways to increase blood supply to the brain. At least such physical treatments were based on a theory as to the cause of the underlying disorders, even if they often evoked fear, terror, or pain, psychological states that Rush also believed could be therapeutic. One psychotherapeutic technique he adopted was to fix his and the patient's eyes and stare him or her out.

The Romantic Brain

Irving Massey summarized the neuroscientific achievements of the romantic era in a review of Richardson's book as follows:

> During the later eighteenth century, psychology moved from a Cartesian mind–body dualism on the one hand, and Lockean *tabula rasa* principles on the other, through what might be called a mechanistic phase among some of the French *ideologues*, to arrive at what Richardson regards as a typically Romantic—and modern—view of the mind . . . [which] unites mind and body, operating as a single activity/entity in which affect and thought are one in

their encounter with experience, so that, in a Wordsworthian phrase, we "half create" what we perceive, or, better still, in Shelley's terms, learn to imagine what we know.[50]

As Vincenzo Ferrone summed it up, "European intellectual life in the last quarter of the century (18th) was strongly affected by the contest between the supporters of a mechanistic and physical-mathematical view of the great chain of being and those who believed in a Renaissance-style *natura naturans* . . . a dynamic view of a nature that existed firmly anchored in time and overflowing with vital energy. . . . Feelings, sensibility, and man's anxieties took their place next to the cult of reason, and deductive and inductive reasoning learnt to coexist with intuition, imagination and reason as a poetic stamp."[51]

The romantic poets, but especially Coleridge, were exploring the creative mind in entirely new ways, which Coleridge found quite uncomfortable. This post-Kantian enterprise involved concessions to an unconscious edifice of uncertainty and an independence of the mind from conscious control. Examining dreams, drugs, and disorders of the mind led a cadre of scientists such as Darwin and Gall to provide an entirely new view of the mind, which, as Glenda Sacks succinctly put it, "foregrounded feeling and emotion at the expense of reason, reassessed the significance of the natural environment on the body, emphasized sensation and sensibility, and postulated an active and creative mind."[52] Meyer Howard Abrams, taking as the title of his masterful book *The Mirror and the Lamp: Romantic Theory and the Critical Tradition*, deliberately emphasized the comparison between the mind as a reflector of external objects and the mind as a radiant projector, which makes a contribution to the perceived object: "The change (of arts) from imitation to expression, and from the mirror to the fountain, the lamp and related analogues . . . [was a feature of the] . . . movement from 18th century to early 19th century schemes of the mind and its place in nature."[53]

Coleridge's contention that thought is motion; Keats's poetic expressions of emotional reactions such as sweating, blushing, and the beating heart; Gall's and others' assertions that some form of thought preceded language, that cognition itself was embodied and preceded spoken language; and the emphasis on instinct and intuition as opposed to logic opened up a whole new understanding of brain–behavior relationships. Attention was directed to the visceral nature of our experiences as opposed to the primacy of the external sensations, with the organs of the body having sympathy with the organs of the mind and electricity as a galvanizing force. It could be said that modern neuroscience really began in the romantic era. But while Gall and others may have clarified how the brain could create order out of chaos by actively participating in constructing the world and suggested a universal model of brain structure across all humans, their critics argued that what was missing was a central organizing principle. What was becoming of the self? Even more disturbingly, since there was no clear division between the human brain and that of other animals, at what point did the moral agent—humanity—emerge?

In Mary Shelley's book, the monster refers to himself as the Adam of Frankenstein's labors. The following lines from Milton appear on the title page, in which Adam laments his fall from grace:

> Did I request thee, Maker, from my clay
> To mould me Man, did I solicit thee
> From darkness to promote me?[54]

The full title of the book is *Frankenstein; or, The Modern Prometheus.*

NOTES

1. Trimble MR, 2000, 276–278. Charles Lloyd's poems were also included in one of the most well-known tracts of romantic poetry, *The Lyrical Ballads*, published along with those of Coleridge and William Wordsworth in 1798. Lloyd was at one point admitted to the York Retreat, and he was also treated at Dr. Willis's institution, then being run by the sons of Francis Willis, whose hospital notes read, "died at age 64 still insane" (282).

2. There was much to this, as the poem, although allegorical, has a vast political backdrop, covering the troubled times of Milton's life span. Milton himself was a republican, who was only saved from punishment at the time of the restoration of the monarchy in 1660 on account of his reputation as a poet and his infirmity and blindness. In the poem, Satan inquires, "Which way shall I fly / Infinite wrath, and infinite despair? / Which way I fly is hell; myself am hell" (*Paradise Lost*, 4.73–75).

3. Letter quoted in Holmes R, 1989, 167. The line "a thousand . . . " is from "The Ancient Mariner." "With music . . . " is from "Kubla Kahn."

4. The relationships between the poets and scientists around this time are superbly laid out in Holmes R, 2008. Nitrous oxide was combined with wine to see whether it would help hangovers (apparently it did). Davy became addicted to it but failed to realize its great potential as an anesthetic.

5. Wordsworth and his sister returned much sooner than Coleridge. Wordsworth was quite miserable while there, but he did begin his autobiographical poem *The Prelude* at that time.

6. Coleridge S, 1817/1971, 64.

7. Coleridge was particularly influenced by Kant, Friedrich Schelling (1775–1854), and Johann Gottlieb Fichte (1762–1814). *Naturphilosophie* was a German philosophical movement

related to German idealism. Fundamentally romantic, *Naturphilosophie* embraced the totality of nature and the human mind as actively interacting with the universe, a philosophy of nature. *Esemplastic* is from the Greek *eis en plattein* ("to shape into one"). Coleridge S, 91.

8. Coleridge S, 152–153.

9. Coleridge S, 1.137.

10. "When I have fears . . ." is from "On Seeing the Elgin Marbles." See Richardson A, 2001, 114–150.

11. Shelley, "Ozymandias," lines 10–12.

12. Darwin E, 2005, sec. 3.3.

13. Darwin E, sec. 3.5.

14. Darwin E, 2008, l.269–280.

15. Bell C, 1868, 154, 157, 164, 167.

16. Wordsworth W, 1995; see lines 109–150.

17. Wordsworth W, *The Excursion: Being a Portion of the Recluse*, X11, an unfinished work, lines 63–66.

18. Bell, C, 1806, 123.

19. Bell C, 5–6.

20. Bell C, 1877, 179.

21. Charcot and Alfred Vulpian (1826–1887), shortly after being appointed to the Salpêtrière, discussed the similarities and differences between this condition and multiple sclerosis. This was clarified in *Lectures on the Diseases of the Nervous System* (1877). In that volume, separate lectures are given on paralysis agitans and disseminated sclerosis. Charcot's descriptions are remarkable for the further observations on such premonitory symptoms as fatigue, neuralgic pains, onset following shock, sudomotor changes, and later alteration of the mental state. He was surprised that the characteristic rigidity of the disorder had not been noted by Parkinson. Thinking that vibration may be therapeutic, he had a harness made that fitted under the patient's head and chin and bounced them up and down. Neither this nor electricity helped, and the disorder remained untreatable until well into the twentieth century. For the original descriptions, see Parkinson J, 1817.

22. Parkinson J, 3, 1. This observation has proved quite false, as later investigators and clinicians have described in great detail the personality and behavioral manifestations of the disorder. These pose an important clinical challenge for neuropsychiatrists at the present time.

Another important movement disorder with marked psychopathology is Huntington's chorea. George Huntington (1850–1916) was a general practitioner in East Hampton, New York, and was 22 when he published his report that brought him unsought fame. As with Parkinson's descriptions, there is so much included in his four-page communication that little was added from a clinical perspective for many years. He referred to its onset in adulthood, the hereditary nature of the disease, how it could skip a generation, and its potential confinement to a geographical area on the east end of Long Island. He noted the tendency toward insanity and the high frequency of suicide. The history of the disorder in America has been traced back to an Englishman called Jeffers, who married a woman with chorea and then emigrated. It is possible that many females with the illness were identified as witches and burned. See Huntington G, 1872.

23. Spurzheim split from Gall in 1813, in part because of his development of their ideas in ways that did not agree with Gall's theories (Gall later referred to him as a quack). It was Spurzheim who most popularized *phrenology*, a word not used by Gall, who preferred the terms *organology, organoscopy*, and *cranioscopy*.

24. Hunter R, Macalpine I, 1963, 714.

25. Combe G, 1841, 339.

26. For a fuller discussion of Swedenborg and other religious mystics, see Trimble MR, 2007, chap. 7.

27. His works and very original findings also had much influence in literary and artistic circles, impacting those with an interest in spiritualism and transcendentalism, such as William Blake (1757–1827), Charles Baudelaire (1821–1867), and, in America, Ralph Waldo Emerson (1803–1882) and Walt Whitman (1819–1892). See Gross C, 1997.

28. Lawrence W, 1819, 7, 110–114. He was a friend of the Shelleys, his views were equated with those of the radical atheist poet Lord George Byron (1788–1824), and, like others putting forward materialistic views, he was attacked on either political grounds or psychological principles.

29. Richardson A, 2001, 29.

30. George Combe was a philosopher and barrister. The Phrenological Society, founded in Edinburgh with his brother Andrew, became one of the largest such associations in the world. He considered Queen Victoria, with her forehead features, to be an admirable leader; he had not examined her but had seen her at an opera.

31. See discussions in Finger S, 2000, 119–136; Peacock A, 1982.

32. The story may well have been apocryphal, even made up by Coleridge as an excuse for not completing one of his now most famous poems. Even the dream origin has been disputed. He delayed publishing it for some 20 years after it was written.

33. Genesis 2:21–22. Milton J, 1977, *Paradise Lost*, bk. 8, lines 463–473.

34. Milton J, *Paradise Lost*, bk. 5, lines 26–94.

35. Constantinople became the capital of the Byzantine Empire for a millennium. The image was a Greek Chi (X) traversed by a Rho (P), together forming a symbol representing the first two letters of the Greek spelling of the word *Christos*. In 325 he summoned the first Ecumenical Council (the Council of Nicaea), which, among other things, instituted the Nicene Creed.

36. Richardson A, 2001, 51. Richardson also notes the dreamlike images and the subconscious in action in some of Coleridge's other well-known poems, such as "Christabel" and "The Pains of Sleep."

37. For more details see Sunstein FW, 1989, 117–131. Claire Clairmont (1798–1879) was Mary Shelley's stepsister and Byron's lover and mother of his child Allegra. Polidori was Byron's personal physician and author of *The Vampyre*.

38. The experiments that led to the invention of the microscope included the work of Galileo, but it was Antonie van Leeuwenhoek (1632–1723) who made the first practical microscopes, and he failed to confirm the observations of Galen that the optic nerve of a cow was hollow.

39. Apparently several of those present thought that the deceased was being brought back to life. Although Mary Shelley had knowledge of galvanism, there is no reference to Victor Frankenstein using "electricity" to bring his monster to life. However, Victor does use the phrase "spark of being" to describe the process of infusing life into his creature.

40. For an excellent account of the development of ideas and investigations about electricity and the nervous system, see Brazier MAB, 1973.

41. For example, admirers imitated Werther by dressing in yellow pants, blue jackets, and tall boots. There were decorated fans and porcelain pieces, silhouettes of Lotte, and a perfume called *eau de Werther*. After the publication of the book, there were many reports of young men committing suicide by shooting themselves. This book was banned in some places. Copycat suicides are still frequently reported. The story was partly autobiographical: Goethe's friend Karl Wilhelm Jerusalem shot himself, and Goethe had fallen in love with

Maximiliane von La Roche, who married another. The combination of blue and yellow in Goethe's *Theory of Colours* gave green—a color often associated with instability and rebelliousness. Blue was identified with German romanticism via the deep blue flower seen in a dream by the philosopher-poet Novalis (1772–1801), which embodied beauty, poetry, and love.

42. For a full analysis of the authors who were splitters and those who were joiners, see Jackson SW, 1986, 249–273.

43. Johnson's psychiatric disorder has been much discussed, and his tics, compulsions, motor hesitations, and speech disturbances all point to a contemporary diagnosis of Gilles de la Tourette syndrome.

44. Chatterton poisoned himself with arsenic in 1770. He was eulogized by all the major romantic poets, and the play *Chatterton* by Alfred de Vigny (1797–1863) was said to have led to a spate of suicides in France. Wordsworth gives us the following eulogy: "I thought of Chatterton, the marvellous Boy, / The sleepless Soul that perished in his pride; / Of Him who walked in glory and in joy / Following his plough, along the mountain-side: / By our own spirits are we deified: / We Poets in our youth begin in gladness; / But thereof come in the end despondency and madness." Wordsworth, "Resolution and Independence," lines 43–49; see Wu D, 2006, 530.

45. She had a fall, hit her head, and at first was lifeless. She remained unconscious for some time. Although later she was said to be much recovered, she was altered, as indicated by Captain Harville: "there is no running or jumping about, no laughing or dancing; it is quite different." Austen J, 1993, chap. 12, and 248.

46. For notes and quotes of Mead, see Hunter R, Macalpine I, 1963, 385–388; for those of Ferriar, see Hunter R, Macalpine I, 543–544.

47. The poem was dedicated to the English collector Richard Heber, who filled eight houses with his collection of over 150,000 books. It describes many other diseased bibliomaniacs.

48. See Hunter R, Macalpine I, 1963, 662.

49. For further information on Cox, see Hunter R, Macalpine I, 594–598. See also Freeman W, 1968, 145–152.

50. Massey I, 2002, 78.

51. Ferrone V, 2015, 137, 138, 144.

52. Sacks G, 2004, 553. Kant admired the sciences but not romanticism. For him, man was not a machine, but neither was he bound to nature. He was a freely acting being who used reason and is an end unto himself. The moral law is within; man is free to choose right or wrong. A problem with some of the theories of mind discussed in this chapter was the lack of the "homunculus" within.

53. Abrams MH, 1953, 57.

54. Milton J, 1977, *Paradise Lost*, bk. 10, lines 743–745. In a version of the Prometheus myth, he creates mankind from clay.

REFERENCES

Abrams MH. *The Mirror and the Lamp: Romantic Theory and the Critical Tradition.* Oxford University Press, Oxford; 1953.

Austen J. *Persuasion.* Wordsworth Editions, Hertfordshire; 1993.

Bell C. *Essays on the Anatomy of Expression in Painting.* Longman, Hurst, Rees, and Orme, London; 1806.

Bell C. Reprint of the "Idea of a New Anatomy of the Brain," with letters, &c. *Journal of Anatomy and Physiology* 1868;3(1):147–182.

Bell C. *The Anatomy and Physiology of Expression as Connected with the Fine Arts.* 7th ed. George Bell, London; 1877.

Brazier MAB. The evolution of concepts relating to the electrical activity of the nervous system, 1600–1800. In: *The History and Philosophy of Knowledge of the Brain and Its Functions.* B. M. Israel, Amsterdam; 1973.

Coleridge S. *Biographia Literaria.* J. M. Dent and Sons, London; 1817/1971.

Combe G. *Notes on the United States of North America during a Phrenological Visit.* Maclachlan, Stewart, Edinburgh; 1841.

Darwin E. *Zoonomia; or, The Laws of Organic Life.* Project Gutenberg, EBook #15707. April 25, 2005.

Darwin E. *The Temple of Nature; or, The Origin of Society. A Poem, with Philosophical Notes.* Project Gutenberg, EBook #26861. October 9, 2008.

Ferrone V. *The Enlightenment History of an Idea.* Translated Tanantino E. Princeton University Press, Princeton, NJ; 2015.

Finger S. *Minds behind the Brain.* Oxford University Press, Oxford; 2000.

Freeman W. *The Psychiatrist: Personalities and Patterns.* Grune and Stratton, New York; 1968.

Gross C. Emanuel Swedenborg, a neuroscientist before his time. *Neuroscientist* 1997;3:142–147.

Holmes R. *Coleridge: Early Visions.* Hodder and Stoughton, London; 1989.

Holmes R. *The Age of Wonder: How the Romantic Generation Discovered the Beauty and Terror of Science.* Harper, London; 2008.

Hunter R, Macalpine I. *Three Hundred Years of Psychiatry, 1535–1860.* Oxford University Press, Oxford; 1963.

Huntington G. On chorea. *Medical and Surgical Reporter* 1872;26:317–321.

Jackson SW. *Melancholia and Depression.* Yale University Press, New Haven, CT; 1986.

Lawrence W. *Lectures on Physiology, Zoology, and the Natural History of Man, Delivered at the Royal College of Surgeons.* J. Callow, London; 1819.

Massey I. Review: Alan Richardson, *British Romanticism and the Science of the Mind. Criticism* 2002;44(1):77–80.

Milton J. *The Portable Milton.* Ed. Bush D. Penguin Books, London; 1977.

Parkinson J. *An Essay on the Shaking Palsy.* Sherwood, Neely and Jones, Paternoster Row, London; 1817.

Peacock A. The relationship between the soul and the brain. In: Clifford Rose F, Bynum WE, eds. *Historical Aspects of the Neurosciences.* Raven Press, New York; 1982.

Richardson A. *British Romanticism and the Science of the Mind.* Cambridge University Press, Cambridge; 2001.

Sacks G. Review: Alan Richardson, *British Romanticism and the Science of the Mind. Poetics Today* 2004;25(3):553–554.

Sunstein EW. *Mary Shelley: Romance and Reality.* Johns Hopkins University Press, Baltimore; 1989.

Trimble MR. Charles Lloyd: Epilepsy and poetry. *History of Psychiatry* 2000;11:273–289.

Trimble MR. *The Soul in the Brain: The Cerebral Basis of Language, Art and Belief.* Johns Hopkins University Press, Baltimore; 2007.

Wordsworth W. *The Prelude.* Penguin Classics, London; 1995.

Wu D. *Romanticism: An Anthology.* 3rd ed. Blackwell, Oxford; 2006.

CHAPTER SIX

Late Romanticism

Strange Times

There is a crack in everything: that's how the light gets in.

Leonard Cohen, "Anthem"

In February 1860, the following note was written: "What I felt is indescribable, and if you will deign me not to laugh, I will try and translate it for you. . . . To be immediately felt carried away, under a spell . . . I often experienced quite a strange feeling, the pride and enjoyment of understanding, of being engulfed, overcome, a really voluptuous sensual pleasure, like riding into the air or being rocked on the sea. In general those deep harmonies remind me of those stimulants which accelerate the pulse of the imagination. . . . There is everywhere something elevated and everlasting, something reached out beyond, something excessive, something superlative . . . this would be if you like, the final paroxysm of the soul."

The Sound of Music

Charles Baudelaire (1821–1867), poet and art critic, described by fellow poet Edgar Allan Poe (1809–1849) as "the writer of nerves," had attended a performance of Richard Wagner's (1813–1883) *Tannhäuser* in Paris, and he sent the above description of his feelings to the composer.[1] He even included physiology in his poetry, describing the brain as a "cramped and mysterious laboratory . . . there is a nervous shock that makes itself felt in the cerebellum"—which has an echo in Emily Dickinson's (1830–1886) "I felt a funeral in my brain."[2] Yet Baudelaire was describing the immense feelings that music could arouse in him and its connections to some epiphanic experience.

Throughout the ages, music has been known to evoke quite profound emotions, across all known cultures. But something happened to musical style around 1800, especially in Germany, linked to the romantic revolution. That something was Beethoven's Third Symphony.

Allow me a digression and a return to chapter 1. What else happened around the campfires all those hundreds of thousands years ago, even before the telling of stories? *Homo sapiens* is a primate species of the class Mammalia, and the only living representative of the genus *Homo*. The defining feature of mammals is the need for and the presence of a mother in early development. The comparative neuroanatomist and physician Paul MacLean (1913–2007) put the point forcefully with the neuroanatomical imperative "The history of the evolution of the limbic system is the history of the evolution of the mammals, and the history of the evolution of the mammals is the history of the evolution of the family."[3] We will have much more to say about the limbic system later, but MacLean's point is that the history of our evolution is integral to the growing complexity of family and social activities, as well as the close infant–mother interrelationships with attachment and bonding so relevant in *Homo sapiens*. Somewhere along the evolutionary way, our ancestors, with considerable emotional expression, began to articulate and gesticulate feelings—denotation before connotation. But, as the philosopher Susanne Langer (1895–1985) noted, "The most highly developed type of such purely connotational semantic is music."[4] In other words, meaning in music came to us before meaning given by words.

The first musical instrument was the human voice. As soon as our ancestors' vocalizations developed the frequency range they eventually did, greater incidentally than needed for our speech, and tones were heard and appreciated for their expressive potential, the basis for the music we appreciate today was born. While the hitting of one object on another gave the beat of integrated elementary rhythms, harmonies were audible within a single sung tone, since the first sounds of a definite pitch would be heard as a chord; the tonic was discovered. An original perpendicular harmony soon became horizontal, sung in time with rhythm. From the rhythmic beating within the mother's body for the fetus to the primitive drumlike beating of sticks on wood and hand clapping of our adolescent and adult non- or proto-speaking ancestors, the growing infant is surrounded by and responds to rhythm. But, "being more variable than the drum, voices soon made patterns and the long endearing melodies of primitive song became a part of communal celebration."[5]

The ear is always open, and unlike the visible gaze, sound cannot readily be averted. There are certain defining features of being human, one of which I have discussed in some detail in my book *Why Humans Like to Cry*, namely, the ability to cry emotionally, and especially in response to music. Another is the facility to respond to and entrain to rhythm, namely, to dance.[6] The Greeks both revered music, as in the festival plays, and feared it, in the sense of Plato's reservations, since it evoked the passions and hence undermined *logos*. He would not allow music in his model republic, aware, no doubt, of the half-bird beautiful women sirens, who sang so enchantingly that they lured sailors to their island and hence to their death.

Before the so-called romantic era, there had been a steady development of the regulated form and style of Western music, with the gradual permeation into the music of the emotions of the composer: self-expression, with the development of counterpoints and harmonies. While not suggesting that there is a language of music akin to our spoken vocal language, Langer noted that to be "keyed up" means more than just to tune the piano, since "music is not the cause or the cure of feelings, but their logical expression."[7] Richard Wagner put it thus: "What music expresses, is eternal, infinite and ideal; it does not express the passion, love or longing of such-and-such an individual, on such-and-such an occasion, but passion, love or longing in itself . . . the exclusive and particular characteristic of music, foreign and inexpressible to any other language."[8] Music is a language of feeling, and "just as words can describe events we have not witnessed, places and things we have not seen, so music can present emotions and moods we have not felt, passions we did not know before."[9] Musical rhythms are life rhythms, and music—with its tensions, resolutions, crescendos and diminuendos, major and minor keys, and silent interludes—does not present us with a logical language but "*reveals* the nature of feelings with a detail and truth that language cannot approach."[10]

This idea seems difficult for a philosophical mind to follow, namely, that there can be knowledge without words. Indeed, the problem of describing a "language" of feeling permeates the whole area of philosophy and neuroscience research and highlights the relative futility of trying to classify our emotions—"Music is revealing, where words are obscuring."[11]

The point cannot be more strongly put, since there are those commentators who have asserted that music in and of itself is meaningless, "auditory cheesecake, an exquisite confection crafted to tickle the sensitive spots of at least six of our mental faculties."[12] Yet any preliterate ancestors, living in small groups, who were capable of uttering meaningful gestures around the campfire, in unison, harmonizing together, perhaps echoing the ambient sounds of the nearby savannah, perhaps even seeking out spirits, would surely have had an evolutionary advantage. What better to do after a meal than huddle together, move together, incant together, in short, to dance—the first *com-unions*. Males and females together, perhaps aided by the warm moonlight, may have bonded together in ways that were never possible before music penetrated aural space, for by now our hominid forefathers' brain had developed significantly, giving more space to the areas of cortex devoted to hearing and meaning.[13]

Paris: Capital of the Nineteenth Century[14]

The University of Paris, which was founded in the twelfth century, was an important center of learning in the Middle Ages. Like all French institutions, it was radically influenced by the French Revolution. So was the growing discipline of *psychiatrie*. Pinel released the insane and others from their chains, favoring a less organic view of mental illness than had Cullen, even though he embraced the latter's views on classification. Pinel failed to find consistent head sizes or skull shapes linked to psychiatric disorders—one purview that led him to favor "moral" treatments. The therapeutic move was away from asylum and confinement and toward psychotherapeutic and social models. The philosopher Jean-Jacques Rousseau (1712–1778), who espoused the idea that man is born free but is everywhere in chains, and whose views influenced the promulgators of the French Revolution, was a considerable influence on Pinel, as were the theories of Locke.[15] Enter onto the stage Pinel's pupil Jean Étienne Dominique Esquirol (1772–1840). His book *Des maladies mentales considérées sous les rapports medical, hygiénique et medico-légal* (1838) is regarded as one of the classics of psychiatry. Like Pinel, he was an acute observer of his patients at the Salpêtrière, later at the Charenton in Paris, and in other hospitals around France. He continued the trend toward improving psychopathological classification. He recognized four forms of insanity (monomania, mania, dementia, and idiocy), distinguished hallucinations from illusions (a feat that still evades many psychiatrists today), and was part of a developing group of psychiatrists referred to by Jan Goldstein as the Esquirol Circle. He established the study of the insane as a medical specialty, requiring special training. He further opined that the center of such learning was in Paris, from which knowledge should flow centrifugally. He became concerned with the architectural structure of mental hospitals, calling for, for example, separate wards for different kinds of patients, and he wanted these specialist hospitals to be called asylums.

The circle embraced a large group of acolytes, some of whom are of importance for contributions to what would become today's neuropsychiatry.[16]

We here approach the unraveling of the pathology of general paralysis of the insane (GPI), further investigations of epilepsy, and clarified descriptions of these disorders, with a greater emphasis on the neurological and emotional underpinnings of psychopathology. As we will see, many of the circulating ideas were influenced by and influenced literature and other arts of the times. Two pathological constructs need first to be discussed: monomania and degeneration.

Monomania

Honoré Daumier (1808–1879), the republican French satirist, published a coruscating set of lithographs in 1832, depicting 16 politicians or magistrates with a parliamentary mandate, each absorbed in their singular obsession. They were published in *La Caricature* and entitled *The Ministerial Charenton: Monomaniacal Varieties of Political Lunatics*. Like many fads in the realm of mental disorders, monomania was by then already on the lips of the general public and in fashionable use. The term is attributed to Esquirol. He commissioned artists to draw or paint the facial features of psychiatric patients, a venture followed by many others after him, before the development of photography.

Monomania was a condition referred to as a partial lesion of the intelligence, the mind of the afflicted being concerned mostly only with a single object. This was seized upon and pursued contrary to any logic, and in the absence of any fever. It was not the full-blown picture of mania, and neither was it akin to melancholia. It was a partial delirium, an *idée fixe* with subtypes: the intellectual, the instinctive, and the affective; lypemania with sad delusions was a part of the first.

Esquirol and his school added an important discipline to the study of mental disorder: they collected statistics. By the late 1830s, monomania was diagnosed in 45 percent of patients at the Charenton and, according to Goldstein, accounted for 10 percent of admissions to the Salpêtrière and Bicêtre.[17]

Étienne-Jean Georget (1795–1828), who commissioned the painter Théodore Géricault (1791–1824) to paint patients with monomania, published a series of forensic case histories, which included *monomanie homicide*, a compulsion to commit murder. The insanity defense became the subject of much debate, within and outside the medical profession, and, as Goldstein explained, monomania became politicized. The disorder was imbricated with the politics of the Restoration monarchy, arguments for law reform in relation to mental illness, and the advancement of psychiatry as an independent, but necessary, discipline—*médecin des aliénés*.

Degeneration: A European Disorder[18]

To anyone not familiar with European history, its culturally rich background (hence the continued adoration of the Greek heritage), and the progressive, but not necessarily successive, cultural developments over the past 2,000 years or so, the ideas behind the concept of degeneration will seem bizarre. Those who espouse the idea that all men are born equal, who believe that historical progression is not only teleological but always progressing toward "higher" and better goals, and who consider that the life trajectory of *Homo sapiens* is only one of improvement will have to falter at this stage. The influence the idea of *dégénérescence* (degeneration) had over European neurology and psychiatry was profound.

As with monomania, "degeneration" permeated all forms of life, especially law, literature, poetry, and the arts, and it entered the debates about evolution. Degeneration related not only to societies, social groups, and races but also to families, bridging generations, as the sins of the fathers are borne by their sons and grandsons. It fostered concerns about the growing numbers and causes of the insane and criminal in society, as well as the effects of "civilization" and the Industrial Revolution on these statistics. As Pinel's influence waned and experimental laboratory work combined with more careful patient observations followed, an emphasis on biology transfigured the perspectives of the romantic era toward a more materialist view of the embodied mind, one that emphasized organic influences on human nature and art.

Causation of these ills became a biological imperative, and not the outcome of Rousseauean social pressures and chains. Man was not born free. Victorian optimism, encouraged by geographical exploration, anthropological inquiry, scientific discovery, and economic successes, had to face up to the body of medical and scientific facts of degeneration. It became indeed "the condition of conditions, the ultimate signifier of pathology."[19]

At around the time that Coleridge and his friends were experiencing the delights of inhaling gases in Bristol and Thomas de Quincy (1785–1859) was writing his *Confessions of an English Opium-Eater* (1821), Jacques-Joseph Moreau (1804–1884), referred to as Moreau de Tours, was extolling the virtues of hashish. Student of Esquirol, but no great advocate of moral treatment, he traveled widely, and in the Orient he encountered the widespread use of hashish. He was impressed by its power to alter the mental state, and he speculated on its potential for exploring the field of mental pathogenesis. It might reveal the mysteries of alienation, revealing the source of delusions.

He was a member of the Hachischins Club in Paris and began to use hashish combined with rose or jasmine essence to treat patients.[20] Moreau was led to the conclusion, not accepted at that time, not only that delusions were always pathological but also that there was continuity between normality and abnormality. This could be explored artificially, revealing the primordial state underlying hallucinations and delusions. His work may be considered as the beginning of psychotropic exploration of the effects of drugs on the brain.

One physician who made degeneration central to his theories was a friend of Moreau de Tours. Bénédict Augustin Morel (1809–1873) was not part of the Esquirol Circle, and except for his student days, he lived and worked entirely outside Paris. Moreau de Tours had suggested the importance of the law of inheritance in many medical disorders, including insanity, a theme taken up by Morel: "If we now move away for a moment from the observation of facts in the individual, and study the successive evolution of pathological facts at the level of the species, we will be permitted to observe a similar phenomenon, that is the successive and progressive evolution of hereditarily transmitted pathological phenomena following and influencing each other over a series of generations." The first generation may just show a "nervous temperament," but in the second some "pathological disposition of the nervous system . . . epilepsy, hysteria, hypochondria." Then with progression of these "hereditary predispositions" came "degeneration of an intellectual and moral character," leading finally to "imbecility, idiocy and cretinous degeneration."[21]

Morel described the stigmata of degeneration, often detected by the physician's eye or crude measurement (small stature, small head, or pointed ears, for example); its links to crime, prostitution, anarchy, and sexual deviancy were coterminous. Syphilis, goiters, rickets, infertility, and madness were within a web of pathologies that wove threads through generations. This was yet another push that established psychiatry as a discipline essential to the legal system and for protection of society from the aberrant. To be fair to Morel, he was considered by his contemporaries to be concerned as much with amelioration or even prevention of these conditions as with understanding their causes. In treatment he was an advocate of nonrestraint. Earlier in life he had intended to become a priest and was a creationist, degeneration thus being for him a deviation from the perfect human type. Degeneration had its natural limits. As the progeny dies out, there can be no more progression in that family line. Nature is restored to balance.

The theory of degeneration was taken further in France by Valentin Magnan (1835–1916). Interested in the effects of cocaine, morphine, and alcohol on the brain, he applied Morel's theories to alcoholics. He classified four groups of hereditary degenerates, namely, idiocy and mental deficiency, cerebral abnormalities, episodic disorders, and delusions proper. He noted one form of delusion which appeared quite suddenly, which he only saw in degenerates and in people with epilepsy, which he distinguished from chronic delusions. With regard to the latter, he suggested that "all the brain centres are in a state of complete erethism; the anterior brain creates the delusional ideas and evokes the images in the posterior centres. These images are then represented in the anterior centres, with such vividness, that they are interpreted as realities. Thus, the hallucination has its direct cause in a series of delusional ideas which brings it into being, and it arises like a form of reflex. The centripetal route begins from the anterior centres." This is quite an insight for 1891.[22]

While this observation is later in the century, it does emphasize the prolonged and continued influence of the degeneration concept. As Daniel Pick summed it up, "It was only from the 1870's onwards that *dégénérescence* was taken to be of undisputed importance in clinical psychiatry. It took national defeat by Prussia and the Paris Commune to seal the importance of this word in historiography, social diagnosis, and cultural critique."[23]

The monomania concept was criticized and fell apart. Another member of the Esquirol Circle, Jean-Pierre Falret (1794–1870), working at the Salpêtrière from 1822 to 1867, believed that careful longitudinal observation of patients over time was essential for diagnosis. In 1854 he published an essay titled "Of the Nonexistence of Monomania." He pointed out that the frequency of monomania varied considerably between authors, and there was variability and often lack of clinical detail in diagnosed cases. Intellectually, there was a problem with subcategories of the disorder, which implied subdivisions of the mind which could be separately diseased, and too much emphasis was given to a prominent delusion, whereas the mental derangement went much beyond such a limitation. He also noted the failure to distinguish excessive sentiments (e.g., religious) from corresponding monomanias, thus overdiagnosing mental illness.

Falret's own observations led him in 1854 to describe *la folie circulaire* (circular madness). In the same year, another member of the circle, Jules Baillarger, described what he referred to as the dual form of insanity (1854), identifying alternating periods of excitation and depression. These descriptions became one of the cornerstones of Emil Kraepelin's (1856–1926) diagnostic pillars, manic-depressive psychosis.

But by 1870 the diagnosis of degeneration was still on the rise, and the ideas spread and were adopted in other countries. In Italy the main exponent was Cesare Lombroso; in England, Henry Maudsley (1835–1918), who we will discuss in the next chapter. In the meantime, monomania had all but disappeared.

Anatomical and Pathological Advances

From the beginning of the nineteenth century there were considerable developments in embryology, comparative anatomy, and the power of the microscope, and the alienists were interested in examining the brains of their patients. Pinel found little that added to his theories, and Esquirol found no pathological changes he could link to melancholy. Baillarger examined thin slices of the human cerebral cortex with the naked eye and described its six layers, noting how the gray matter lay outside the white. But it was not until 1867 that Theodor Meynert (1833–1892) revealed the different distribution of neurons in the layers. The psychiatrist François Leurat (1797–1851), examining the brains of several mammals, revealed the increasing complexity and patterning of the cerebral convolutions, linking these to increasing intelligence. Many others were parceling the mammalian brain, identifying nuclei and other brain structures, beyond discussion here. However, psychopathologists were by now also neuropathologists, a trend that persisted up to the end of the century.

Antoine Laurent Bayle (1799–1859) studied three disorders that were linked to mental illness: chronic arachnoiditis, gastroenteritis, and gout. The outcome was the real beginning of "organic psychiatry," the development of psychopathology secondarily arising from medical conditions. Most significant was his thesis that GPI was a symptomatic disorder secondary to chronic arachnoiditis. He described three phases of the condition: monomania, mania, and dementia; he even identified ambitious delusions, the grandiose delusions that are now well associated with this disorder. Although its syphilitic nature would not be discovered for another 50 years, Bayle's work is credited with the tendency of French psychiatrists for the rest of the century to pursue a materialist, organic perspective to mental illness.

It was the view, following on from the work of Flourens, that the cerebral cortex was not irritable, this being a property confined to the spinal cord. On these grounds, the origin of epilepsy was attributed to the medulla, and alteration of consciousness was thought to be due to vascular causes. There was little advance in the understanding of the various forms of seizures, although Esquirol had used the terms *le grand* and *le petit mal*, the former referring to generalized seizures. But he had some quite harsh things to state about epilepsy, especially with regard to his work on asylum design. He followed on from Pinel, insisting on separation of those with epilepsy from other psychiatric patients on account of the adverse influence of witnessing seizures. Many clinical observations of epilepsy in the latter part of the century were reported from asylums. Degeneration alongside violent passions, GPI, or even lunar influences were explored statistically, the latter examining the association between seizures and the phases of the moon.

The number of cases in which mental disorders were seen in epilepsy was published. This was the beginning of what is now referred to as comorbidity of the two conditions, and statistics were provided by, among others, Esquirol and his pupils Camille Bouchet (1801–1854) and Jean-Baptiste Cazauvieilh (1801–1849). The last two drew attention to the high incidence of seizures in the insane and wanted to understand the relationship between them. Attempts to locate the cerebral origin of epilepsy had to that date proved as fruitless as those seeking a seat of insanity. Bouchet and Cazauvieilh made insightful pathological observations, namely, that the mental disorder was related to the gray substance of the brain, while the epilepsy was due to underlying white matter. Temkin considered that attributing mental activity to the cortex in 1825 was a tour de force, supporting the ideas of Gall and Bell. However, their additional suggestion that pathology in the hippocampus (Ammon's horn) was important was groundbreaking. They identified at autopsy similar inflammatory pathology in this area in those who were insane and in those with epilepsy—this later became referred to as hippocampal sclerosis. The concept of temporal lobe epilepsy was born, as was the link to psychiatric disorder.[24]

The aura or "cold breeze," *état de mal* (status epilepticus), and *absence* all became familiar terms. *Furor epilepticus*, epileptic mania, *furor épileptique*, and epileptic delirium—all terms that reflect the sudden onset of violent rages noted in association with epilepsy—were variously described as occurring before or after seizures, but also in the absence of attacks. The link of these outbursts to actual seizures attracted much attention. The episodes remitted after a day or two, but during them, interpersonal aggression—even murder—was recorded. Another association was with somnambulism and automatic fugue-like states, these disorders reviving arguments about criminal responsibility. It was only a short step to the suggestion that episodes of violence without epilepsy were really *formes frustes*, epileptic equivalents; the link to degeneration remained.

Interest in epilepsy was sharply brought to the public attention by Émile Zola (1840–1902). His 20-volume cycle of novels *Les Rougon-Macquart* traces the destiny of a degenerate family, setting the decline of France in the Second Empire, played out through the family tree. He put his project thus: "My aim is to explain how a family, a small group of human beings, behaves in a given society after blossoming forth and giving birth to ten or twenty individuals who, though they may seem at first glance totally dissimilar from each other, are, as analysis shows, linked together in the most profound ways. Heredity, like gravity, has its laws."[25] Zola was well versed in the scientific developments of his times and was attracted to the realism of such painters as Édouard Manet (1832–1883). He visited the Paris Morgue to obtain images of murder and natural death, and he emphasized physical characteristics and psychological aberrations in his characters. His own observations of Paris life, with its underbelly of squalor and crime, came from the streets of old Paris and not Haussmann's boulevards: his clinical eye analyzed the lives of people dominated by their nerves and blood.

Zola's *La Bête humaine* is filled with crimes and murders, and the protagonist, Jacques Lantier, ponders the origins of his obsession, which is to stab a beautiful woman to death. He wonders whether this relates to an illness from an earlier generation in which his race was infected by a woman and passed down through generations of men. This heredity taint he felt as episodes of instability in his being, which led to times when he had no control over himself. His urge to kill seemed a legacy of an atavistic past. The beast within him was widely considered to be epilepsy.[26]

Heroic Triumph

The determinism underlying the concept of degeneration, as a biological imperative, which, by a natural force, led to madness and imbecility, had huge consequences, not the least being an acceptance of the relative incurability of mental illness. On the horizon was a growing interest in unconscious forces that affect behavior. Reason was clearly under attack. In his writings, Kant used the analogy of revolution, a Copernican revolution that he thought was needed in philosophy. Knowledge was not just passively received and imprinted on the mind, hence a posteriori, but knowledge of the external world of objects was a priori. This he referred to as the categorical imperative, something not given only by experience. The mind has a structure and rules that are universal and guaranteed by God. Kant gave us the distinction between the noumena, things in themselves, and phenomena. The former are beyond our grasp, since we experience only phenomena, but the latter, our synthetic a priori judgments, of necessity imply a creative principle, the result of the powers of the imagination, which were unconscious. The thing in itself (*Ding an sich*) remains unknown to us as it is independent of our experience of it.

Kant set limits to human understanding and access to knowledge, while emphasizing the active synthetic properties of the mind. Such views were well received by romantic authors, as already discussed, and altered the course of Western philosophy—Kant's Copernican revolution.[27]

As the historian Peter Gay, noting the extremely long life of romanticism and telling us why the romantics matter, wrote, "What is forgotten, unjustly, is the rich romantic past that was the first to question age-old traditions, and to teach novelists and poets, composers, painters, and dramatists, to say nothing of architects, where its creators' passions originated and how much they still live in that world"—amen to that, since ideas in neuroscience should surely also be included.[28]

Beethoven wrote in 1802 what has become referred to as the Heiligenstadt Testament, a letter to his brothers about his encroaching deafness, his despair, and his suicidal ideation. In 1803 he composed his Third Symphony in E-flat major, a composition of struggle and triumph, music being his way of overcoming human adversities. At first it was dedicated to Napoleon (when he heard that Napoleon was to be crowned emperor, he altered this), but it was inspired by Prometheus. He had composed a ballet titled *The Creatures of Prometheus* in 1801, and the music is echoed in the symphony. Beethoven's Promethean act of overcoming, the apotheosis of his own heroic struggle as Prometheus, changed Western music forever.[29]

NOTES

1. Pohanka JJ, 2010, 1.

2. Calasso R, 2012, 23. Emily Dickinson wrote her poem around the same time. The full first verse is, "I felt a Funeral, in my Brain, / and Mourners to and fro / Kept treading—treading—till it seemed / That Sense was breaking through—."

3. MacLean P, 1990, 247.

4. Langer SK, 1951, 93.

5. Langer SK, 116.

6. Trimble M, 2012.

7. Langer SK, 1951, 185.

8. Quoted in Langer SK, 188.

9. Langer SK, 189.

10. Langer SK, 199; italics in the original.

11. Langer SK, 206.

12. Pinker S, 1998, 534.

13. The psychiatrist Stephen Porges associates the emergence of empathy in primates with the increasing complexity of the autonomic nervous system and an increased sophistication of what he refers to as the Social Engagement System. He notes another important mammalian anatomical feature, namely, the shift to audio-vocal communication with the evolution of the mammalian middle ear, which developed from the jawbones of earlier reptiles. The human middle ear carries sound at only specific frequencies. It is naturally a-tuned to the sound of the human voice, although it has a range greater than that required for speech. Further, the frequency band that mothers use to sing to their babies, so-called motherese or child-directed speech, with exaggerated intonation and rhythm, corresponds to that which composers have traditionally used to compose melodies. See Porges SW, 2011.

14. Taken from Walter Benjamin's (1892–1940) title for the Arcades Project, an unfinished work on the city life of Paris in the nineteenth century.

15. Rousseau's account of the ideal education in his *Emile* was echoed by Pinel with the latter's use of the term *alienation*, better than the earlier, more popular *folie*, since the purpose of treatment was to reverse the alienation.

16. Included are Étienne-Jean Georget (1795–1828), Jean-Pierre Falret (1794–1870), Louis Florentin Calmeil (1798–1895), Alexandre Brierre de Boismont (1797–1881), Jean-Baptiste Cazauvieilh (1801–1849), Jacques-Joseph Moreau de Tours (1804–1884), and Jules Baillarger (1809–1890). There is a complete list with a diagram of their links in Goldstein J, 1987, 140.

17. See Goldstein J, 154. The Charenton became known as the Esquirol Hospital.

18. Taken from the title of the book by Pick D, 1989.

19. Pick D, 8.

20. For those now aware of the growing literature on the negative effects of marijuana on the human brain and its link to psychosis, Moreau de Tours was there a long time before, observing that hashish could lead to a wide range of abnormal mental states, from the mildest morbid impulse to manic excitation and furious delirium.

21. Morel B, 1860/1999, 176–178.

22. Magnan V, 1999, 275–276.

23. Pick D, 1989, 50. The Second Empire, under Louis-Napoleon (Napoleon's nephew), had provoked war with Bismarck and Prussia and was defeated. In 1871, a new legislature was declared—the Third Republic. The Parisians started to organize an alternative legislature in the capital, and a left-wing Communard government seized power. The royalists who had fled to Versailles invaded Paris. The Communard's rise was brief and their defeat brutal.

24. Bouchet C, Cazauvieilh M, 1825.

25. Zola E, 1871/2012, 3.

26. In the book there is no actual seizure written about, but the doom of the hereditary past permeates the whole story. He gets his woman in the end, but not in the way this may occur in a romantic tale. Nana is a prostitute and is the cipher of another novel named after her. She is descended from generations of drunkards, a plant nurtured on a dung heap, a force of nature and destruction corrupting Paris with her body, with thighs as smooth and white as marble. Cesare Lombroso conjectured that Zola had epilepsy; his case was written up by Dr. Édouard Toulouse (1865–?), who considered Zola a neuropath with nervous complaints, a view that Zola seemed happy with. Magnan considered Zola a degenerate.

27. Kant does not refer to his Copernican revolution; it was a term utilized by others. Although he was interested in medical matters and mental disorder and wrote an essay on illnesses of the head, Kant opined that the causes of them lay not in the brain but in the "digestive parts." He was woken from his dogmatic slumber by the Enlightenment philosopher Hume, who declared that all that bound cause to effect was that certain objects (sensations) are always conjoined, the associationist perspective that knowledge comes only from experience and is not a priori.

28. Gay P, 2015, 117.

29. The importance of art in overcoming adversity is found in the work of Milton, who lost a wife, two children, his eyesight, and his faith in Cromwell before writing *Paradise Lost*: his mind wanted to make a heaven of hell, salvation through art. Johann Sebastian Bach (1685–1750) is another example, who, like Milton, suffered terrible bereavements (of 20 children, only 10 reached adulthood, and his first wife died), yet he produced some of the most sublime music ever written.

Beethoven continually emphasized his individuality. Beethoven's patron Prince Lichnowsky (1756–1814) invited some French officers to dinner, with Beethoven present. Napoleon had made himself emperor, much upsetting the composer, who refused to play the piano, and he stormed out of the dinner party, locked himself in his room, and refused to come out. The door was broken down, and Beethoven threatened to hit Prince Lichnowsky with a chair. Later he wrote, "Prince, what you are, you are by accident of birth. What I am, I am through myself. There have been and will always be thousands of princes. There is only one Beethoven." Suchet J, 2012, 133.

REFERENCES

Bouchet C, Cazauvieilh M. De l'epilepsie consideree dans ses raports avec l'alienation mentale. Recherche sur la nature et le siege de ces deux maladies. *Archives de medicine générale et tropicale* 1825;9:510–542.

Calasso R. *La folie Baudelaire.* Trans. McEwan A. Allen Lane, London; 2012.

Gay P. *Why the Romantics Matter.* Yale University Press, New Haven, CT; 2015.

Goldstein J. *Console and Classify: The French Psychiatric Profession in the Nineteenth Century.* Cambridge University Press, Cambridge; 1987.

Langer SK. *Philosophy in a New Key.* Mentor, New York; 1951.

MacLean PD. *The Triune Brain in Evolution: Role in Paleocerebral Functions.* Plenum, New York; 1990.

Magnan V. Clinical lessons on mental diseases. In: *Anthology of French Language Psychiatric Texts.* Trans. Crisp J. Instititut D'edition: Institut Synthélabo, Paris; 1999.

Morel B. Treatise on mental disorders. In: *Anthology of French Language Psychiatric Texts.* Trans. Crisp J. Instititut D'edition: Institut Synthélabo, Paris; 1860/1999.

Pick D. *Faces of Degeneration: A European Disorder, c. 1848–1918*. Cambridge University Press, Cambridge; 1989.

Pinker S. *How the Mind Works*. Penguin, London; 1998.

Pohanka JJ. *Wagner the Mystic*. Wagner Society of Washington, Washington, DC; 2010.

Porges SW. *The Polyvagal Theory*. W. W. Norton, New York; 2011.

Suchet J. *Beethoven: The Man Revealed*. Elliott and Thompson, London; 2012.

Trimble M. *Why Humans Like to Cry: Tragedy, Evolution, and the Brain*. Oxford University Press, Oxford; 2012.

Zola E. *The Fortune of the Rougons*. Trans. Nelson B. Oxford World Classics, Oxford; 1871/2012.

CHAPTER SEVEN

Charcot's Joints

Since then, the degeneration of this ancient house had clearly followed a regular course, with the men becoming progressively unmanly; and over the last 200 years, as if to complete the ruinous process, the Des Esseintes had taken to intermarrying among themselves, thus using up what little vigour they had left.

Joris-Karl Huysmans, 1884/2003, 3

On November 21, 1907, a marble bust, made by one of London's most fashionable sculptors, Herbert Hampton, was unveiled in a ceremony at the National Hospital for Nervous Diseases, Queen Square. The subject was asked by Samuel Kinnier Wilson (1878–1937) whether it was tiring to sit for such a period of time while the work was created. "It is hard work doing nothing. Sitting still is being under restraint, and that is an active inhibitory process," was the reply. The presentation was made by Sir William Gowers (1845–1915), who finished his address with the words "let us look upon this bust and then turn to the living counterpart—our Master." The "Master" was John Hughlings Jackson (1835–1911).[1]

Unfinished Business—The Mad Genius

The French were defeated by the Prussians, which marked another turning point in French culture and society. The axis of our exploration of the romantic brain soon swings to Germany, and then to England and the writings of Hughlings Jackson. But the legacy of degeneration lingered and spread to infiltrate the arts and ideas about creativity.

In *Against Nature* (*À rebours*), Huysmans portrays Duc Jean Des Esseintes, the last of the line of descendants of a noble but degenerate family, and who bore a striking resemblance to his distant ancestors. Zola had written about the decline of families, while Huysmans, in the style of naturalism, gave us the prototype of "degeneration." His lonely Des Esseintes more and more seeks out the ornamental and artificially extravagant things around him. He had a need to provide for all his senses, ending with his impotence in what is a plotless novel that signaled the end of creativity itself.[2]

Yet there was still considerable creative life in a decadent France, and ideas about creativity, the brain, and its disorders began to flourish. The link between genius and madness echoed the Greek alignment of poetic genius with inspiration from the gods, the singing poet even considered as a god, signified by Orpheus, whose head continued to sing while floating down the river Hebrus after he was dismembered by the Bacchae. The divine frenzy of poetic inspiration is discussed in the *Phaedrus* of Plato. As John Dryden (1631–1700) put it, "Great wits are sure to madness near allied / And thin partitions do their bounds divide."[3]

Post-Kantian philosophies and romanticism were concerned with inner creative forces and the concept of genius. Over time, views have varied from the idea that creativity is evidence of mental disorder, and thus that the creations themselves are a manifestation of mental illness, to views that the act of creation actually provides a means of escape from or mollification of breakdowns, as we have noted with Beethoven and others. The ideas of Moreau de Tours, who referred to creativity as a *névrose*, were taken up by Lombroso. He was particularly interested in criminal anthropology but classified genius as a variety of insanity. He was influenced by current theories of evolution, positivism, materialism, phrenology, and "craniometry," but degeneration was pivotal.[4] In a similar way that the French literary world and some scientists had portrayed a decline and fall of French families as an analogy for the decline of French society, Lombroso, with cranial measurements, photographs, and statistics, plotted the atavistic nature of his degenerate times. In the skull of the criminal he saw the basic nature of their personality, revealing "ferocious instincts of primitive humanity and the inferior animals."[5] In *The Man of Genius* Lombroso discussed the relevant clinical signs of degeneration in people of outstanding creativity, ranging from the small stature of Aristotle, Plato, Mozart, Beethoven, Charles Lamb, De Quincey, and Blake to those who were taller than average, such as Petrarch, Goethe, Schiller, Tennyson, and Walt Whitman. A "cretin-like" physiognomy was accredited to Socrates, Rembrandt, Pope, and Erasmus Darwin, and left-handedness affected Michelangelo and Leonardo da Vinci.[6] Writers in whose works he identified stigmata included Baudelaire, Poe, Zola, and Dostoyevsky.[7]

As the anecdote goes, Lombroso thought he would test his theories by meeting and observing Leo Tolstoy (1828–1910), widely regarded as a genius. He traveled to Tolstoy's home, Yasnaya Polyana, thinking he was going to meet a degenerate-looking cretin. He found an elderly, well-preserved, but irritable man, who barked that Lombroso's ideas were all nonsense—leading Lombroso to conclude that Tolstoy's views reflected a sickly, eccentric mind. After the visit, Tolstoy recorded in his diary that Lombroso was a "limited old man." Later, Tolstoy finished his novel *Resurrection*, a story of a trial of a prostitute falsely accused of murder. In the story, Lombroso's anthropological ideas were put before the court by the public prosecutor and rejected. The court president gave his views that Lombroso was "a very stupid fellow."[8]

In his book *The Insanity of Genius* (1893), journalist John Nisbet (1851–1899) put forward neuroanatomical and neurophysiological evidence to give Lombroso's ideas some scientific standing. He cited examples of men of genius who showed stigmata of mental illness, or in whose families these could be revealed. He thought that the ideas of men of genius were overabundant, series of facts, tones, and colors combining in ways not available to the normal mind. Both the genius and the madman suffered from excessive stimulation, depression, or excitability of certain regions of the brain. There was much phrenology in his ideas, especially related to head size and the frontal parts of the skull. Yet another skull that had made the rounds was that of Jonathan Swift, which had been exhumed and examined and was reported to reveal a frontal development likened to that of a fool, not that of a great wit. But helpfully the areas devoted to motor and sensory capacity were large.

Fin de Siècle

The most remarkable book on links between personality, mental states, and art was *Degeneration* (1892), written by Max Nordau (1849–1923), also a journalist, who qualified in medicine in Budapest in 1876. He was enthralled by Lombroso and dedicated this book to him.[9] Nordau thought that degenerates revealed themselves through their art, and so he analyzed the works of artists for the stigmata of degeneration. A long list of people was thus compiled. These included poets who used extraneous refrains, the stringing together of disconnected words, repetitions, and senseless meanderings with a lack of focus and incoherent thoughts, as well as those who exhibited graphomania. He identified painters such as the pre-Raphaelites, with their cluttered, overly elaborate, often mystical themes; musicians such as Wagner (reflecting a general crisis of German hysteria); and the metaphors of

Friedrich Nietzsche (1844–1900), which, like Wagner's music, were powerful and had dangerous appeal. The works of such degenerates, he wrote, appealed to sexual psychopaths.

The book caused a sensation and is still in print today. It has to be noted that Nordau was living through a time of many emerging artistic isms, such as impressionism, post-impressionism, naturalism, symbolism, and synthetism, and it is not surprising that he wrote that "the spasmodic search for new forms (of art and poetry) is nothing but hysterical vanity. . . . Those that have talent can create within the limitations of ancient forms."[10]

Fin de siècle was a term Nordau introduced for what he was describing, and the first part of his book uses that title. States of fatigue and exhaustion, related to the vertigo and whirl of a frenzied lifestyle, explained degeneration and hysteria, these being consequences of excessive organic wear and tear, which he said particularly affected those living in France.

The influence of Nordau's book lingered on, and the concept of *entartete Kunst*, degenerate art, was central to the constriction and destruction of much art in the time of Nazi Germany a generation later.[11] But it also spilled over into ideas of the brain of the genius being different in structure and function from others, with geniuses thus being more prone to suffer from various diseases than the general population. Having examined the Registrar General's mortality statistics for 1888, Nordeau highlighted gout, paralysis, and epilepsy. He thought that excessive or blunted excitability of the brain was implicated, states of ecstasy being related to violent excitation, brain centers possessing powerful explosive potential.

The influence of Lombroso lives on today in the continued search for differences in the brains of criminals or the highly talented with brain imaging techniques and DNA analyses to identify various behavioral predispositions. Pathological examination of brains of the intellectually superior has included those of Einstein, Stalin, and Carl Friedrich Gauss. Unsurprisingly, as little has emerged from these studies as from the earlier cranioscopy.[12]

Progress

Several neuroscientific conceptual threads of relevance for the later development of neuropsychiatry came together in the mid-nineteenth century, especially advances in neuroanatomy and chemistry, laying humoral theories finally to rest. Rudolf Virchow (1821–1902) enunciated his theory *omnis cellula e cellula*, which emphasized the cellular nature of the normal and the pathological. In physiology, Johannes Müller differentiated nervous from electrical conduction. His law of specific nerve energies stated that for experience the pathway that carried sensory information was important, not the origin of the sensation itself.

His pupil Hermann von Helmholtz opened our eyes to the subjective nature of our sensations: "The nature of the sensation depends primarily on the peculiar characteristics of the (receptor) nervous mechanism; the characteristics of the perceived object being only a secondary consideration. . . . The quality of the sensation is thus in no way identical with the quality of the object that arouses it. Physically, it is merely an effect of the external quality on a particular nervous apparatus. The quality of the sensation is, so to speak, merely a symbol for our imagination."[13]

Helmholtz discovered the rods and three color receptors in the retina, with different light wavelengths of sensitivity. The resulting conclusion that our response to color is dependent on our eyes and unconscious influences rather than a property of a viewed object explained such phenomena as visual afterimages and influenced the future of neuroscience and art—impressionism was born. Subjective responses to light, such as Monet's fleeting impressions, were painted as opposed to representations of the world or objects.

The scientist and artist Ernst Haeckel (1834–1919), most famous for his apothegm "ontogeny recapitulates phylogeny" and who also gave us the word *Darwinism*, showed that the eye was an extension of the brain, and Helmholtz's invention of the ophthalmoscope gave neurology for the first time an ability to see, via the eye, a part of the brain in vivo.[14] Together with Charles Darwin's evolutionary ideas, these scientific revelations provided an additional shift of focus to the dissolution of form of the impressionist schools. This was, after all, a partial capitulation of the idea of the eye as a passive receptor of sensations, incidentally supported by Charcot. The creative nature of our visual system was advanced by Paul Cezanne (1839–1906), who imposed his own structure on the world as he saw it. Painting nature was more than capturing light; Cezanne famously revealed the cylinders, spheres, and cones inherent in nature as imaginative reconstructions of the painter.

Helmholtz estimated the speed of nerve conduction, which was much slower than electricity, and his work *On the Sensations of Tone* influenced musicology as much as his theories of vision had done for the visual arts. His research proved that visual perceptions are not simply external imprints on the retina but require a priori unconscious inferences, for how else could the retina sort out all the incoming sense information into a meaningful whole? A new philosophy and physiological understanding of perception was needed, since perceptions of the present require knowledge of the past. The importance of this had to wait nearly 100 years before a fuller explication.

Hughlings Jackson and the Beginnings of Modern Neuropsychiatry

The first patients were admitted to the National Hospital for the Relief and Cure of the Paralysed and Epileptic, situated in Queen Square, London, in 1860. This was the first hospital dedicated to the care of those with nervous diseases, and its roll call of physicians and surgeons from its founding up to the present represents a history of elegant neuroscience talent. The remarkably skilled Charles-Édouard Brown-Séquard (1817–1894) was one of the first. He performed many neurophysiological investigations, some on himself. He considered epilepsy to be a reflex disorder, linked to vasomotor changes consequent on activity of sympathetic nerve fibers. Theories of epilepsy as disturbances of vascular and/or nutritional changes were becoming accepted by midcentury.

Whereas Gowers referred to Hughlings Jackson as "the Master," Macdonald Critchley (1900–1997), in his biography, called him "the Father of English Neurology."[15] Born in Yorkshire, poorly educated, and uninterested in the arts—prone to leave the theater after the first act—Hughlings Jackson's contributions to neurology and neuropsychiatry are substantial. He misused books and liked penny dreadfuls, often tearing out pages that he had read and discarding them into the street from his carriage window. He had few books at his home, and it is said that if he borrowed a medical book, he would rarely give it back, or if it came back, it was with pages removed.[16]

Hughlings Jackson never wrote a textbook, and perhaps because of this he was less popular than Gowers, whose books on neurology were and still are regarded as some of the finest clinical books on neurology ever written. Gowers's *Manual of the Diseases of the Nervous System* (1886/1888) became the "Bible" of neurology, based almost entirely on his own observations, largely using his own illustrations, written in an elegant and readable prose.[17] In contrast, Hughlings Jackson was a theorist and much interested in philosophy and psychology. Indeed, he considered philosophy as central to neurology. He took careful, detailed notes on patients, but his published writings were unclear, with footnotes, brackets, and qualifications making his texts difficult to read. Neither was he a brilliant lecturer, and consequently his colleagues did not quite comprehend him. He was simply ahead of his time, as we shall see.

Hughlings Jackson was a resident medical officer at the York Dispensary, and there he was influenced by the ideas of Thomas Laycock (1812–1876). Laycock was a physician in York and became professor of medicine at Edinburgh. He was well acquainted with French and German ideas, and as far as he was concerned, insanity was a physical disorder of the brain. He was interested in hysteria and put forward a view, revolutionary at the time, that the encephalon (cortex), like the spinal cord, was endowed with excito-motor phenomena (reflexes). There could therefore be unconscious reflex activity via the cortex which lay at the basis of hysteria and other twilight mental states (such as somnambulism). Although a devoted Christian, he was interested in the mind–body problem and how to bring together mental and cerebral physiology. He argued that the brain was a continuum of the spinal cord and would therefore be subject to the same laws, but he had a teleological perspective, physical forces bringing about an ordering of things most manifest in consciousness and intellectual activity.

Laycock described two cases of "hysterical" ischuria in teenage girls. In addition to weakness and abdominal pains, there was oliguria. One of the girls vomited urine and had discharges of urine from her navel and ears. Eventually, a catheter was found in her bed, and she admitted to self-catheterization and drinking the urine. The second one was also shown to be faking her symptoms and signs. Charcot read Laycock's papers on hysteria and congratulated him on not being deceived by the two young girls.[18]

Hughlings Jackson became acquainted with mental illness while at York via Daniel Hack Tuke (1827–1895) of "The Retreat," the psychiatric hospital founded by Tuke's great-grandfather. Hughlings Jackson was accompanied to London by his friend and fellow graduate Jonathan Hutchinson (1828–1913), who described what became known as Hutchinson's teeth in cases of congenital syphilis. He persuaded Hughlings Jackson to continue with medicine, but to combine that with his interests in philosophy. Central to Hughlings Jackson's ideas are his four principles of nervous action (table 7.1) and his four factors of insanity. First, we must briefly discuss Darwinian evolution, Herbert Spencer, and a return to trilogies and hierarchies.

Evolution

Many events in history have been occasioned by psychiatric or psychological disorders. On August 12, 1828, Pringle Stokes (1793–1828), the first captain of

TABLE 7.1
Hughlings Jackson's Four Principles of Nervous Action

Evolution of nervous functions
Hierarchy of those functions
Negative and positive symptoms of dissolution
Local and uniform dissolution

the ship *Beagle*, died from the effects of a self-inflicted gunshot wound. The newly appointed captain, Robert FitzRoy (1805–1865), whose own family was beset with mental illness, was concerned that a long sea journey may not only upset his own emotional stability but also lead him to the same fate as that of Stokes.[19] He put out a request for an educated scientist of equal social standing to accompany him on his next voyage, whose mission was to continue surveying the South American coast. He invited the young Charles Darwin to accompany him. Darwin was notoriously prone to seasickness and therefore spent as little time on the ship as possible. This allowed him long land explorations and much time not only to collect specimens but also to think about ideas that were currently exciting the world of natural philosophy.

The biblical account of creation states that in the very first days of the universe, God created animals, birds, fish, and the rest, in their own kind. There was no possibility of one becoming another; evolution of life-forms was not possible. Indeed, this view remained fixed well into the seventeenth century. There were a few exceptions. The Greeks were the first to promote an idea of evolution, as in the concept of becoming beyond being discussed by Heraclitus, the weeping philosopher.[20] But for Aristotle, kinds were fixed. During the Enlightenment, notions of evolution were debated, especially in France, as Napoleonic explorers were bringing home fossil specimens that seemed to prove that there was no change in species' forms over at least 6,000 years. But they also provided paradoxical evidence that species had become extinct.

The story of Darwin's discoveries on the Galápagos Islands, born from the fire of volcanoes, is well known, especially how, on the basis of finding different species of mockingbirds and finches on different islands, he explored ideas of geographical separation as one potential explanation of new life-forms. His book *On The Origin of Species by Means of Natural Selection, or The Preservation of Favoured Races in the Struggle for Life* (1859) finishes with the words "There is a grandeur in this view of life . . . from so simple a beginning endless forms most beautiful and most wonderful have been, and are being, evolved."[21]

When published in 1859, Darwin considered that his theory of evolution better explained his and other naturalist observations than a creationist alternative. The principle of evolution was based on random variation and natural selection. Darwin was familiar with the ideas of Thomas Malthus (1766–1834) but had no idea of Mendelian genetics, patterns of inheritance which neo-Darwinian ideas later embraced. For Darwin, natural selection provided an answer, the principle that favorable variations in animals are preserved while those injurious are not, changes brought about in the "struggle for life" therefore becoming inherited qualities.[22] He conceded that *natural selection* was a bad term, and it is one that has subsequently been confused and abused, as has the term *species*.

Hughlings Jackson on Evolution and Dissolution

Darwinian evolution and its correlative social associations put forward by Herbert Spencer (1820–1903) were the key to Hughlings Jackson's concepts of the organization of the brain, in health and disease. Spencer was a philosopher, anthropologist, and sociologist who attempted to synthesize knowledge in the sciences with evolutionary theory. He was much influenced by the ideas of Coleridge, the philosophy of associationism, and German *Naturphilosophie*—a belief that there was an underlying unity to all nature and a teleological progression of natural development. After reading Darwin's *Origin of the Species*, Spencer coined the term "survival of the fittest," which he then applied to social structures, referred to as social Darwinism. However, what was very important to Hughlings Jackson was Spencer's idea that structures evolved from the undifferentiated, simple, and homogenous to the differentiated, complex, and heterogeneous with integration of differentiated parts. For Spencer this was a universal law, one that for Hughlings Jackson could be applied to the evolution of the functions of the brain.[23] The progression was from the most automatic to the most voluntary actions, and units of constitution were re-represented at different levels in the brain, becoming more complex in form. Hughlings Jackson conceptualized the structure and function of the brain in a hierarchical manner, with interactions between levels, the highest level being in the prefrontal cortex. His ideas were better applied to the motor than the sensory systems, and he was concerned with movements rather than muscles.

An important principle was that based on inhibition and release, namely, that clinical signs involved both processes simultaneously. This led to concepts of negative and positive symptoms, which he opined were present in every neuropsychiatric case—"a double symptomatic condition, one negative and one positive."[24] Lesions could never localize a function, but only degrade a system, and the effects reflected the continued level of activity of the parts of the brain spared by the lesion. Hughlings Jackson was clearly straying away from Darwin's concepts, in which the idea of dissolution is not found, as this was reversing evolution.

His interest in philosophy brought him to the royal gates of neurophilosophy, namely, to consciousness and mind–brain interaction. He associated mind with consciousness, rejected panpsychism, and was reluctant to accept the idea of the unconscious mind: "but to be thoroughly materialistic as to the nervous system, we must not be materialistic at all as to mind. . . . There is no physiology of the mind any more than there is a psychology of the nervous system."[25] He advocated the Doctrine of Concomitance, a parallelism of mental states and neural states. This is based on three main premises: that mental states are different from neural states (a dualist position); that there is a correlative nervous state for every mental state, but the relationship is totally unknown; and that although mental and nervous states occur in parallel, there is no interference of one with the other.

Normal mentation was a rhythm and combination of what he called subjective and objective consciousness, ideas that he developed from his careful studies of aphasia. The contributions of Broca are discussed below, but suffice it to say that while Hughlings Jackson initially accepted the concept that speech may be localized to the left anterior hemisphere, he soon shifted his position, so well analyzed by Anne Harrington in *Medicine, Mind and the Double Brain*.[26] This too was related to the mind–brain problem. He struggled with the question of how a mental activity (speech) could be linked to a physical one (hemiplegia), and he concluded that aphasia was in reality also a physical disorder. He was impressed that aphasics retained some language, albeit emotional and with very limited expression, and he considered that normal propositional speech was "voluntary" while the latter was "involuntary." These words had special references for him, *voluntary* implying activity in the less perfectly organized complex cortex, a physiological concept with no reference to volition.

Hughlings Jackson thought that the left hemisphere was the leading one, giving rise to propositionizing, acts that were accompanied by consciousness. There was less awareness of more automatic actions, the right hemisphere having no capacity for consciousness in words. He took notice of a finding relating to the developing brain by the French anatomist and zoologist

Louis Pierre Gratiolet (1815–1865). He worked with Pierre Paul Broca (1824–1880) on aphasia and divided the brain into five lobes (frontal, temporal, parietal, occipital, and insular). He reported that the left frontal lobe developed in advance of the right, and the right posterior lobe in advance of the left. Harrington concludes that "Jackson's discovery of the French anatomical studies led to the revelation that the left anterior and the right posterior lobe each were slightly more advanced evolutionarily than their twins." This not only counteracted a prevailing Continental view that the left hemisphere was superior to the right but also suggested that the posterior lobes were the seat for most mental operations (i.e., in the visual area). The less evolutionary advanced right frontal and left posterior areas were more automatic. This was reflected in the emotional utterances of aphasics, but also perhaps in normal speech aberrations. "Jackson believed, in fact, that they played a vital role in the sensory-motor activities underlying propositional speech and perception. The energising of the lower arrangements, he argued, 'although unattended by any sort of consciousness', was essential for activating those higher, more voluntary arrangements in the left anterior and right posterior parts of the brains, arrangements whose activities *were* attended by consciousness. In short, *mentation was a dual process*, played out between the two hemispheres of the brain."[27] Further, evolution was involved, even if pre-Darwinian, since the fittest and the best images survive to become conscious. These ideas become revived in some current theories of Gerald Edelman, whose theory of consciousness is based on re-entrant neuronal circuits and what he called "neural Darwinism."[28]

There has been much debate about some of Hughlings Jackson's theories and how they may be relevant to late twentieth-century neurology. His hierarchical organization of brain functions, in keeping with the Victorian social structure of his times, and his insistence that neurological signs cannot be only the result of the effects of damage of a lesion continue to have resonances to this day. These views implied a subordination of the lower automatic levels to less organized, more complex, higher ones, as well as a passage from the most automatic to the most voluntary, all important for integration as a whole. Ideas related to positive and negative symptoms become even more relevant for today's neuropsychiatry, as do his discussions of hallucinations, delusions, illusions, and disturbed conduct with dissolution.[29] In his paper "The Factors of the Insanities" he states that "in every insanity there is morbid affection of more or less the highest cerebral centres . . . [they] are out of function, temporally or permanently from some pathological process."[30]

However, the subsequent clinical picture related to the depth of the dissolution, the rate at which it had taken place, the influence of the state of the body and external circumstances, and the personality of the person affected (referred to as the four factors of insanity).

Hughlings Jackson's incorporation of evolutionary theory into neuropsychiatry emphasized that the brain is not the static organ of the pathologist's table, but the product of millions of years of development in space and time. With his adherence to the Doctrine of Concomitance, Hughlings Jackson thought he could free neurology from the influences of faculty psychology and its psychologico-materialistic theories, using only physiological ones.[31]

Hughlings Jackson had a considerable interest in epilepsy. He introduced the concept of "dreamy states." He had observed these abnormal alterations of consciousness in patients with seizures and noted an association with episodes of déjà vu and jamais vu. He took a lead from James Crichton-Browne (1840–1938), who noted that in cases of epilepsy such events always involved damage to the right side of the brain, and as Harrington summed this up, Hughlings Jackson identified "the left brain with active, objective (conscious, voluntary, intelligent) processes, and the right brain with more passive, subjective (unconscious, involuntary, visceral/emotional) ones."[32] As we progress through our journey into the next century, these insights will be seen as being simply remarkable.

Hughlings Jackson and His Friends

Hughlings Jackson made many other contributions to neurology, beyond the scope of this book. He wrote much about epilepsy and was also interested in associated states of psychosis and dementia. He married his cousin, who, following septic thromboencephalitis, developed seizures of the form now referred to as Jacksonian, with a motor march. He described a case of multiple tics with coprolalia in 1865, 20 years before Gilles de la Tourette. He maintained his interest in mental illness, visiting both the Bethlem and Guy's Hospitals and discussing cases with George Savage (1842–1921); he also visited the Wakefield Lunatic Asylum, where he conducted experiments with Crichton-Browne and David Ferrier (1843–1928).

Crichton-Browne deserves more than a brief mention in the history of modern neuropsychiatry. Referred to by one obituarist as a "pioneer neurologist and a scientific drop-out," he served as director of the West Riding Lunatic Asylum, Wakefield, and was a Lord Chancellor's Visitor in Lunacy. He was highly influenced by Laycock, establishing autopsies on deceased patients as a routine, with an associated laboratory for neuropathology. He published the annual *West Riding Lunatic Asylum Reports*, containing many articles about mental illness (including data related to 1,500 autopsies). Hughlings Jackson published papers in this journal, including some on epilepsy and related mental disorders, but together with Ferrier, professor of forensic medicine at Kings College, London, he went to Wakefield to use the laboratory that Crichton-Browne had established there, in part to conduct vivisection investigations. Ferrier needed to find a suitable location for his experiments as he was attempting to study localization of function in the brain using electrical stimulation and lesions in animals. In 1876 Parliament passed the Cruelty to Animals Act, and antivivisectionists targeted Ferrier and took him to court for operating without a proper license. He won his case, and his research helped confirm some of Hughlings Jackson's views derived from clinical experience on the localization and hierarchical nature of the sensorimotor functions of the brain.

The *Asylum Reports* stopped when Crichton-Browne left the hospital, but he cofounded the journal *Brain*, along with John Bucknill (1817–1897), Ferrier, and Hughlings Jackson. This became   and probably still is—the world's most influential neurological journal. Crichton-Browne was proposed as a fellow for the Royal Society by Charles Darwin and admitted. He was knighted in 1866.[33]

In contrast to Gowers, Hughlings Jackson was much admired for his dignified, commanding appearance; for his intellect; and, in spite of a relative lack of warmth, for his kindness. His place in the history of today's neuropsychiatry is secure. As Frederick Golla (1878–1968), in a review of *Selected Writings of Hughlings Jackson*, wrote, "His fundamental conception of the real significance of symptoms as the evidence of release of control is never lost sight of, and provides the clue to much that has been hopelessly entangled by the irrational attribution of positive symptoms to nervous matter that has undergone destruction."[34] This simple lesson has been lost to present-day psychiatrists, with their muddles over the terms *positive* and *negative*.

Hughlings Jackson's influence was worldwide and still remains so, at least for those who need to dwell on the signs and symptoms of neurological and psychiatric disorders and those who aspire to be neuropsychiatrists. A list of those who were influenced by his ideas is given in table 7.2. His importance for Freud will be encountered in the next chapter. A number of American neurologists were influenced by him, notably Silas Weir Mitchell (1829–1914), who dedicated his book *Lectures on Diseases of the Nervous System* (1881) to

TABLE 7.2
List of Scholars Influenced by Hughlings Jackson

Country	Those Influenced
England	Henry Head (1861–1940); Russell Brain (1895–1966); Macdonald Critchley (1900–1997)*
Germany/Austria	Sigmund Freud (1856–1939); Eugene Bleuler (1857–1939); Kurt Goldstein (1878–1965)
France	Théodule-Armand Ribot (1839–1916); Pierre Janet (1859–1947); Henri Ey (1900–1977)
Czechoslovakia	Arnold Pick (1851–1924)
America	Adolf Meyer (1866–1950); Silas Weir Mitchell (1829–1914); James Jackson Putnam (1846–1918); William James (1842–1910)
Russia	Constantin von Monakow (1853–1930); Alexander Luria (1902–1977)

* The author, although not listed in the table, must be included among those influenced by Hughlings Jackson.

Hughlings Jackson, and Adolf Meyer, whose writings on comparative anatomy and neuropathology aligned with his own attempts to unite the biological with the psychological in psychiatric disorders.

James Jackson Putnam (1846–1918) was a distinguished neurologist who visited the National Hospital and the London Hospital and discussed with Hughlings Jackson their mutual interests, especially his concepts of hierarchies of nervous function and positive and negative symptoms. However, Hughlings Jackson's writings on aphasia caused a storm of neurological protest, so first we must consider events in France, as well as the relationship between Hughlings Jackson, Broca, and Charcot.

The Hysterical French Revolution

"There is an enigma that remains unexplained; the word 'hysteria' resolves nothing; it may be enough to define a physical condition, to denote the uncontrollable turbulence of the senses, but it does not get at the spiritual consequences that fasten upon it, or especially the sins of duplicity and falsehood which nearly always takes root in it. What are the ins and outs of this sin-laden malady? . . . Nobody knows: on this subject medicine talks nonsense and theology remains silent."[35] Admittedly, these views were written by Huysmans, a novelist whose stories were much bound into theories of degeneration, but they nevertheless emphasize ongoing conceptual enigmas surrounding this age-old disorder.

In Paris in the mid- to late nineteenth century, the number of cases so diagnosed was on the rise, and many were admitted to the Salpêtrière, mostly working-class women. Pinel had advanced a scientific, Enlightenment perspective to psychiatry and a put a nosological structure around disorders of the body and mind. Taking a lead from Cullen, but moving away from his view that the *neuroses* were physical conditions, Pinel thought that these were disorders of mental alienation, best viewed as abnormalities of sense and movement, with moral (i.e., psychological) causes, and without pathological findings in the brain. Hysteria was listed under genital neuroses of women, even though he recognized a similar condition in men (with satyriasis as opposed to nymphomania as one variety). The "sympathetic" nature of the neuroses reflected various influences of parts of the body (e.g., stomach, genitals) on the brain.

Gradually, a number of Pinel's neuroses were shown to have an organic basis (such as epilepsy), and it was Georget who separated out hysteria, hypochondriasis, and a number of other conditions that were characterized by being chronic, not dangerous, and intermittent, which gave an impression that there was some serious underlying disease, but which yielded no pathology at postmortem. But this view sacrificed a growing trend, namely, a search for localized anatomical lesions in mental disorders. Into the discussions entered Achille Louis Foville (1799–1878), who argued that there could be "functional localization" in the brain, and Paul Briquet (1796–1881). In his book *Traité Clinique et Thérapeutique de l'Hystérie* (1859) Briquet described in some detail 450 patients personally examined, identifying cases of hysteria in males, and noting that patients' symptoms could last many years.[36] Briquet, today famous for the eponymous Briquet's hysteria (sadly rechristened several times), had, like many others by his time, abandoned any uterine theories, and he considered that the parts of the brain that received affective impressions and sensations were involved, although he recognized many other influences, including heredity, trauma, and physical abuse. He felt that the term *hysteria* should not be abandoned, as it had been in use so long that everyone understood its meaning.

The last quarter of the nineteenth century saw a rise in the status of and societal interest in psychiatry in Paris, and the doyen who came to dominate the area was Charcot. By neurologists considered a neurologist, he is often referred to as a psychiatrist, but perhaps better embraced here as a neuropsychiatrist.[37] He was appointed as chief of the Salpêtrière in 1882,

Figure 7.1. A Clinical Lesson at the Salpêtrière (Une leçon clinique à la Salpêtrière),
by Pierre Aristide André Brouillet. The image shows patient Blanche Wittman
falling, to be caught by a nurse; Babinski is the male figure between the nurse and
Charcot. An all-male Parisian audience is filling the back rows of the "theater,"
and behind them is the image by Richer suggesting the *arc en cercle* that Blanche
will adopt. Other notable medical figures include Gilles de la Tourette (front
row, far right), Richer (sitting to the right of Charcot), and Marie (two over from
Richer). Charcot's theories, their weaknesses, and the sociopolitics of the time
are all encapsulated here. His medical approach is suggested by his dress and his
approach to the patient being hypnotized, which he considered a special mental
state. The theatrical surround and the image on the back wall of the posture that
Blanche Wittman is expected to adopt implied that "suggestion" was the more
important component of her "seizure."

having worked there since 1862. Administrative changes at the hospital led
to patients with hysteria being transferred to the wards for patients with
chronic epilepsy. Like Briquet before him, Charcot had fallen into the subject. As we have seen, at this time, hysteria was considered within the
provenance of neurology, there was no psychiatry as such, and the so-called
alienists were mainly interested in the psychoses, not the neuroses.

With regard to hysteria, Charcot considered that it should be studied in
the same way as any other medical condition, and that a sufficient number of
observations would establish its course and causes, the latter of which should

of necessity be sought in the brain. The famous painting of him demonstrating a case of hysteria at a Friday lecture to a distinguished crowd of males, *Une Leçon Clinique à la Salpêtrière* by André Brouillet (1857–1914), is a picture worth a thousand words (fig. 7.1). In the audience there are some doctors, including Joseph Babinski (1857–1932), Pierre Marie (1853–1940), and Georges Gilles de la Tourette (1857–1904), but also Paris dignitaries.[38] The patient, Blanche Wittman, about to have a seizure under hypnotic suggestion, already displays a hand dystonia. She will be caught by one of only two other females present, a nurse. On the wall opposite her is an image from Charcot's artist, Paul Richer (1849–1933), showing the position she should adopt, the *arc en cercle*, one of the phases of the attack which Charcot recognized. Thus, he distinguished two forms of hysteria: *grand hystèrie*, with convulsions, and *petite hystèrie*. He documented many of the stigmata that can be identified on examination, such as tubular vision or hemianesthesia. He parsed out the progression of the attacks into different recognizable periods or segments, from prodromes, through to *attitudes passionnelles*, to the period of delusion. He noted "hysterogenic zones" and ovarian tenderness, at which sites pressure could bring on an attack. This link to the feminine did not mean that he did not recognize hysteria in men, but it echoed to the past, as did his views of hereditary predisposition in the disorder, as a feature of degeneration.

Arc en Cercle

The painting displayed and contrasted two views of hysteria. There was a rival school to that of Charcot at Nancy (under Hippolyte Bernheim [1837–1919] and Ambrose-August Liébeault [1823–1904]), which believed that there was nothing more to the state of hypnotism, which was one of Charcot's methods of inducing symptoms, than suggestion. Proponents of the latter school were unimpressed by the revealed stigmata and the idea that there was something special about the brains of people with hysteria and when in a state of hypnosis; the hysterical diathesis was not pathological but a condition that could be found in everyone to some degree. Charcot, seeking a neurologically based theory, believed that there was a hereditary neuropathy giving rise to personality predispositions; essentially, hysterics were neuropaths.

Hypnosis became one battleground, and the academic and less academic arguments were acrimonious. The implication was that Charcot was suggesting the form of the attack that his patients should adopt; his patients were following his overt and covert instructions, in effect deceiving him. The painting reveals this debate, as the grand master is seen in a theater setting, seemingly conducting the play that is about to commence for an all-male audience.

Three Graces

Into this arena stepped two other female graces, referred to as Augustine and Geneviève. Their stories have been well told in other texts, and their images well portrayed, since they all fell under the captivating gaze of Charcot's photographer Paul-Marie-Léon Regnard (1850–1927). He convinced Charcot to publish a journal with his images in it and an accompanying text written by Désiré-Magloire Bourneville (1840–1909). The latter was an alienist, a neurologist, and a journalist who became an assistant to Charcot. In addition to founding the *Iconographie Photographique de la Salpêtrière* (1876/1877), later to be succeeded by the *Nouvelle Iconographie de la Salpêtrière* (1888–1918), he also started *Progrés Médical* and the *Revue photographique des hôpitaux de Paris.*[39] Charcot and his team thus introduced the visual photographic image to clinical medicine and gave us the real faces of hysteria. According to Georges Didi-Huberman, the camera was an instrument that framed and fashioned the signs of hysteria, the women posing for Charcot in postures that he desired; these images became iconic of hysteria, and the camera literally reinvented the condition. The faces, gestures, and postures of the women represented a *tableaux vivant,* one that was fixed and memorialized for future generations. Labels were attached to their states, such as "ecstasy," "crucifixion," "eroticism," and "catalepsy," identifying the supposed affects and passions of the hysterics.[40]

Although the image of Blanche may be the most recognized (because of her appearance in the Brouillet painting), it was Augustine who filled many of the illustrations; she has become Charcot's most celebrated hysteric. Later generations turned her into an icon, especially as a feminist victim of misogyny. There are those who maintain that hysteria was and is a social construct, imposed by male doctors on women, yet the signs and symptoms of the protagonists described so well in Asti Hustvedt's *Medical Muses* have a regularity that can only echo from the clinics of the past to today. Convulsions, fainting spells, paralyses and contractures, complex and often florid hallucinations, mutism, sensory stigmata, eating disorders, intense religiosity, self-mutilation, somnambulism, and so on, are still seen in certain medical settings. Case histories often contain evidence of early separation from parents, either abandoned or fostered, and sexual or physical abuse. Temper outbursts and labile moods pepper some of the anamneses.[41]

Of the three, only Blanche leaves us her tale. According to Hustvedt, after the death of Charcot in 1893, she never had any more convulsions, paralyses, or deliria. She became an employee of the Salpêtrière and eventually worked as a technician in the radiology laboratory. As a consequence of radiation exposure, she first had the fingers of one hand amputated but then lost the whole arm before the process started in the other limb. Her story, including the eventual amputation of three limbs, has been novelized by Per Olov Enquist, in an engaging tale of her relationship with Charcot and Marie Curie. Blanche was interviewed later in life and asked the inevitable question whether she had simulated her attacks. This she denied and stated, "Simulation! Do you think it would have been easy to fool Monsieur Charcot?"

Charcot and Hughlings Jackson

Charcot's main concerns were with theories of cerebral activity and how the signs and symptoms of a damaged nervous system emerged, an enterprise not dissimilar to that of Hughlings Jackson. Yet these two Rhadamanthine champions of the art of clinical observation of the neurologically damaged could scarcely have been more different. Charcot spoke excellent English and visited England on several occasions; Hughlings Jackson spoke limited French but had extensive knowledge of French medical literature. It is known that they met at a British Medical Association annual meeting in Brighton in 1886. What is also known is that Charcot had a very high opinion of Hughlings Jackson's work on epilepsy. He even gave the name Jacksonian epilepsy to the focal unilateral convulsions he described, though a similar pattern had been described before by a Frenchman, Louis François Bravais (1801–1843). It is also known that Charcot had a portrait of Hughlings Jackson in his office. Charcot had an interest in art and music (although the impressionist school was too imprecise for his taste; he preferred the Dutch old masters), but apparently Hughlings Jackson did not. Charcot loved animals and was against animal experiments, whereas Hughlings Jackson needed animal experimentation to prove his theories, hence his collaboration with Ferrier and his excursions to Wakefield.[42]

Unlike "the master" Hughlings Jackson, whose name remained vibrant in English neurology and today is well known in neurological circles all over the world, after his death Charcot's reputation virtually collapsed. For the last 15 years of his life his interest in hysteria may have made him famous and a household name, coinciding with a time when hypnosis was becoming a fad in French social and medical circles. Studies of transient alterations of the mental state with apparent therapeutic benefits had run an earlier course, with techniques such as Franz Anton Mesmer's (1734–1815) use of "animal magnetism" and his treatment known as mesmerism, which coincided with an ever-growing interest in somnambulism, cataplexy, and automatic writing.

Mesmer was interested in planetary influences and the power of a universal fluid with quasi-magnetic properties which could be used to restore any disturbed energy balance in the body. At first he used magnets, then his own charisma, with fixed gazes and hand passes over the body to redistribute such forces. He attracted so many patients that he developed a "baquet," a large circular vessel with holes in it which contained ground glass and iron filings. Iron rods protruded from it, which people held and through which "magnetism" was transferred to them. These were placed at different angles and so could be applied to different parts of the body. Music was played, and Mesmer, wearing a silk cloak and carrying an iron wand, would "magnetise" the participants, of which over 20 could be accommodated at a time. The whole business provoked an investigation, instigated by King Louis XVI, who appointed four members of the Faculty of Medicine to evaluate "magnetism." They were unable to find any evidence of the fluid substance that was the basis of the treatment.

Mesmer was undone, at least in Paris, and Charcot's work on hysteria dented Charcot's reputation. There was a backlash against his theories, notoriously led by Babinski, who sided with those who considered that the mechanism of hysteria was suggestion, and who was one of the first to abandon the term *hysteria*. He unsuccessfully tried to replace it with *pithiatism* (curable by persuasion). It may be that the school from Nancy won out in the end.

Charcot is perhaps best known for the eponymous Charcot's joints, the neuropathic joints associated with syphilis, and for his descriptions of diseases such as disseminated sclerosis and paralysis agitans (Parkinson's disease); in the latter disorder he noted mental symptoms as a part of the condition.

But perhaps a major difference between Charcot and Hughlings Jackson was in relation to cerebral localization. The story of localization will be discussed in the next chapter, but before Charcot's investigations there was much less interest in or acceptance of links between focal lesions and cortical signs and symptoms. But Charcot was a localizer, whose experimental method was to carve nature at its joints. He followed patients from the clinic to autopsy, seeking circumscribed lesions in the cerebral hemispheres, even in cases of hysteria.

Hughlings Jackson sat on the fence. George York and David Steinberg note that he realized that his clinical observations and theories could not support either a localization or a universalization theory (the idea that brain functions were widely distributed in the brain). As noted, he could not accept that language was the property of only one hemisphere, and by insisting on the exclusive sensorimotor nature of nervous functions and the re-representation of one level in the next highest level, and so on, he avoided a strict localization perspective.[43]

English physicians, in contrast to the French, took much less interest in the neuroses, and even less in hysteria.[44] There were few articles on hysteria in the early editions of *Brain*, but there were publications by Sir Benjamin Brodie (1783–1862), Sir John Russell Reynolds (1828–1898), and Henry Charlton Bastian (1837–1915). Kinnier Wilson saw cases at the National Hospital "in abundance."[45] Hughlings Jackson seems to have had little or no interest in hysteria and disliked the word "functional." He rarely mentions hysteria presenting as a seizure disorder, although working with so many patients with seizures, he must have formulated some views on associations with hysteria. It is recorded that he occasionally used hypnosis.

Charcot's Ark

There were many in the circle of Charcot who made magnificent contributions to what became the neurology of the twentieth century, but here only a couple are noted who contributed importantly to the history of neuropsychiatry as we know it today.

Gilles de la Tourette is a name that has now become more famous than that of Charcot, thanks to Charcot's habit of giving eponyms to diseases. The story is briefly told. He was one of Charcot's favored students, joining the staff at the Salpêtrière in 1881. Charcot had read Gilles de la Tourette's translation of a paper about the "Jumping Frenchmen of Maine," a movement disorder described by George Beard (1839–1893), characterized by exaggerated startle responses, imitated movements and sounds, and unusual noises. He reckoned that if there were Frenchmen that behaved so in Maine, then they should be found in Paris, and Gilles de la Tourette set out to find them. He found no equivalent patients but something else, even more extraordinary, namely, a condition characterized by compulsive gestures, tics, and swearing. The first publication on this was in 1884, and a more extended paper was written in 1885. He identified one patient who had been described by Jean Itard (1775–1838) some 60 years before. The uttering of obscenities was a common feature, present in five of nine cases, but they all made involuntary jerks and tics, different from chorea, and had associated obsessions, compulsions, and episodes of bizarre behavior. Charcot gave the name for this disorder as *la maladie des tics de Gilles de la Tourette*. It is a neuropsychiatric disorder par excellence, with such a unique combination of motor abnormalities, obsessional rituals, and some serious psychopathological links, including self-mutilation.

Gilles de la Tourette also showed a great interest in hysteria and wrote about the links between anesthetic patches and the earlier-described witches' patches, examining in detail the case of Soeur Jeanne des Anges. Her account of her hysterical illness, provoked by her unrequited love for the Loudun priest Urbain Grandier, led the latter to be burned at the stake for witchcraft. Poor Gilles de la Tourette died unhappily. After his mentor Charcot and his own son had died, he was shot in the head by a patient who claimed he hypnotized her. Although he survived, his behavior became quite erratic, and he later became quite insane, although this was not due to the head injury. He had published a paper on the progression of syphilis into neurosyphilis in 1899, noting the development of paralysis and insanity. He recognized such symptoms in himself and died at an asylum near Lucerne.

Pierre Janet (1859–1947) was at the center of the hysteria debates between the Nancy and the Paris schools, and he wrote extensively on the disorder and its symptoms. He demonstrated that the clinical signs in hysteria did not follow known anatomical patterns, and he explored psychological mechanisms of hysteria. Failure of psychological synthesis, amnesia, and constriction of the field of consciousness were the elements of his theories. Subconscious parasitic ideas, concealed in the deeper layers of the mind and related to traumatic events, acted as *idées fixes* and were a target for treatment.

His concept of psychological automatisms was very influential, leading to later ideas about dissociation, hysterical amnesia, multiple personalities, and the subconscious, the latter being able to influence motor somatic expression. Janet did not use the term *unconscious*, was critical of psychoanalysis, and was quite disliked by Freud, in part because of Janet's insistence that he had discovered the cathartic method of treating hysteria before Freud.

The division between madness and sanity was becoming blurred, and there was a call for nervous disorders to be recognized in their own right;

the *demi-fous* needed their own specialists. What became the later psychiatry emerged from the study of the neuroses (*névroses*) and madness (the *aliéné*), the neurotic and the psychotic, the former being physiological rather than a structural disorder of the nervous system. But there was another spin-off that was to have huge consequences, and that was the possibility of seeing and treating patients as outpatients. The doorway was open to private offices and the possibility to conduct psychological treatments, away from onlookers, which shifted the scope of neurological and psychiatric practice forever.[46]

Charcot's ideas were captured in America by George Beard's neurasthenia, a term that not only had Charcot's approval but characterized a neurosis he considered comparable to hysteria. He disputed Beard's suggestion that neurasthenia was only an American disease, arguing that the pressures that Beard suggested led to the disorder in America also existed in France.

The increased emphasis on the psychological as opposed to the neurological as a basis for hysteria heralded a geographical shift from France to Austria, as well as a conceptual shift with significant consequences for neuropsychiatry, as we shall see with Freud's theories.

But it is also the time to acknowledge Charcot, sometimes referred to as the Napoleon of Neurosis, as the neuroscientist who changed our views of the brain and its functions and who, along with Hughlings Jackson, must be viewed as a founder of today's neuropsychiatry.

Hughlings Jackson, philosophically of Aristotelian persuasion, could not commit to any kind of mind–brain identity and set a challenge to those who espoused a strict localization of functions in the brain. The greater challenge emerged from Vienna, and the battleground became aphasia. Neurology and psychiatry were about to emerge as quite independent disciplines, a situation that cut up the brain and curdled the mind.

NOTES

1. Critchley M, Critchley EA, 1998. For the full story of the curious history of what happened to the bust, see Trimble M, 1997. For many years it stood in the front hall of the hospital, but it was later moved to the Institute of Neurology, from whence it was stolen. Its whereabouts are known to a Canadian neurologist, who to this day declines to reveal this information. However, a copy of the bust was made and displayed at the Montreal Neurological Institute, which, under the auspices of Dr. Fred Andermann, had a copy of the copy made, which was presented back to Queen Square on July 18, 1996, where it now stands in the library, safely chained among the books. PS: If the conscience of the neurologist who knows

where the original is may be aroused (as not before) by reading this, the author would be delighted to hear from him.

2. Des Esseintes also, incidentally, had a book fetish, yet he had given up reading; it was the lush paper and bindings that absorbed him.

3. Dryden, *Absalom and Achitophel* (1681), pt. 1, lines 163–164; see Fowler A, 1992, 679.

4. For a lucid account of Lombroso's influences and ideas, see Pick D, 1989, 111–152.

5. The so-called *Risorgimento*, the Italian political movement that led to eventual Italian unification, was an ongoing process up to 1871 and occasioned much social unrest. Lombroso was a progressive socialist and believed that his theories would lead to identification of criminality and to preventative actions that might progress to avoidance of prison detention and punishment. His views of degeneration were more positive in their intentions than the all-pervasive degeneration theories of Morel, Magnan, and Zola, with Lombroso advocating social reforms to lessen poverty, crime, and incarceration.

6. Lombroso C, 1891.

7. He even identified the practice of tattooing as the trace of a primitive language of the "lower orders." Pick D, 1989, 117.

8. Mazzarello P, 2001, 983.

9. The tendency for journalists to pick up scientific (especially medical) ideas and popularize them, often getting the facts and certainly the details wrong, but making much money publicizing their own embellishments, is rampant today, especially when it comes to popular neuroscience.

10. Nordau A, Nordau M, 1943, 406.

11. The exhibition of "Degenerate Art" opened in Munich in July 1937 and was taken around to 11 German and Austrian towns. It contrasted the "degenerate" style of many contemporary artists (now much sought after, such as Emil Nolde, Max Beckmann, Otto Dix, and Ernst Ludwig Kirchner) with the ideals of classic German styles, such as the paintings of Adolf Ziegler.

12. Einstein's brain was preserved in formalin, photographed, and then cut into 240 blocks for microscopic examination. These have been analyzed by several pathologists, looking at neuronal and glial patterns, or gross anatomical features, notably in the frontal and parietal cortex. A recent paper has examined the cell microcolumn morphometry of "three distinguished scientists." Significant differences between their brains and comparison specimens were noted in minicolumnar width and cell spacing, the former bearing "similarity to that described for both autism and Asberger's syndrome." There is speculation about whose brains may have been examined, although case 1 bears a striking resemblance to the behavioral neurologist Geschwind. See Casanova MF, et al., 2007, 557.

13. Gamwell L, 2002, 57.

14. The expression "ontogeny recapitulates phylogeny" implied that the development of the embryo resembled the evolution of the species. Some credit the invention of the ophthalmoscope to Charles Babbage (1791–1871), whose prototype was never used.

15. Macdonald Critchley was appointed to the National Hospital in 1923 and took an interest in organic mental disorders, but he was quite put off by states of neurosis and the area of psychoanalysis. He visited patients at many mental hospitals, including the Maudsley Hospital, attending the ward rounds of Edward Mapother (1881–1940). Critchley traveled widely in the Second World War and observed shipwreck survivors, writing a monograph on what is now referred to as posttraumatic stress disorder. Some early twentieth-century neurologists, such as Gordon Holmes (1878–1965) and Kinnier Wilson, were most hostile toward psychiatry, while Francis Walshe (1885–1973) was at least indifferent. Risien Russell (1863–1939), a close friend of Hughlings Jackson and colleague of Gowers, had a huge private practice of psychoneurotics and tried to keep psychotic patients out of asylums as long as

possible, recognizing the fate of those so admitted. He served as a chairman of the National Society for Lunacy Reform.

16. Penny dreadfuls were popular stories published serially, printed on cheap paper and sold weekly at a penny a piece.

17. Gowers was of humble origins but was considered a brilliant student, who taught himself painting and engraving. He was an eloquent teacher, and he examined patients with great care. He made notes in shorthand and would insist that his house officers learn this method. Not a warm person and highly obsessional, he was not empathetic toward nurses or patients, and he fell out with colleagues. He developed pseudobulbar palsy, from which he died (Critchley M, personal communication).

18. For more on Laycock, see James FE, 1998.

19. He later cut his throat in a state of depression.

20. As opposed to the pre-Socratic Democritus (460–370 BC), the laughing philosopher, who was a strong advocate of an atomic theory of the universe.

21. Darwin C, 1859/2006, 760.

22. Darwin C, 501–533.

23. Spencer and Hughlings Jackson corresponded with each other for 40 years, but it is unclear whether they ever met. Hughlings Jackson was a friend of Darwin. Spencer was the best-known philosopher of his day, whose works sold over a million copies in his lifetime.

24. Taylor J, 1958, 2:414.

25. Taylor J, 1:367, 417.

26. Harrington A, 1989.

27. Harrington A, 223, 226, 227; italics in the original.

28. See, e.g., Edelman G, 1989.

29. Taylor J, 1958, 2:46–47.

30. Taylor J, 2:411.

31. Faculty psychology is the view that the mind is composed of separate faculties or modules, still adopted by many psychologists.

32. See Harrington A, 1989, 233. She also notes that Arnold Pick (1851–1924) made the same observations.

33. Jellinek EH, 2005. Anyone who thinks that attacks on today's neuroscientists for animal experiments reflect something new simply ignores history.

34. Frederick Lucian Golla was a physician at the Maida Vale Hospital in 1913, and later professor of mental pathology at the University of London. He then became director of the Burden Neurological Institute in Bristol. He was accredited as the "grand old man of electro-encephalography," being in 1929 the first neurologist to take Hans Berger's claims seriously (see chap. 10 for more on Berger). Golla F, 1932, 204–6.

35. Huysmans J-K, 1884/2003, 201–211. This was written as an appendix 20 years after his novel was first published.

36. Briquet P, 1859. In 59 patients symptoms lasted more than 20 years; in five patients, 55 years. For an account of the development of the neurosis concept in France at this time, see López Piñero JM, 1983, 44–63.

37. Simple designation of these terms is difficult, a point made in the preface to this book. It must be clear that the ideas that have been discussed up to this point, but covering in particular the late eighteenth and nineteenth centuries, have culminated in a body of knowledge which, in today's terms, was neither strictly neurology nor psychiatry. The term *neuropsychiatry* was not used but surely could be attributed to the growing number of physicians who were interested in clinical problems that had a basis in neurological signs and symptoms and embraced a wider perspective that included emotional states, as well as a background philosophy that was not wedded to a strict empiricism.

38. Many people came to know of Charcot through the writings of the Swedish psychiatrist/physician Axel Munthe (1857–1949). He wrote an autobiography, *The Story of San Michele* (1929). He attended Charcot's lectures and was hardly complimentary: "short of stature, with the chest of an athlete and the neck of a bull . . . a white clean shaven face, a low forehead, cold penetrating eyes, an aquiline nose, sensitive cruel lips, the mask of a Roman emperor. [He was] a tyrant who was feared by his patients and his assistants, for whom he seldom had a kind word of encouragement . . . indifferent to the sufferings of his patients." However, this was written some 40 years after his time with Charcot and he had been expelled from the Salpêtrière, in part for having taken a young female hysteric away from the hospital to live with him, to escape from Charcot's clutches. Some consider the description given rather biased and harsh. Hierons R, 1993, 1589–1590.

39. The three graces are the *femme incomprise* or the *femme fatale* of Asti Hustvedt's *Medical Muses*.

40. Didi-Huberman G, 2003; see also Enquist PO, 2004. Hustvedt A, 2011, is more avowedly feminist in its approach.

41. It remains paradoxical that there are historians and clinicians who persist in writing that hysteria has disappeared, as would be expected if it were a social construct. Patients with such symptoms are still seen, for example, at the National Hospital for Neurology and Neurosurgery. The complex array of signs and symptoms remain a puzzle as they move from one specialist opinion to another. When seizures invade the picture, they come to a seizure clinic. When the seizures are revealed as non-epileptic, the clinical picture unravels. By this time they may well have had several unnecessary operations and have been dosed with multiple medications, including opiates. The name of the condition simply changes, from Briquet's hysteria, to somatization disorder, to whatever the latest fad in the American Psychiatric Association's diagnostic manuals (DSM series) suggests.

42. For an excellent account of Hughlings Jackson's influences on French neuropsychiatry and psychiatry, see Dewhurst K, 1982.

43. York GK, Steinberg DA, 1994. They explain Hughlings Jackson's theory of representation within the brain as follows: "Because the hierarchical motor and sensory centres contain a nested representation of the body, Hughlings Jackson's concept can be viewed as ordinal representation. In ordinal representation, each element of one level consists of the entire preceding level. Thus each element of the middle and highest level contains a complete representation of the body and can control the function of the entire body, at least theoretically" (163).

44. A link between the French and English may be Brown-Séquard, a Mauritian with a French mother, who had shown that stimulation of a peripheral nerve could lead to vasoconstriction in the spinal cord—a functional lesion.

45. Wilson SAK, 1910, 298.

46. For a fuller exploration of this, see Goldstein J, 1987, 336–338.

REFERENCES

Briquet P. *Traité Clinique et Thérapeutique de l'Hystérie.* J.-B. Baillière, Paris; 1859.

Casanova MF, Switala AE, Trippe J, Fitzgerald M. Comparative minicolumnar morphometry of three distinguished scientists. *Autism* 2007;11:557–569.

Critchley M, Critchley EA. *John Hughlings Jackson: Father of English Neurology.* Oxford University Press, Oxford; 1998.

Darwin C. *On The Origin of Species by Means of Natural Selection, or The Preservation of Favoured Races in the Struggle for Life.* In: Wilson EO, ed. *From So Simple a Beginning: The Four Great Books of Charles Darwin.* W. W. Norton, New York; 1859/2006.

Dewhurst K. *Hughlings Jackson on Psychiatry.* Sandford, Oxford; 1982.

Didi-Huberman G. *Invention of Hysteria.* Trans. Hartz A. MIT Press, London; 2003.

Edelman G. *The Remembered Present: A Biological Theory of Consciousness.* Basic Books, New York; 1989.

Enquist P O. *The Story of Blanche and Marie.* Vintage, New York; 2004.

Fowler A. *Seventeenth Century Verse.* Oxford University Press, Oxford; 1992.

Gamwell L. *Exploring the Invisible: Art, Science and the Spiritual.* Princeton University Press, Princeton, NJ; 2002.

Goldstein J. *Console and Classify: The French Psychiatric Profession in the Nineteenth Century.* Cambridge University Press, Cambridge; 1987.

Golla F. Review of *Selected Writings of John Hughlings Jackson*, ed. Taylor J. *Journal of Mental Science* 1932;78:204–206.

Harrington A. *Medicine, Mind, and the Double Brain.* Princeton University Press, Princeton, NJ; 1989.

Hierons R. Charcot and his visits to Britain. *British Medical Journal* 1993;307:1589–1591.

Hustvedt A. *Medical Muses. Hysteria in Nineteenth-Century Paris.* W. W. Norton, New York; 2011.

Huysmans J-K. *Against Nature (A Rebours).* Trans. Baldick R. Penguin Classics, London; 1884/2003.

James FE. Thomas Laycock, psychiatry and neurology. *History of Psychiatry* 1998;9:491–502.

Jellinek EH. Sir James Crichton-Browne (1840–1938): Pioneer neurologist and scientific drop-out. *Journal of the Royal Society of Medicine* 2005;98:428–430.

Lombroso C. *The Man of Genius.* Walter Scott, London; 1891.

López Piñero JM. *Historical Origins of the Concept of Neurosis.* Trans. Berrios D. Cambridge University Press, Cambridge; 1983.

Mazzarello P. Lombroso and Tolstoy. *Nature* 2001;409:983.

Nordau A, Nordeau M. *Max Nordau, a Biography.* Nordau Committee, New York; 1943.

Pick D. *Faces of Degeneration: A European Disorder, c. 1848–1918.* Cambridge University Press, Cambridge; 1989.

Taylor J, ed. *Selected Writings of John Hughlings Jackson.* Staples Press, London; 1958.

Trimble M. Hughlings Jackson comes home. *Journal of the Royal Society of Medicine* 1997; 90:350–351.

Wilson SAK. Some modern French conceptions of hysteria. *Brain* 1910;33:293–338.

York GK, Steinberg DA. Hughlings Jackson's theory of cerebral localisation. *Journal of the History of Neuroscience* 1994;3:153–168.

CHAPTER EIGHT

The Division of the Hemispheres

Habit of seeing opposites.—The general imprecise way of observing sees everywhere in nature opposites (as, e.g. "warm and cold") where there are, not opposites, but differences of degree. This bad habit has led us into wanting to comprehend and analyse the inner world, too, the spiritual-moral world, in terms of such opposites. An unspeakable amount of painfulness, arrogance, harshness, estrangement, frigidity has entered human feelings because we see opposites instead of transitions.

Friedrich Nietzsche, 1878/1986, "The Wanderer and His Shadow,"
vol. 2, aphorism 67

"Who's there?"

This is the opening line of one of the world's most famous tragedies—Shakespeare's Hamlet*—but may well have been the question a passerby would have asked the puzzled landlord Mr. Fino, as he heard the disturbed pounding piano sounds rising above the morning air, day breaking above the crimson dawn of Turin.[1] Mr. Fino's visitor could be heard improvising on the piano, often striking chords with long silent intervals in between. Yes, he always seemed somewhat eccentric, and it was true that the landlord's wife had espied him through a keyhole dancing naked, prancing and singing as if performing some religious rite. And he had asked for his room to be adorned like a temple. As the days passed, the sounds were more frenetic, more energetic, more erethitic. This frequent visitor to the Finos' house was playing again, but wilder than before and singing old Italian gondolier's songs.*

Yes, it is known who was there, but perhaps the guest—the shortsighted, slight man, with a large forehead and a bushy, drooping Prussian moustache, small ears, and a scar on his nose—was not "all there"? Did he or did he not later go out into the street and tearfully embrace a horse, itself battered by life's unkindness? Did they both fall?

Friedrich Nietzsche (1844–1900) became one of the most interesting, illuminating, and inspiring of all nineteenth-century philosophers, said not only to have overthrown two millennia of Platonic hegemony but to have skewed—some would say skewered—Western philosophy ever since. He danced on a rhythmic verbal trampoline, falling into madness in January 1889, at age 44.

After having been brought back to his lodgings by his landlord Mr. Fino, he apparently stayed on his sofa for two days without moving or speaking a word. His friend Franz Overbeck, with the help of Dr. Bettman, persuaded him to accompany them to Basel, holding out to him the prospect of a great reception and celebrations in his honor. At the station he tried to embrace everyone, but he calmed down when he was told that such behavior was not worthy of such a distinguished gentleman.

He was admitted first to the Basel mental asylum and then transferred to the asylum at Jena. A diagnosis of GPI was made, and this is still widely accepted. However, even some of Nietzsche's contemporaries doubted this conclusion.[2]

Where? The Rise of Localization

It is time to consider in more detail one of the most divisive developments in clinical neuroscience, the doctrine (for that is what it became) of localization. Some of the early background has been touched on. The brain itself became a focus of the mind in Greco-Roman times; animal spirits or vital spirits were moving agents, which localized there, but the ventricles held pride of place rather than the substance of the brain. The shift of interest to the solid parts of the brain took centuries. The seventeenth and eighteenth centuries saw the contributions of Descartes and Willis, and Swedenborg alone seems to have recognized the importance of the cortex, introducing the idea of a motor homunculus. The second half of the eighteenth century saw an anti-localization bias, with some scholars, such as Samuel von Sömmerring (1755–1830), who classified the 12 cranial nerves, reviving ventricular theories. However, it is Gall and Spurzheim who are largely credited with attempts to localize functions in the brain, leading to a quest for topographical locations of psychological and other faculties. Each of their modules in one hemisphere had a corresponding one on the opposite side.

Gall's attempts to seek unity among diversity are often overlooked, and in spite of his parcellation of the cortex, he never considered any hemispheric imbalance. There was a backlash against their views, as phrenology was quickly discredited, and some authors, such as Magendie and Flourens, strongly supported the view that there was equivalence of mental functions throughout the brain. Both Magendie and Flourens were vigorous vivisectionists, allowing for the effects of destructive brain lesions to be studied, while others such as Galvani used electrical stimulations. Flourens, for example, removed the cerebral hemispheres of pigeons and noted that they could still fly if thrown in the air, concluding from this that the hemispheres had nothing to do with movement. He was the first to suggest a "field theory" of equipotentiality.[3]

By the mid-eighteenth century the cerebral convolutions had been accurately described. Histological contributions revealed individual layers of cells and fibers, and Gratiolet had noted a plan of the cerebral convolutions which could be traced in mammals up to man and the different rates of development between the anterior and posterior parts of the two hemispheres. The modern era of localization began when Gustav Theodor Fritsch and Eduard Hitzig, zoologist and psychiatrist, respectively, explored the exposed cortex with electrical stimuli and noted different areas that provoked motor and sensory events. They then proved this association with ablation experiments, and along the way they reported that excessive stimuli caused convulsions. However, it was the anthropologist and comparative anatomist Broca, much influenced by Geoffroy Saint-Hilaire (1772–1844), and his studies of aphasia which caused a kerfuffle in academic and religious circles, as well as putting the localizers and anti-localizers once again in opposition. He noted with some surprise that by 1863 he had collected eight cases of language disturbance, all of which had lesions in the left hemisphere.

Aphasia

The researcher who first reported that speech was located in the frontal areas of the brain remains in debate. François Pourfour du Petit (1664–1741) had noted the contralateral motor effects of soldiers' head wounds, and in 1766 he described a man with a complete right-sided hemiplegia who was unable to speak. He dissected the brain at postmortem and noted the damaged left cortex and "corpora striata." But it was the work of Marc Dax (1771–1837) which caused what became known as the Broca–Dax controversy.[4]

Jean Baptiste Bouillard (1796–1881) at the Charité in Paris had made extensive clinical and pathological studies of speech disorders and offered a wager of 500 francs to anyone who could demonstrate that they were not associated with the frontal lobes. On April 4, 1864, Ernest Auburtin (1825–1893), Bouillard's son-in-law, presented a paper to the Anthropological Society of Paris stating that language was localized in the frontal lobes. In the audience was Broca. Broca had just admitted to the Bicêtre Hospital a speechless hemiplegic who had died and had a lesion in the left second and third frontal convolutions of the brain. This man, called "Tan"—the sound he made when he tried to pronounce his name—has become perhaps the most famous neurological patient in history. His illness turned the localization debate firmly to the left, literally and metaphorically. Broca collected 22 brains from similar cases, and at a meeting of the society in 1867 he stated categorically that in aphemia (his term for aphasia) the lesion is always localized to the left side of the brain. Further, patients with lesions in the same area on the right side did not develop speech abnormalities.

At the next meeting (April 21), Broca reported that a Mr. Gustave Dax (1815–1898) had complained to the Academy of Medicine that his late father had known this for a long time and had actually already written on the topic, for a congress in Montpellier in 1836. Broca declared ignorance of this and said that he was unable to trace the article. It turns out that Dax *père* (Marc) had indeed written a memorandum to his colleagues in 1858, repeated in 1860, entitled "Observation Directed Towards Proving the Association between Derangement of Speech and a Lesion of the Left Cerebral Hemisphere." Further, the son Gustave discovered many of his father's written manuscripts, which revealed that his father had collected 40 cases, all implicating the left hemisphere with speech disorders. The arguments over priority continued.[5]

Broca, a Protestant in a largely Catholic France, and one of the most respected scientists in Paris at the time, implied not only that speech was localized but that other, higher cerebral functions, today referred to as executive, could also be localized. He went further. After an autopsy on a woman with epilepsy, who had no cortex at the language area but no language deficit while alive, he suggested that her right hemisphere had adopted the role of the missing left hemispheric tissue.

A police spy had been present at the meetings of the Anthropological Society, and Broca, who supported some of Darwin's theories, was accused by the Catholic Church of being a materialist and a corrupter of youth. Perhaps even more subversive was Broca's assertion that faculties related to man's highest sentiments could be localized in the brain, which, it seemed, was turning out to be imperfect, asymmetrical, and unbalanced.

Broca versus Hughlings Jackson

In 1868, Broca attended a meeting of the British Association for the Advancement of Science in Norwich, England. The notification for the meeting suggested that there would be a discussion on the paper to be read by Hughlings Jackson on aphasia: "It will be no small attraction to the meeting that visitors will thus have the opportunity of immediately comparing the best English and the best French views on the pathology of this remarkable disease."

Broca was by now the acknowledged champion of localization, supporting the view, along with Gratiolet, that the left hemisphere was more advanced than the right. Hughlings Jackson had already stated his reservations about such precise localization, saying he was neither a universalizer nor a localizer, and he rejected the concept of localization of a faculty. Praising Broca for bringing his findings to the attention of the scientific community, he stated that he had come to believe less in Broca's views than at first. He went on to say, "I think that both sides are probably educated, but the left one is the one that begins to act," and he then referred to his views about the right side and its association with "involuntary educated utterances." As the neurologist Robert Joynt (1925–2012) reported, unfortunately no minutes of the "very animated and very spirited" discussions and no comments about what Hughlings Jackson may have said to Broca and vice versa were recorded.[6]

The localization theme was further examined in Britain, especially by Hughlings Jackson and his coworker Ferrier. At Wakefield, the latter continued the observations of Fritsch and Hitzig, using apes and other animals, and implementing ablation and stimulation techniques. He minutely mapped the motor cortex, and at the 1881 International Medical Congress in London he presented a hemiplegic monkey whose motor area was ablated on one side. It is reported that Charcot, who was in the audience, exclaimed, "It's a patient!" Ferrier went on to confirm the views of Hughlings Jackson on the cerebral cause of unilateral epileptic seizures. According to Lawrence McHenry, Ferrier was "the link between Jackson and the modern work on the cerebral cortex of Sherrington and others."[7] However, this did not end the debates with those who believed in a theory of brain equipotentiality.

On to Berlin! On to Berlin! On to Berlin![8]

The decline of the second French Empire under Napoleon III, nephew of Napoleon I, coincided with the rise of a unified Germany under Otto von Bismarck (1815–1898). This shifted the balance of power in Europe.[9] Germany had always had more universities than England, and by the middle of the nineteenth century they were all research orientated, with a technical infrastructure. This engendered the rise of independent disciplines of study, as well as competition between them. Of 29 psychiatric professorships in Germany, 21 were created before 1880. Dedicated journals were established, and research was seen not only as a contribution to human knowledge but also as part of an individual's moral development: *Wissenschaft* (a research-based development of science and knowledge) was integral to *Bildung* ("a process of fulfilment through education and knowledge . . . an amalgam of wisdom and self-realisation").[10]

Wilhelm Griesinger (1817–1868) was appointed as professor to the *Klinik für Nerven- und Geistkrankheiten* (Clinic for Nervous and Mental Diseases) in 1865 at the Charité Hospital, Berlin. His aim was to relieve psychiatry of its romantic associations, and he famously stated that "we therefore primarily, and in every case of mental disease, recognise a morbid action of that organ." He wanted to keep the study of mental diseases within the general corpus of medicine: "Insanity being a disease, and that disease being an affection of the brain, it can only be studied in a proper manner from the medical point of view." His *Pathologie und Therapie der psychischen Krankheiten* (Mental pathology and therapeutics), published in 1845, was a bombshell. By this time, lesions had been found in the brains of people with GPI, and lack of iodine in the diet was associated with cretinism. His review of postmortems of the insane led him to conclude that "the majority . . . show anatomical changes to exist within the cranium."[11]

In addition to establishing two psychiatric journals and advocating an organic approach to the subject, he recognized both idiopathic and symptomatic causes of mental illness and promoted somatic and psychological treatment methods. Hitzig was one of his students.

The idea that mental diseases were related to brain pathology was taken in a different direction by Theodor Meynert and his student Carl Wernicke (1848–1905). Wernicke practiced neurology in Berlin and became professor of psychiatry in Halle. He was an outstanding neuroanatomist and continued the ideas of Broca with the publication of his text on aphasia in 1874

(*Der aphasische Symptomenkomplex*). His later three-volume book (*Lehr-buch der Gehirnkrankheiten*, 1881–1883) was the first comprehensive account of cerebral localization. He noted that not all cases of aphasia had lesions in the areas outlined by Broca, and he described a different form that occurred after lesions situated more posteriorly and affecting the left superior temporal areas. This language disturbance, now referred to as Wernicke's aphasia, contrasted with Broca's aphasia, but once again it concerned only the left hemisphere. He also described cases of alexia and agraphia.

Wernicke's classification of aphasic disorders became widely accepted, as other cortical syndromes were described. These included alexia without agraphia (Joseph Jules Dejerine, 1849–1917), ideomotor and other forms of apraxia (Hugo Liepmann, 1863–1925), and visual agnosia (Heinrich Lissauer, 1861–1891).[12]

Wernicke's reduction of cortical functions to discrete brain areas depicted on box and wire-line diagrams was not the first attempt at this kind of depiction, but by this time it held authority and the backing of empirical evidence. This trend evoked a backlash; for example, the neuroanatomist and psychiatrist Auguste Forel (1848–1931) wrote in his memoirs that he was disappointed when he worked with Meynert because he realized that many of the brain tracts Meynert said he had discovered were creations of his imagination. However, this attracted little attention, especially because of the simplicity of such drawings depicting how the brain works (see fig. 8.1); their naivety meant they took a stronger and stronger hold on theorizing well up to the present day.[13]

The parcellation of the cortex into different anatomical divisions proceeded apace, as did the identification of different cortical signature syndromes. Baillarger made his observations of the six cortical laminations without a microscope in 1840. Meynert divided the brain into the neopallium (nonolfactory cortex) and the archipallium on the basis of cell structure.[14] The detailed histological images of the brain by Santiago Ramón y Cajal (1852–1934), using newly developed staining techniques (such as those of Golgi and Nissl), have become legendary.[15] Paul Flechsig (1847–1929) noted that every cortical area possessed a special anatomical position and functional significance although he was concerned that this was close to the older phrenology. He noted that what he referred to as projection areas and association centers—the former being related to sensation, the latter including areas of the frontal, parietal, temporal, and insular cortex—were in direct relation to several or all of the sensory areas. He noted, "Following their bilateral destruction, the intellect

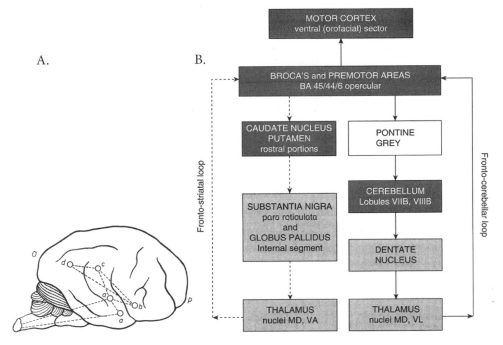

Figure 8.1. A, Wernicke's original wire diagram (1874), showing the auditory input (a), the auditory language center (a), the motor speech area (b), and their connections: c = concept formation; d – visual language area. B, a modern equivalent. Such diagrams are supposed to help us understand how our brains work, in this case explaining the complexities of human speech. Panel A reproduced from Benson DF, 1997, 93. Panel B reproduced from Morgan A, Liegeois F, Vargha-Khadem F, 2010, 103.

appears to be diminished; the association of ideas is especially disturbed."[16] The splitting up of the cortex progressed as Korbinian Brodmann (1868–1918) published his map of 52 cerebral cortical areas in 1907.

These were all considerable advances that form a setting for our current understanding of the underlying neuroanatomy of many neuropsychiatric syndromes. The early origins of the limbic system are in view, as are the descriptions of some specific cortical signature disorders. The whole concept of the latter was revisited nearly a century later in a more refined way by the neurologist Norman Geschwind (1926–1984). He revived many of Wernicke's ideas, especially noting the importance of disconnection between various brain areas caused by white matter lesions, later recognized as an essential element of Wernicke's initial theories.[17]

But this triumph of localization of function in the cortex had consequences as the split between neurology and psychiatry intensified.

What Freud Was Not

Sigmund Freud was not a psychiatrist, he was not the discoverer of the unconscious, he did not originate the sexual theories of neuroses, he was not the founder of psychotherapy, he was not a philosopher, and he was not an enthusiast of the arts, even if he enlisted several artists to his psychoanalytic pantheon. In fact, he was a neurologist and psychologist, a profound admirer of Charcot, and a great storyteller, and he enjoyed cocaine. Further, he believed that he was a scientist.

There can be no better history of the unconscious than Henri Ellenberger's (1905–1993) superb book *The Discovery of the Unconscious.* He does not get around to discussing Freud and psychoanalysis until approximately halfway through this almost 900-page-long book.

But let's backtrack. In 1876, Freud joined the Physiologisches Institut in Vienna as a research assistant to Ernst Wilhelm von Brücke (1819–1892) and von Helmholtz. The latter encouraged researchers to apply scientific methods as used in physics to mental phenomena. The intellectual environment was full of discussion about physicochemical forces active in an organism; there was a need to abolish vitalism, with the adoption of von Helmholtz's doctrine of the principle of the conservation of energy, namely, that the sum of forces remains constant in every isolated system. Forces acted together or could inhibit each other, ideas that gave the young Freud a physiological perspective and concern with understanding nervous forces and energy. Helmholtz, as we have noted, had shifted the emphasis on apperceptions from the sole influence of exteriority to interiority.

In 1883 Freud was working in Meynert's department, doing histological work. He learned about cocaine in articles from Lima, Peru, suggesting its effects as a stimulant and its possible use with psychiatric patients. It seemed not to be addictive, and so he used it on himself, finding a positive effect against fatigue, but also noting that it caused numbness of the mouth. He reported this anesthetic effect to his colleagues, and the Austrian ophthalmologist Karl Koller (1857–1944) stole a march on him by publishing on its use for eye surgery while Freud was away visiting his then fiancée Martha Bernays.[18]

Having missed an opportunity to make an important scientific discovery, Freud was at the threshold of another when he first met Charcot in Paris on October 20, 1885. He was there only until February 23, 1886, but in this short time he came to view Charcot and his theories regarding hypnosis and hysteria as groundbreaking. Freud was enamored by him. He translated Charcot's lectures into German, he named his first-born son after him, and he hung a copy of a painting of Charcot demonstrating a patient's examination in his consulting room in Vienna, which he took to London when he had to flee from Vienna.

But for Charcot, psychology was not part of medical science, and there were deep suspicions in France regarding supposed therapies for hysteria, especially after Mesmer had been exposed as a charlatan by the investigation of the French Royal Academy of Science. After all, the development of animal magnetism and electrical therapies undermined the developing Enlightenment values of a mechanistic universe—the Newtonian physico-mathematical model was becoming strained and drained. For Charcot, hypnosis was a method of inducing symptoms and signs but not a treatment.

One evening at Charcot's house, after listening to a discussion about a patient who had come to Paris with her husband on account of her neurotic disorder, Charcot implied that the husband's impotence had something to do with her illness. Charcot is reported to have said, "Mais, dans des cas pareils c'est toujours la chose génitale, toujours . . . toujours" (Always a question of the genitals, always . . . always). This struck a chord with Freud, since his Viennese colleague Josef Breuer (1842–1925), with whom he was developing ideas about hypnosis, had said of hysteria that "these cases are always secrets of the alcove, the marriage bed." The next day, when he had reflected on what the great man had said, he told Charcot the story of his conversations with Breuer and asked him whether there was any potential for hypnosis to be helpful therapeutically. Charcot replied, "No, no, there is nothing of interest there"—and the great split between psychiatry and neurology was born.[19]

Railway Lines

The story of how Freud returned to Vienna and incurred the wrath of his colleagues over a paper on male hysteria presented at the Academy of Sciences on October 15, 1886, is well recounted.[20] The associations surrounding the displeasure are not. Freud described a case of Charcot's, and then five weeks later one of his own cases, emphasizing the stigmata that Charcot relied on for a diagnosis of hysteria. As Ellenberger outlined the story, two main problems were the dislike of the Viennese for the introduction of French ideas and the confusion over hysteria in males. The latter was well described, but what was not accepted was Charcot's linking traumatic neurosis—in other words, posttraumatic hysteria—with male hysteria, as well as his use of hypnosis.

This has to be seen in the context of discussions over another condition, namely, railway spine. Europe and America were witnessing the rapid development of the railway system and with it railway accidents. With them came an interest from lawyers and rising compensation claims. The main battle in England was with John Eric Erichsen (1818–1896), who had claimed in his book *On Concussion of the Spine* (1882) that, following accidents, especially in trains, the symptoms that developed were due to spinal damage, with the organic structure of the cord being deranged as a result of "molecular changes." This "organic" injury was even referred to as Erichsen's disease, and it necessitated compensation. The alternative view was held by Herbert Page, who advised railway companies. He dismissed the whole concept of concussion of the spine, preferring to view many cases as examples of "nervous shock."

"Railway spine" was a hot topic at the Imperial Society of Physicians in Vienna, since Hermann Oppenheim (1858–1919), the leading German neurologist of the day, had claimed that traumatic neuroses, including the symptoms and signs of "railway spine," were caused by minute, undetectable lesions—a condition that was causally and pathologically different from hysteria. Oppenheim rejected the French position, as espoused by Charcot, arguing that it blurred the distinction between illness and simulation.[21] Oppenheim eventually lost this fight and his reputation, but Freud's split from academic neurology in Vienna had begun.

Freud remained under the influence of Charcot and was very impressed by his clinical methods and research techniques, for Freud believed himself to be a scientist and sought empirical foundations for his theories. From Charcot he had received the concept of "unconscious fixed ideas," which Janet had developed further with his *idées fixes* (fixed ideas), images or thoughts linked to feelings, physical postures, and/or bodily movements, due to psychological trauma, which became encapsulated and unavailable to consciousness.

From the Brain to the Mind

While in Paris, Freud worked on the pathology of the brains of children, some of whom had been subject to physical abuse, giving him ideas about early childhood trauma and psychological development; he published papers on cerebral palsy, which were well received. But his other very important influence was Hughlings Jackson, and Freud's book *Zur Auffassung der Aphasien. Eine kritische Studie* (On aphasia), written in his pre-psychoanalytic period and now largely neglected, was hugely influenced by Hughlings Jackson's ideas.[22]

Hughlings Jackson did not accept a "faculty" of language, and neither did Freud. The latter credited Hughlings Jackson as his inspiration toward an anti-Wernicke understanding of speech disorders and a rejection of localization of speech or of individual muscle movement in the brain. Freud initially looked for theories that explained psychological processes as localizable to anatomical brain areas, but he supported Hughlings Jackson's view on the Doctrine of Concomitance and noted the recurrent emotional utterances of aphasic patients, linking stress and consequent emotion to existing cerebral processes. Freud rejected the principle of hereditary degeneration but—influenced by Darwin and Spencer, as well as Hughlings Jackson—incorporated ideas of "dissolution" (the opposite of evolution) and of hierarchies of neurological and psychological processes into his theories. The ability of higher aspects of psychological function to be overwhelmed by lower ones became the idea of regression, fundamental to psychoanalysis.[23]

In Freud's book we find ideas such as "overdetermination" and words such as *projection* and *representation* as physiological concepts. *Besetzung*, which became translated as "cathexis," became a central psychoanalytic theme for the way that the libido becomes invested in objects.[24]

However, at this stage in his thinking Freud had not abandoned his attempts to consider his knowledge of neuroscience; in his *Project for a Scientific Psychology*, the recently developed concept of the neuron, introduced by Heinrich Wilhelm Gottfried von Waldeyer-Hartz (1836–1921), was central.[25] The project, as referred to by Karl Pribram and Merton Gill in *Freud's Project Reassessed*, was a "psychological document cloaked in neurological terms."[26] It deals with energy transfer within and between neurons and the external world using reflex arcs, and it was an attempt to correlate psychological events with the distribution of energy in the brain. The theories outlined were based on the properties of different types of neurons, their excitability, and principles of inhibition and excitation. The physiological was always imbedded in the psychical, reflecting a reciprocal causality.

Differing types of neurons were involved with different energy quantities (such as phi [Φ] or psi [Ψ]), but there were also omega (ω) ones, which were linked to the quality in the psychic energy, such as associated with hunger and pleasure. Some neurons were permeable, allowing for new information to flow through them (Φ), and some were impermeable (Ψ); in the latter resistance was set up, leading to permanent changes. The project was linked to Jacksonian principles since Freud linked symbolic verbal representations to consciousness, speech associations linking Ψ neurons to neurons associated with motor speech images and discharge. For Hughlings Jackson, the emotional outbursts in health and in aphasia were "on the physical side a process during which the equilibrium of a greatly disturbed nervous system is restored ... by expenditure of energy"; swearing was a safety valve to feelings and a substitute for aggressive muscle action. There were then in both Hughlings Jackson's and Freud's models centers/neurons of resistances. These links in which "psychic processes must be 'cloaked' in words in order to be perceived by consciousness" were perhaps "the initial rationale for the role of insight in therapy—with speech essential to permit thoughts to become conscious." Through language, and in psychoanalysis, energy could be redistributed and released and pathological processes resolved.[27]

As Stengel pointed out, both Hughlings Jackson and Freud were interested in nervous functions as "manifestations of nervous energies." As an important etymological note, both used the term *functional* adjectivally, to refer to the function of the nervous system to store up and expend energy. It refers to the physiology of the nervous system and was used by Freud in this same sense in his book on aphasia. Jacksonian neurodynamics were transformed into Freudian psychodynamics, neuroses into psychoneuroses, and Hughlings Jackson's three divisions of the brain transmuted into the id, ego, and superego—the *Project* simply failed and was abandoned as Freud moved on to different pastures.

Another Yorkshireman

Born in 1835 in Yorkshire and living in London, with an acute interest in the relationship between the mind and brain, and having spent some time at the Wakefield Asylum, Henry Maudsley (1835–1918) might have been expected to have met with Hughlings Jackson, but there is no evidence that he ever did.[28] Famous for funding the construction of the hospital that bears his name, Maudsley was appointed as medical superintendent to the Manchester Royal Lunatic Asylum at the age of 24.

When he moved to London, his reputation grew rapidly, in part on account of the books he published; he became one of the most eminent psychiatrists in England. He laid down some very special criteria for the Maudsley

Hospital, including that it would take only cases that were certified as insane or convalescent after cure of insanity. Most important to his enterprise was that it should have due provision for clinical and pathological research.[29] British psychiatrists at that time (sadly, as is much the case today) were isolated from their medical colleagues and from universities, unlike the situation in Germany, where academic psychiatrists had university posts and clinics in general hospitals: this gave them facilities for teaching and research, as well as prestige. There was a pathological laboratory at Claybury Hospital (the London County Lunatic Asylum, Ilford), which was run by Sir Frederick Mott (1853–1926), and the plan was for this to move to the Maudsley site (which it eventually did in 1916). Edward Mapother (1881–1940) was the first medical superintendent, and Frederick Golla (1878–1968) became the director of the pathological laboratory.[30]

Maudsley's first important book, *The Physiology and Pathology of the Mind* (1867, revised in 1895), was a compendium of information on the role of evolution, brain structure, and brain function related to mental illness. He emphasized the links between the study of mental disorders and general medicine and affirmed the integrity of the mind and the body. Maudsley had a brief correspondence with Charles Darwin; the latter had studied his books. However, his evolutionary ideas were not based on Darwinian principles but more on Continental degeneracy theories; he was well acquainted with the French and German authors—the book begins with a quote, in German, from Goethe's *Faust*. Maudsley wanted to understand how, with specialization of structure in the human cortex, there was "the highest and the most complex manifestations of mental function . . . for in the functions of man we observe, as in a microcosm, an integration and harmonious coordination of different vital actions"—a Goethean enterprise.[31] Not surprisingly in a paper on aphasia, he caviled about Broca's ideas of localization.

The book brought him instant fame and was soon translated into German, French, Italian, and Japanese; it was even read by Tolstoy. Maudsley became a coeditor and then senior editor of the influential *Journal of Mental Science*, published by the Medico-Psychological Association. This was aimed largely at medical superintendents of asylums, but it gave him access to current ideas, as well as opportunities to influence research and practice. Maudsley was interested in the links between madness and "genius," as well as what psychopathology could learn from a study of literature and the arts.

He published a paper on Edgar Allan Poe, and later came an article on *Hamlet*. He described the hereditary traits that influenced Poe, a man destined by his constitution to follow the wrong turnings he took in his life, yet an outstanding poet. The journal for which he served as editor contained many papers on medical jurisprudence, and Maudsley was critical of the poor ability of judges to understand medical matters and the courts' inability to accept that criminal actions could be involuntary. Patients should not be treated as criminals, since the matter was a scientific one and not a moral one—ideas expounded in some detail in his *Responsibility in Mental Disease* (1864).

Maudsley also discussed the "double brain." In this he was influenced by Arthur Ladbroke Wigan's (1785–1847) book *The Duality of the Mind* (1844). The symmetry between the two sides of the brain had been taken for granted since ancient times, but there were growing speculations, even before Broca and Dax, that inharmonious actions of the two hemispheres might reflect some kind of double consciousness. Wigan considered the two hemispheres of the brain independent, each being a "distinct and perfect whole" capable of separate processes of thinking, one perhaps being rational and the other irrational. He discussed his own mental experiences, feeling himself to have a double identity, two minds in two brains. Maudsley contended that despite the duality of the brain, there was a unity of action, although he speculated on how discordant action between the two sides of the brain may be linked to personality and mental illness through degeneration. Feelings he attributed to the "splanchnic distribution of the cerebrospinal system," the life of feeling being fundamental to the life of thought.[32]

Maudsley wrote about links between epilepsy and insanity and became interested in electrical charges in the brain, even suggesting that generalized convulsions might be artificially induced as a cure for mental disorders. He also observed that intercurrent infections could alter the course of chronic mental conditions. He recognized that epilepsy could be associated with short episodes of dementia or violent dangerous mania and, if continuous, could lead to complete dementia. Transitory manias could occur, taking the place of seizures—so-called masked epilepsy—which may lead to profound "moral disturbance," hence his interest in the forensic implications of brain disorders. He wrote about the insanity of Swedenborg, a man whom he considered wonderfully original, with elements of greatness, yet who was a monomaniac with probable epilepsy.

Maudsley's philosophical views tended toward Lockean ideas of association, and his overall perspective was very similar to that of Griesinger, namely, that "mental disorders are neither more nor less than nervous diseases in which mental symptoms predominate, and their entire separation from other nervous diseases has been a sad hindrance to progress."[33] He was critical of introspection and did not accept the concept of the unconscious, or at least of the Freudian variety. Maudsley was well ahead of his time, and he resigned from the journal after some nine years, his colleagues being unhappy with his positivism, his irreligious materialism, and his ideas' lack of practical application in treatments.[34] Like Hughlings Jackson, he tended toward a philosophical orientation and was quite a pessimist. It is sad that before he died he instructed his publisher to destroy all the remaining copies of his books, apparently concluding that they were out of date.

Epilepsy, Energy, and Nerve Power

The predominant theories about epilepsy in the nineteenth century were vascular, and perhaps one of the most quoted historical definitions given of epilepsy was that of Hughlings Jackson: "Epilepsy is the name for occasional, sudden excessive, rapid and localized discharges of grey matter."[35] This definition was linked to ideas of instability of the gray matter, the strength of the discharge being associated with different patterns of seizures. Hughlings Jackson invoked altered activity in arteries and their vasomotor nerves and, as Edward Reynolds has noted, did not imply that electrical activity was involved. This idea originated from Robert Bentley Todd (1809–1860), who was influenced by the work of Michael Faraday (1791–1867). Todd used the analogy of the electrical forces in a galvanic battery to describe the tensions that derived from abnormal cellular nutrition in the brain. "These periodic evolutions of the nervous force may be compared to the electrical phenomena described by Faraday under the name of disruptive discharge."[36]

The nosology of epilepsy was changing. The French physicians had introduced us to *grand mal* and *petit mal*. The aura had taken on a more modern meaning as a sensory event, but others described motor auras. Todd used the term *epileptiform* for focal motor seizures, whereas Hughlings Jackson introduced *partial seizures* and used *epilepsy proper* or *genuine epilepsy* for attacks in which loss of consciousness is the first or nearly the first manifestation.[37]

Hughlings Jackson, as we have seen, was very interested in the associated psychological alterations seen with epilepsy. This topic, as already discussed, goes back to ancient times, but the asylum doctors, especially in France and Germany in the nineteenth century, had the opportunity to observe many patients with intractable seizures and reported on the associated mental disorders. Hughlings Jackson was not an alienist, but he saw many patients in an outpatient setting, describing several cases as "a peculiar variety of epilepsy ('intellectual aura')." Of singular importance was Arthur Thomas Meyers (aka Dr. Z), a doctor with "dreamy states," in which he became "bemazed" briefly, although he could continue talking. He also had "reminiscences." He kept careful, extensively written notes of his episodes and developed an indexing system. He was very interested in mystical and psychical research and was said to be melancholic; he eventually committed suicide.[38] At postmortem his brain was shown to have a small lesion in the left uncinate gyrus. Hughlings Jackson made the point that loss of consciousness was not essential in making a diagnosis of epilepsy; there could be a "defect" only. Some of the cases he described experienced clear déjà vu episodes and crude smell sensations. Dr. Z may have been a typical case of temporal lobe epilepsy with personality changes, which we meet in a later chapter.

Transformations

The term *epileptic equivalent*, implying seizures that were only revealed by alteration of the mental state, was first used by the psychiatrist Friedrich Hoffmann (1660–1742). However, Carl Friedrich Flemming (1799–1880), director of the Sachsenberg, the first psychiatric asylum in Germany, claimed in his textbook *Pathology and Therapy of Psychoses* (1859) that epilepsy was one of the more frequent causes of psychosis. He wrote, "A seizure of habitual epilepsy fails to appear and is replaced by a paroxysm of insanity . . . epilepsy and mental disorder are two states of illness of the very closest relationship; they represent identical pathological conditions in two different areas of the nervous system."[39] Morel used the designations *larval* and *masked* epilepsy for a form interlinked with character changes. Jules Falret (1824–1902) noted the *troubles intellectuels*, ranging from brief episodes linked to the time of the attack (including auras) to the *folie épileptique*. The latter he divided into two types, *petit mal* and *grand mal intellectual*. He had also observed cases in which the delirium substituted for the seizure. For Falret, the mental symptoms were more important than the convulsions, and he was very interested in a growing literature on automatisms. In postictal delirium he had observed extreme violence in some patients, including suicide and homicide. Griesinger had noted many cases of epilepsy in which the patient had chronic mental disorders in between their attacks (now referred to as interictal psychoses), but it was his assistant Paul Samt (1844–1875) who developed the concept of equivalents, describing over 40 cases. His classification of epileptic insanity recognized two major forms: psychic equivalents and postepileptic insanity. The former included episodes of violence, religious ecstasy (noting the patients' God nomenclature of their surroundings), anxious delirium, and stupor. Another Dr. Hoffman, Heinrich Hoffmann (1809–1894), most known today for his children's book *Der Struwwelpeter*, in 1859 described the course of epilepsy over a seven-year period in 45 patients, noting mental symptoms that occurred, some of which were "an equivalent" of an epileptic seizure.

It would seem that by the end of the century mental state changes in preictal, ictal, postictal, and interictal phases and as equivalents were well described. It is interesting to note here how little attention was paid to these clinical states until the second half of the twentieth century.[40]

Nietzsche's Fall

It may appear astonishing today, but a psychiatrist working in an asylum at the end of the nineteenth century would be looking after many cases we would now designate as neuropsychiatric. The concept was emergent, even if the designation had to wait until the next century. As Kraepelin stated, "The greatest step towards understanding the aetiology of mental disease was the discovery that paresis resulted from syphilis."[41]

GPI was common, recognized as a disease entity by the French since the 1820s, and was more fatal than epilepsy, life expectancy being very limited once the psychosis emerged. In 1905, the zoologist Fritz Schaudinn (1871–1906) and his assistant Erich Hoffmann (1868–1959), using serum from a genital lesion and new staining methods, identified the spirochete (*Treponema pallidum*). Then in 1906, August Paul von Wassermann (1866–1925) developed a test for the identification of antibodies in the blood and cerebrospinal fluid. Hideyo Noguchi (1876–1928) in 1913 confirmed the presence of the spirochete in sections taken from brains of patients. Mott confirmed the pathological link between GPI and syphilis.

Diagnoses of GPI before these times gave great emphasis to the mental state, and the margin for error was wide. The neuropsychiatrist Richard Hunter and his mother, Ida Macalpine, noted in their book *Psychiatry for the Poor* that no less a person than Kraepelin had diagnosed GPI in 30 percent of his patients in the period 1895–1897, but the subsequent course of the illness revealed that he was wrong in most cases, the true incidence being around 5 percent. They noted, after the tests were available, that at the National Hospital for Nervous Diseases some 10 percent of admissions had cerebrospinal syphilis, and most had either tabes dorsalis or GPI.[42]

When precisely Nietzsche first developed the symptoms of his final fatal illness remains unclear. Two old friends of his, Erwin Rohde and Paul Deussen, commented on changes in his personality in 1886 and 1887, when he was in his early forties.[43] These personal observations suggest that the onset of his decline may have been in the mid-1880s, but by 1888 he clearly was disturbed. He became confused, and his previous smart appearance became somewhat disheveled. Ideas of grandiosity became obvious. He began to sign his letters with names such as "Phoenix," "Antichrist," and "the Crucified," and by December he was drafting letters to the Kaiser and to Bismarck. He sent invitations to a congress he was organizing to the Italian king Umberto II, Moriani, the secretary to the pope, and the "Haus Baden."

He was transferred to the Jena mental asylum on January 18, 1889. The psychiatric presentation was as follows. He entered his room with majestic strides and thanked those present for the "great reception." He bowed repeatedly. He did not know where he was and had lost insight. He gesticulated, spoke in a lofty tone, and mixed Italian and French words. He confused the simplest Italian words.

A Doubtful Diagnosis?

The diagnosis of syphilis in the case of Nietzsche has always been doubted.[44] However, to be diagnosed as such was the fate of many well-known and talented people at the time; rumor was rife, but the diagnosis was often insecure.[45] Two of Nietzsche's maternal aunts had psychiatric illnesses, one committing suicide, and one maternal uncle developed mental illness in his midsixties. Another maternal uncle probably died in an asylum. Perhaps more relevant and often overlooked was the fact that Nietzsche's father died at the age of 35 of a neurological illness, of which it was said that at postmortem softening of a quarter of the brain was observed. An alternative diagnosis must have been frontotemporal dementia, which is a common hereditary cause of intellectual and behavioral decline, and the signs and symptoms fit well Nietzsche's pathography.[46]

Of special interest are the reports of the emergence of artistic talents in patients with frontotemporal dementia, as well as a temporary increase in verbal skills. Obviously, as the disorder progresses, artistic and creative talent wanes, but such productivity that Nietzsche had at the end of his active life is hardly compatible with the development of a chronic progressive brain infection such as syphilis and subsequent GPI.[47]

Nietzsche, embowered in his own tragedy, signed a letter to Overbeck with the name Dionysus. In the end, endlessly without words, expressing himself only through music, improvising phrases of a *Tristan*-like inspiration on the piano, this philosopher of Dionysus was shattered by Promethean forces. In Turin he fell to the ground, pitying an injured horse with which he knew he had symbiotic altruistic feelings, and cried.[48]

NOTES

1. The First Folio of 1623.

2. For example, see the discussion in Julian Young's recent excellent biography (Young J, 2010).

3. These ideas of the supposed equipotentiality stress patterns of CNS activity rather than segregated loci of excitation, prefiguring later theories of Gestalt psychology, gradients of neural activity, and the mutual influence of neighboring parts of the brain on each other.

4. There are several accounts of this, and that of Critchley M, 1979, is quoted here.

5. Nobody who had attended the congress where Dax claimed his father had read his paper could recall a Dr. Dax.

6. The details and quotations are given in Joynt R, 1982.

7. McHenry LC, 1969, 219, 223.

8. This is the refrain and the last line of Zola's *Nana*. The second French Empire collapsed and decayed (in Zola's view), as did Nana's body, as she became disfigured and died from smallpox. The French government had just declared war on Prussia, as the name of Bismarck is repeated in the story. The French lost.

9. In 1870 Bismarck started a war against France in a continuation of the old conflicts between the two countries. The Second Reich came into being as German unification proceeded.

10. For a very thorough discussion of the German genius, there can be no better book than Watson P, 2010; quote on 53–54.

11. Griesinger W, 1867, 1, 9, 413.

12. Hughlings Jackson had described apraxia before Liepmann in 1861, but Liepmann gave more detailed descriptions and subtypes.

13. On Auguste Forel, see Ellenberger H, 1970, 434.

14. These later became also referred to as *isocortex* and *allocortex* (the classification of Oskar Vogt [1870–1959] and Cécile Vogt [1875–1962]).

15. The Golgi method of silver staining revealed axonic nets and neurons. Nissl's methylene blue method gave even better delineation of cell bodies.

16. Clarke E, O'Malley CD, 1968, 552.

17. See Benson DF, 1997.

18. There are debates as to whether or not Freud became addicted to cocaine. His English analyst collaborator Ernest Jones (1879–1958) estimated that Freud used the drug on and off for some 15 years. The way this may have influenced his creative thinking and his dreams is a matter of speculation.

19. Stone I, 1971, 183–184.

20. See Ellenberger H, 1970.

21. For an account of the battle between Erichsen and Page and the history of posttraumatic neurosis, see Trimble MR, 1981. Other good references are found in Micale MS, Lerner P, 2001.

22. There is much written about this, but the translation by Stengel E, 1953, as well as the excellent small book by Dewhurst K, 1982, covers the important ground. The book was not well received, and only 142 copies were sold in the first year and 115 in the following nine.

23. Freud went on to clarify his view that Charcot's attribution of hereditary neuropathic influences to both organic disorders and the neuroses failed to understand the etiological and nosological differentiation of the neuroses. Obsessions and phobias were true neuroses, with different etiologies, and he distinguished anxiety neuroses from neurasthenia. The causes were revealed by psychological analysis, leading to his later insistence that the common pathological source of the neuroses lay in the sexual life of the individual, either in the current life or in a significant past event (as expounded in his paper of 1896 in *Revue Neurologique*).

24. Stengel E, 1954, 86.

25. Von Waldeyer-Hartz used silver nitrate staining to reveal the small cells of the brain, with an axon and dendrites, clarifying the neuron doctrine of brain structure, although this was clearly a development of ideas hinted at by others before him, notably Cajal.

26. Pribram K, Gill MM, 1976, 14.

27. Pribram K, Gill MM, 14, 127. Hughlings Jackson quotes from Stengel E, 1963, 350. See also Harrington A, 1989, 240–241. "Any impression which the nervous system has difficulty in disposing of by means of associative thinking or of motor reaction becomes a psychical trauma." Thus, neurotic symptoms became explicable in terms of symbols, reminiscences, defenses, and pathogenic ideas. Freud further stated that "the cyclical experiences forming the content of hysterical attacks . . . are all of them impressions which have failed to find adequate discharge, either because the patient refuses to deal with them for fear of distressing mental conflicts, or because (as in the case of sexual impressions) he is forbidden to do so by modesty or social conditions, or, lastly, because he received these impressions in a state in which his nervous system was incapable of fulfilling the task of disposing of them" (Freud S, 1966, 1:147–154).

28. Later in his career he moved to Hanover Square, London, and then to Queen Street, hardly a stone's throw away from Hughlings Jackson's house in Manchester Square. Like Hughlings Jackson, he had no children and preferred solitude. Unlike Hughlings Jackson, he was not popular with his colleagues.

29. In 1907, Maudsley offered £30,000 to the London County Council to establish a hospital. Other conditions that he stipulated were that it was for early and acute cases only, it was to have an outpatient department, and it was to take paying and pauper patients. It was to be situated within 3–4 miles of Trafalgar Square. It was built in Denmark Hill, London, where it is still situated. The board that was constituted to view the application consisted of several prominent neurologists: Gowers, Ferrier, and Sir Edward Farquhar Buzzard (1871–1945).

30. The University of London created a chair in the Pathology of Mental Disease (which later became Neuropathology) in 1936. Mapother held this first, followed by Golla. Mapother did much research into neuropsychiatric disorders, including the war neuroses and following head injury.

31. Maudsley H, 1879, 58. Goethe is quoted several times in the book.

32. On Wigan, see Hunter R, Macalpine I, 1963, 933–938. Quote is from Maudsley H, 1889. The best biography of Maudsley is Collie M, 1988.

33. Maudsley H, 1879, 41–42.

34. After his resignation, Maudsley did not publish in the journal for the next 10 years.

35. Taylor J, 1958, 2:100.

36. Reynolds EH, Trimble MR, 2009, 52. Reynolds makes the point that the contributions of Todd (of Todd's paralysis) to the electrical underpinnings of epilepsy seem to have been ignored, even by Hughlings Jackson.

37. Taylor J, 1958, 2:276–286.

38. Taylor J, 2:385–405. For a full description of the case, see Taylor D, Marsh SM, 1980. Links between epilepsy and suicide have been hinted at for years, but they have become an important part of today's clinical epileptology, as have a spectrum of psychiatric comorbidities.

39. Schmitz B, 1998, 8.

40. For more on this historical background, see Trimble MR, 1991.

41. Kraepelin E, 1962, 130.

42. Richard Hunter was appointed to the National Hospital for Nervous Diseases in 1960. He was a convinced psychoanalyst, and his mother Ida Macalpine was a psychoanalyst. At ward rounds he would always ask about families and anniversaries, which went down poorly with Eliot Slater (more later about him), even though he had appointed Hunter as a psychotherapist. He fell out badly with a number of his neurological colleagues and eventually went to Friern Hospital, at one time the Colney Hatch Asylum. He did a complete volte-face in his views, considering mental illness to be neurological illness, especially encephalitis. Although overemphasizing the organic aspects of psychiatric disorders, he was a good representative of the developing field of neuropsychiatry after the Second World War. See also note 45, p. 67.

43. Nietzsche was appointed as professor of philology at the University of Basel in 1869 at the age of 24. Toward the end of his life, he spent much of his time as a *Wanderphilosoph*, moving regularly around various places, including Genoa, Sils Maria, Nice, and Turin. For more on the final illness of Nietzsche, see Orth M, Trimble MR, 2006.

44. One of the staunch supporters of the syphilitic theory for Nietzsche's decline was the eminent Leipzig neuropsychiatrist Paul Möbius (1853–1907), but he could only make the diagnosis from the clinical notes and observations of those still alive who knew Nietzsche. He opined that the signs of the disorder were visible as early as 1881, but he had to confess that he had never encountered or heard of any cases similar to Nietzsche's, nor had any of the colleagues he consulted, and he had not found similar descriptions in the literature or the relevant books.

45. See Hayden D, 2003, for a good account of the supposed and the confirmed sufferers.

46. The disease duration of frontotemporal dementia can be long, up to 20 years. It has an insidious onset and gradual progression, the patient revealing early decline in social interpersonal conduct and regulation of self-conduct. There is emotional blunting, loss of insight, and a decline in personal hygiene and grooming. There are other recorded aspects of Nietzsche's behavior which fit with this diagnosis, including perseverative and stereotyped behaviors and eventual loss of language skills.

47. In the last year of his sentient life he produced six books.

48. His sister, Elizabeth, did her best to suppress the slur of syphilis, and in reality Nietzsche's sexual history will forever remain a mystery. In March 1890, he was discharged into the care of his mother, and when she died, it was Elizabeth who cared for him and carefully managed his reputation. His sister stated that from 1897 he sat mutely in his chair—silenced forever. He was a vegetarian.

REFERENCES

Benson DF. Disconnection syndromes revisited. In: Trimble MR, Cummings JL, eds. *Contemporary Behavioural Neurology*. Butterworth-Heinemann, Oxford; 1997.

Clarke E, O'Mallay CD. *The Human Brain and Spinal Cord*. University of California Press, Los Angeles; 1968.

Collie M. *Henry Maudsley: Victorian Psychiatrist: A Bibliographical Study*. St. Pauls Bibliographies, Winchester; 1988.

Critchley M. The Broca–Dax controversy. In: *The Divine Banquet of the Brain*. Raven Press, New York; 1979.

Dewhurst K. *Hughlings Jackson on Psychiatry*. Sandford, Oxford; 1982.

Ellenberger H. *The Discovery of the Unconscious*. Basic Books, New York; 1970.

Freud S. *Standard Edition of the Complete Psychological Works*. Hogarth Press, London; 1966.

Griesinger W. *Mental Pathology and Therapeutics*. Trans. Lockhart Robertson C, Rutherford J. New Sydenham Society, London; 1867.

Harrington A. *Medicine, Mind, and the Double Brain*. Princeton University Press, Princeton, NJ; 1989.

Hayden D. *Pox: Genius, Madness and the Mysteries of Syphilis*. Basic Books, New York; 2003.

Hunter R, Macalpine I. *Three Hundred Years of Psychiatry, 1535–1860*. Oxford University Press, London; 1963.

Hunter R, Macalpine I. *Psychiatry for the Poor*. Dawsons of Pall Mall, London; 1974.

Joynt R. Historical aspects of the neurosciences. In: Rose FC, Bynum WF, eds. *A Festschrift for Macdonald Critchley*. Raven Press, New York; 1982.

Kraepelin E. *One Hundred Years of Psychiatry*. Philosophical Library, New York; 1962.

Maudsley H. *Body and Mind: An Inquiry into Their Connection and Mutual Influence, Specially in Reference to Mental Disorders*. Appleton, New York; 1879.

Maudsley H. The double brain. *Mind* 1889;14:161–187.

McHenry LC. *Garrison's History of Neurology, Revised and Enlarged*. Charles C. Thomas, Springfield, IL; 1969.

Micale MS, Lerner P. *Traumatic Pasts*. Cambridge University Press, Cambridge; 2001.

Morgan A, Liegeois F, Vargha-Khadem F. Motor speech profile in relation to site brain pathology: A developmental perspective. In: Massen B, van Lieshout P, eds. *Speech Motor Control: New Developments in Basic and Applied Research*. Oxford University Press, Oxford; 2010.

Nietzsche F. *Human, All Too Human: A Book for Free Spirits*. Cambridge University Press, Cambridge; 1878/1986.

Orth M, Trimble MR. Friedrich Nietzsche's mental illness: General paralysis of the insane versus fronto-temporal dementia. *Acta Psychiatrica Scandanavica* 2006;114:439–444.

Pribram KH, Gill, MM. *Freud's "Project" Reassessed*. Basic Books, New York; 1976.

Reynolds EH, Trimble MR. Epilepsy, psychiatry and neurology. *Epilepsia* 2009;50(Suppl. 3):50–55.

Schmitz B. Forced normalization: History of a concept. In: Trimble MR, Schmitz B, eds. *Forced Normalisation and Alternative Psychoses of Epilepsy*. Wrightson Biomedical, Petersfield, UK; 1998.

Stengel E. *On Aphasia: A Critical Study*. Trans. Stengel E. International Universities Press, New York; 1953.

Stengel E. A re-evaluation of Freud's book "On Aphasia": Its significance for psych-analysis. *International Journal of Psychoanalysis* 1954;35:85–89.

Stengel E. Hughlings Jackson's influence in psychiatry. *British Journal of Psychiatry* 1963; 109:348–355.

Stone I. *The Passions of the Mind: A Biographical Novel of Sigmund Freud*. Doubleday, New York; 1971.

Taylor D, Marsh SM. Hughlings Jackson's Dr Z: The paradigm of temporal lobe epilepsy. *Journal of Neurology, Neurosurgery and Psychiatry* 1980;43:758–767.

Taylor J. *Selected Writings of John Hughlings Jackson*. Staples Press, London; 1958.

Trimble MR. *Post-traumatic Neurosis*. John Wiley and Sons, Chichester; 1981.

Trimble MR. *The Psychoses of Epilepsy*. Raven Press, New York; 1991.

Watson P. *The German Genius*. Simon and Schuster, London; 2010.

Young J. *Friedrich Nietzsche: A Philosophical Biography*. Cambridge University Press, Cambridge; 2010.

CHAPTER NINE

Fin de Siècle

That was the end of Lieutenant Carl Joseph, Baron von Trotta.

Joseph Roth, 1932/2002, *The Radetzky March*, 320

In the fall of 1887, Freud and Josef and Mathilde Breuer went to the old Hofburg Theater in Vienna to see a production of the tragedy Oedipus Rex *by Sophocles. There is a moment in the play when Oedipus asks Jocasta, "Yet how can I not fear my mother's bed?" She replies, "So do not fear this marriage with your mother; / Many a man has suffered this before—/ But only in his dreams, Whoever thinks / the least of this, he lives most comfortably."*[1]

Freud, archaeologist and explorer of the depths, found in this play a cornerstone for his developing theories of psychoanalysis. With the failure of the *Project*, his disappointment over the cocaine affair on which others had published before him, and the hostility that his publication on infantile sexuality and the sexual abuse of children brought him from Julius Wagner-Jauregg (1857–1940), he set out with even more determination to become famous. He identified himself with heroes such as Alexander the Great, Hannibal, and Napoleon, and he wrote in a letter to his friend Wilhelm Fleiss in 1897, "The expectation of eternal fame was so beautiful, as was that of certain wealth, complete independence, and lifting the children above the severe worries that robbed me of my youth. Everything depended upon whether or not hysteria would come out right." Wernicke had declared that insanity was incurable; Freud saw himself as Oedipus, "defeating the dark riddling voices of the subconscious," and developing not only new theories about the mind but also new treatments.[2]

Freud was certain that the contents of dreams were a form of sleeping insanity, so he analyzed his own dreams and their memory content, taking the remembered fragments apart, like an anatomist: the interpretation of dreams was the royal road to knowledge of the unconscious mind. Freud's contribution was a watershed in the study of dreams. No one before him had suggested that if they could be deciphered, they would be a source of insight into the waking world of the dreamer, about which the latter was unaware. After Freud, dreams and dreaming became a proper subject for scientific study, rather than simply being used by prophets as a predictor of the future. For Freud the elements of a dream were a complete rupture from the waking experience, particularly with regard to the latent content that had logical and linguistic rules quite different from those of the awake state. The way that philosophers before him had largely ignored the dreaming state is of interest but not discussed further here.[3]

This is not the place to discuss any details of Freud's theories, or his method of psychoanalysis, but Freud's role for the themes developed in this book is very important. There are three main lines, quite divergent. First, he developed ideas about the dynamic workings of the mind and highlighted the fact that much of the driving force of behavior was not conscious. Secondly, he reintroduced a Cartesian framework allowing theories of a brainless mind and a mindless brain to flourish and dominate twentieth-century psychiatry and neurology, respectively. Thirdly, Oedipus was again famous.

The various nonmedical developments underlying Freud's theories are shown in table 9.1. There was a heavy emphasis on philosophers, yet Freud was a "reluctant philosopher," to use the title of Alfred Tauber's book,[4] and very reluctant to acknowledge several authors who, in the latter part of the nineteenth century, had been writing about unconscious mention and its effects on behavior, which must have influenced his own theories. Not only did he fail to reveal the importance of Hughlings Jackson, who incidentally also wrote about the similarities between the contents of dreams and abnormal mental states, but he also minimized the importance of Franz Brentano (1838–1917), Arthur Schopenhauer (1788–1860), and Nietzsche.[5]

During the course of his education, Freud had studied philosophy with Brentano, who used the term *intentionality*, meaning that every mental phenomenon has a content (intentional in-existence), which is directed toward an object (but not necessarily a thing), the intentional object. Mentation is always about something and directed toward something, a basis of Freud's object relations theory (in the development of the individual mental life, someone else is always involved) and hence Freud's ideas of cathexis and libidinal investment. These ideas have important consequences with regard to the developing twentieth-century concept of the romantic brain and are explored further in chapters 10 and 15.

TABLE 9.1
Some Origins of Psychoanalysis

Philosophical	Cultural
Johann Christian Reil (1759–1813): *Rhapsodies on the Application of the Psychic Cure Method to Mental Disorders*	Theology: exorcism
Hegel (1770–1831): life force, consciousness	
Kant: reason as a faculty independent of nature	
Friedrich von Schlegel (1772–1829): *sehnsucht* (longing)	
Friedrich von Schelling (1775–1854): *Naturphilosophie*—seeking underlying spiritual laws, forces, and polarities	Romanticism: Goethe, Schiller, Byron
Arthur Schopenhauer (1788–1860): Will = unconscious; a driving force behind the universe and human individuals; sexual instinct the most important concern of man and the animal	Individualism: Luther
Franz Brentano (1838–1917): *Psychology from an Empirical Point of View* (1874); intentionality	
Eduard von Hartmann (1842–1906): *Philosophy of the Unconscious* (1869); an intelligible blind dynamism underlying the universe; three layers of the unconscious (absolute, physiological, and psychological)	
Nietzsche: see text	

Schopenhauer wrote, "Consciousness is the mere surface of our mind, and of this, as of the globe, we do not know the interior but only the crust."[6] His work in the psychological sphere was markedly deterministic, positing our limited ability to escape the underlying dominance of what he referred to as the *Will*, a universal force, an innermost essence of everything, a force of nature which drove human behavior. According to Tauber, "Schopenhauer placed the subject in its world by articulating an organism that promoted the naturalism that was to penetrate into virtually all areas of *fin de siècle* European thought. He thus initiated a shift in comprehending human nature, which gained momentum with Darwinism and culminates with Nietzsche."[7] This was a reaction to the perceived excesses of positivism, a neo-romanticism, but romanticism with more than a tinge of decay and decadence in the air. Although the underlying philosophy was German, the sentiment was not only evident in medical circles but also reflected in the arts of the time, notably in Paris and Vienna.

The second half of the nineteenth century denoted a time of progressive decline of cultural values, with an inevitable rise of materialism and kitsch. Interest in the mystical and the unconscious led to surrealism. Out of a seeking for beauty came the aesthete, a servant of beauty, exemplified by

Oscar Wilde (1854–1900). The dandy decadentism of his own lifestyle and his fictional *Dorian Gray* extolled beauty as a form of genius: "It is only shallow people who do not judge by appearances, the true mystery of the world is the visible not the invisible." Somehow the brain was involved since "it is in the brain, and in the brain only, that the great sins of the world take place."[8] Neo-romantic feminine beauty came into its own, as in the portrayals of women in the Pre-Raphaelite images.[9] Entwined were dandyism, exoticism, eroticism, and a lot of Satanism. The dark, perverse, fantastic images of Aubrey Beardsley (1872–1898) reminded us that while the times may have seemed all about surfaces, they also asked about depths. There was a veritable spiritual onanism.[10]

The term *id* (*das Es*) originates from Nietzsche, and importantly, the dynamic concept of the mind, with notions of mental energy in latent and inhibited forms and release of energy or transfer of energy from one drive to another, is totally Nietzschean.[11] In Freud's writings, terms such as *nervous energy* and *psychic energy* are interchangeably used; psychic energies struggle with one another, Eros (libido) with Thanatos (the death instinct). "Instinct" is often used as a translation of the German word *Trieb*. In fact, this word more often denotes "drive" than "instinct," and it was a word familiar to and well used by Nietzsche.

Nietzsche's psychological ideas interlinked with contemporary scientific discoveries regarding physical energy, particularly its conservation and transformation. He suggested that dammed-up psychic energy could be released and discharged by minor precipitating causes. The human mind was a system of drives (*Triebe*). He even referred to emotion as "a complex of unconscious representation and states of the will." Nietzsche unveiled the self-deceiving nature of human beings, founded on the unconscious, the latter being an essential part of the individual psyche. One of his more famous aphorisms is, "'I have done that,' says my memory. 'I cannot have done that'—says my pride and remains unshakeable. Finally—memory yields."[12] Consciousness was "a more or less fantastic commentary on an unconscious, perhaps unknowable, but felt text."[13] Nietzsche thought of the unconscious not only as a cauldron of confused emotions and instincts but also as an area of reenactment of past stages of the individual and the species.[14] He was influenced by evolutionary theory (although not a Darwinian), and he associated the physiological rather than the metaphysical with creativity and art, drive and *Rausch* being ingredients. *Rausch* was important for Nietzsche, a German word echoing back to Goethe, meaning an acceleration of movement, a subdued ecstasy, an unconscious primordial force that underpinned Nietzsche's concepts of the Dionysian. *Rausch* affected sensory powers, as the whole affective system was excited and enhanced by it. Silk and Stern refer to these urges as proto-creative processes that are therefore physiological and psychological: "On the psychological level, the Dionysiac and the Apolline are creative human impulses, *Triebe* under which are subsumed modes of perceiving, experiencing, expressing and responding to reality."[15]

Trieb—instinct, meaning a powerful urge to expression, to make an impression on the environment, to attain an end, a momentum without which there could not have been the motion of emotion, that prepared state to act, to move, to prance and dance, to dine, to mate, and to create. We do not obtain knowledge of the environment passively, but by action and motor involvement.

There were fundamental differences between Freud and Nietzsche. One was how the emotional substrata ruling our lives could be transformed by reason (and psychoanalysis) alone. For Freud, *logos* was intact; words and syntax articulated the world and could be mined from the mind. As Tauber put it, "Nietzsche would celebrate the Will, Freud would endeavour to control it."[16] George Steiner expressed it thus:

> The heart of the psychoanalytic method was that Freud, like Aristotle and Descartes took it for granted that syntax relates organically to the realities it segments and articulates, that words speak the world. Only because of their intentional stability, their "truth functions," can words be psychoanalytically excavated, can their vertical concealments and suppressions be unmasked. The deconstructive proposal that language is in arbitrary motion, that meaning itself is a nonverifiable convention, that there are no insured bonds between discourse and that which is naively, ideologically postulated to exist "out there"—an axiom even of uttermost classical skepticism—the claim that "anything goes" would have struck Sigmund Freud as infantile clowning or madness.[17]

It is often thought that Freud's tripartite model of the mind emphasized distinct differences and boundaries between the ego, id, and superego. But as with the theories of Hughlings Jackson, it was not so simple, Freud's psychic systems having only arbitrarily chosen names. The psychical act was unconscious: what was latent and repressed in the unconscious remains capable of action and has retained cathexis. The ego has three adversaries—

the external world, the id, and the superego—and is buffeted, resting on the surface of what is unknown and unconscious while controlling the discharge of excitations, hence motility: "The ego is in the habit of transforming the id's will into action as if it were its own."[18] The ego substitutes the reality principle for the pleasure principle.

The importance of these writers for the theme of the romantic brain is quite understated. The capitulation of our everyday behaviors, let alone psychiatric illness, to unconscious forces was a radical shift of understanding, completely alien not only to traditional Western philosophies but also to neuroscience. The idea of the "self" as a fixed moral entity was collapsing, and the philosophical shift from one of "being" to one of "becoming" was again under way. For Nietzsche there was no identifiable self, there was no identifiable "being" behind doing, only becoming. He argued against any mind–body duality; free will was a theological and philosophical invention. There was no art for art's sake, but art was for life's sake, life itself being an aesthetic phenomenon.[19]

More Trouble at Home

Freud's split from neurology and his rejection by Vienna's medical elite continued, as illuminated by his tussles with Breuer and later with Wagner-Jauregg.

Breuer came from a well-established Viennese family and had done interesting physiological work, including unraveling the role of the vagus nerve in breathing, leading to the eponymous Hering–Breuer reflex. Breuer provided mentorship and financial assistance to Freud; *On Aphasia* was dedicated to him. Very significantly, he introduced Freud to the "hysterical" Bertha Pappenheim, the first case described in *Studies of Hysteria* (1893–1895), referred to as Anna O. She was quite polysymptomatic, with many symptoms we would now recognize as psychogenic with a marked neurological flavor (paralyses, contractures, amnesias, multiple pains, anorexia, and language disturbances, at times not being able to speak her native German and sometimes speaking only English). Breuer was treating her with hypnotism, but she was unsuggestable, so he encouraged her to try a method that she referred to as "the talking cure" or "chimney sweeping."

He described his developing method thus: "Every individual symptom in this complicated case was taken separately in hand; all the occasions on which it had appeared were described in reverse order, starting before the time when the patient became bed ridden and going back to the event which had led to its first appearance. When this had been described the symptom was permanently removed. In this way, her paralytic contractures and anaes-thesias, disorders of vision and hearing of every sort, neuralgias, coughing, tremors, etc., and finally her disturbances of speech were 'talked away.'" In this same communication the word *unconscious* is used in the psychoana-lytic sense for the first time.[20]

Breuer started treatment with Anna O in 1880, but in 1881 he called a halt to the therapy. Freud later told his translator James Strachey that Anna O had made it clear to Breuer that she had pronounced sexual feelings for him, one reason for the delay in publishing the case history (1893).

There were marked differences in approach to treatment and theory between the two men. For some time Freud was content to attribute the discovery of psychoanalysis to Breuer, but he then did a complete volte-face, revealing the problem that Breuer had dealing with Anna O's sexual "transference" (in 1914, in *On the History of the Psychoanalytic Movement*). He considered that Breuer had missed the significance and universal nature of the transference. While Breuer continued to treat psychiatric patients, he no longer used the cathartic method. Freud moved ahead with his theories of the sexual basis of neurotic symptoms, "repression," and the universal nature of the Oedipus complex, but failure to accept the Oedipus complex was the reason for most of the breaks with his disciples.

While hypnosis had revealed the importance of the unconscious for therapy, Breger suggests that just a few years after *Studies* was published, in his quest for fame, Freud turned on Breuer and cut him out of his life entirely. A second edition of the book was published in 1909, with the two authors providing separate prefaces. When Breuer was an old man, he ap-proached Freud in the street, but the latter ignored his open-armed greeting and passed him by.

Off to America

At the end of 1908, Freud received an invitation from G. Stanley Hall, presi-dent of Clark University in Worcester, Massachusetts, to come to America for a series of lectures celebrating the founding of the university. Hall had founded the first American psychological laboratory at Johns Hopkins Uni-versity in 1883 and had been teaching psychoanalysis, freely quoting Freud's papers. Freud at first refused to go, as the quoted fee was too low in terms of the amount he would not earn from his private practice while being away, but when a more generous offer was made, he accepted. Carl Jung (1875–1961) traveled with him. On the journey, after Jung had recounted a story about some prehistoric remains that had been dug up in Germany, Freud retorted that Jung was revealing an unconscious death wish against him—and fainted!

They arrived in New York on August 27, and they were shown around the city by two psychoanalysts, Abraham Brill (1874–1948) of New York, founder of the New York Psychoanalytic Society (1911), and Ernest Jones from Toronto. Freud spent time with James Jackson Putnam of Boston, the founder of both the American Psychoanalytic Association (1911) and the American Neurological Association (1874). Before the trip, he had spent some time polishing up his English, and he was delighted to find receptive audiences, even for some of his theories of child sexuality.

Even so, he was highly critical of America. He thought that Americans lacked culture, referring to them as "savages," who sublimated their sexual impulses into an unhealthy obsession with money. Yet he also said to Jones that "America is useful for nothing else but to supply money."[21] Psychoanalysis became a cure for all ills, particularly impressing a number of neurologists, who were consulted by so many neurotic patients that they did not know how to manage. The popularity for psychoanalysis provided a secure income for the analysts; later, the EEG gave neurologists their own instrument to use and charge for, thus furnishing neurology with money and enhancing it as an independent discipline.

The total separation between neurology and psychiatry was finally achieved in America in the 1930s and during the Second World War. European Jews with psychoanalytic training immigrated to America, and universities with expanding and competing departments of psychiatry appointed a number of them to senior positions, raising the prestige of psychoanalysis as an academic discipline. The theoretical boundaries of Freudian theories were soon breached, not only by dissenters such as Jung and Alfred Adler (1870–1937) in Europe but also by the "ego" psychologies and therapies. In these, the basic tenets of Freud's system were ignored, the ego no longer being understood as a moderator between instincts and value judgments of the superego, but as a conflict-free ego, which was autonomous but struggled to adapt to the environment. There came an encroachment of nondoctors (largely psychologists) developing and selling a profusion of therapies, and the "analytic hour" was soon taken up more by the well-to-do than by those with significant psychopathology.

The descent into brainless mind was propelled by Pavlovian ideas and learning theory, and it was taken to an extreme by the mindless/brainless theories and treatments of John Watson (1878–1958) and the radical behavioralism of B. F. Skinner (1904–1990). The untold damage that this era did to psychiatry is sometimes discussed but always downplayed, as the profusion of quick solutions to life's complexities via brief therapies seduces the media, insurance companies, politicians, and neurotics now as much as ever. In the lay mind, such therapists have become identified with psychiatry, as the psychiatrist now finds his or her decisions regarding patient diagnosis (often only used for payment purposes) and management bunkered in by a "team approach," often alien to any "medical perspective" that may broach the integrity of a patient's mental state. Somewhere along the way the compromise nostrum *biopsychosocial* has snaked into this vocabulary, with arguable consequences.[22]

War Neurosis

Recall that after the death of Emperor Franz Joseph came the outbreak of the First World War, in which the dissolute Austrian state suffered much damage. There was a loss of national identity, incompetence in high places, and huge arguments about the status and treatment of the large number of war neurotics (*Kriegszitterer*—"war shakers"). The battle lines were drawn between those who had used physical treatments and those who condemned them. The authority of the military doctors was challenged, and there was a fear that mass hysteria might break out, leading to revolution. The most prominent psychiatrist in Vienna at this time was Wagner-Jauregg, director of the Clinic for Psychiatry and Nervous Diseases, having succeeded Meynert. He had pursued physical treatments for psychiatric diseases and was interested in pyrotherapy. His idea was to bring a resolution to psychosis by inducing fevers. He first tried but failed with streptococci taken from patients with erysipelas, but he achieved somewhat more promising results with tuberculin. He then used blood from patients with tertian malaria. Using a scientific method for his investigations, unusual at the time, of including a control group of similar size whose members were treated with Salvarsan (arsphenamine, a drug also introduced for syphilis), he claimed considerable success. The treatment was rapidly introduced internationally, and since such psychoses were considered incurable (echoing Wernicke's comments), he was awarded the Nobel Prize in Physiology or Medicine in 1927.

The use of physical treatments at his clinic for psychological war trauma came under scrutiny by the Commission of Enquiry, which investigated wartime misconduct. The case that achieved notoriety was that of Walter

Kauders, a Jewish journalist, who had received a head injury during the war and had developed a neurosis. Wagner-Jueregg considered Kauders to be a malingerer and asked his assistants to treat him with electrical therapy. At the hearings, the professor impressed the commission more than Kauders, but Freud was given an opportunity to provide evidence, and he put forward a case for psychotherapeutic treatment. This brought him into conflict with Wagner-Jauregg, and he lost out, further distancing the developing Freudian psychiatry from any organic or neuropsychiatric leanings.[23]

Freud coveted the Nobel Prize in Physiology or Medicine, but he received the Goethe Prize for Literature. On his 80th birthday, it was not physiologists or psychologists who sent him congratulations; the letter he received was signed by Thomas Mann, Romain Rolland, Jules Romains, H. G. Wells, Virginia Woolf, and Stefan Zweig.[24]

Posttraumatic Neurosis

The war neuroses became a veritable battleground between those favoring an organic interpretation of shell shock and those recognizing the psychological imperatives behind the symptoms, especially in Germany and the United Kingdom. The approaches and types of patients in the two countries differed, in the sense that in Germany, psychiatry and neurology had not progressed to separate as much as they had in England, where asylum psychiatry and neurology were already different realms.

The arguments are well rehearsed in Pat Barker's novel *Regeneration*, the first of a trilogy about British army officers being treated at Craiglockheart Hospital, Edinburgh, for shell shock during the First World War. The physician in charge was the anthropologist and neurologist-psychiatrist W. H. R. Rivers (1864–1922). He had previously worked at the Bethlem Royal Hospital and at the National Hospital for the Paralysed and Epileptic with the neurologist Sir Henry Head (1860–1940). Together they had examined the effect of regeneration of sensation after cutting peripheral cutaneous nerves (Head's own!). The story describes the psychological methods of treatment of the war neuroses, and key patients are the poets Siegfried Sassoon (1886–1967) and Wilfred Owen (1893–1918). This is contrasted with the electrical treatments of Lewis Yealland (1884–1954), who was based at the National Hospital for the Paralysed and Epileptic.

In Germany, Oppenheim (whose views on the causes of these neuroses have been discussed), supporting a view that the signs and symptoms were related to microscopic alteration of the brain substance, lost this argument and was forced to shift his position. The neurologist Karl Kleist (1879–1960) described *Schreckpsychosen* (terror psychosis). These states were caused by the unconscious persistence of the traumatic events, which were acute, in contrast to the war neuroses, which tended to be seen not at the front lines but when soldiers were back at a base hospital. The presentations were chronic, and he reported the war-related nightmares from which they suffered. He also described twilight states (*Dämmerzustände*), with associated posttrauma amnesia and other memory complaints, as well as somatic symptoms such as seizures (which we would now call non-epileptic). Sigbert Ganser (1853–1931) noted a variant twilight state, in the syndrome named after him, with the identifying sign of *Vorbeireden* (approximate answers), and Kurt Schneider (1887–1967) suggested the protective effect of dissociation from intense fear-arousing emotions.

In London, the organic perspective was pursued by Frederick Mott (1853–1926). His earlier ideas were bound up with the denegation era and a belief in a psychopathic constitution, although with greater experience he moved on to emphasize the importance of emotional arousal.[25] There was thus a shift in understanding of the causes of hysteria, precipitated by the war neuroses. Debates arose as to whether or not they were so different from the neuroses seen following civilian accidents—so-called functional disorders. Rivers looked toward a psychoanalytic approach to treatment, although he was highly critical of a Freudian interpretation, with its emphasis on sex—this approach was for him pornographic. But Janet's dissociative-cognitive-reintegration model and the Freudian repression–abreaction one held on as underlying principles. Freudian sexually related complexes as universal underpinnings of the neuroses simply became untenable with so many soldiers succumbing to traumatic war neuroses.

The discussions that were aroused by these arguments about the causes of traumatic neuroses and posttraumatic neurosis have scarcely been resolved. The rechristened posttraumatic stress disorder and the attribution of psychological symptoms of mild head injury to invisible pathologies such as diffuse axonal injury remain highly controversial. The polemics still rebound daily in medicolegal settings, certainly in England and America, in part interlinked with the overtly antagonistic framework of these legal systems.[26]

Reactions

An important figure at the Charité was Karl Bonhoeffer (1868–1948), who worked with Wernicke, introduced the concepts of the exogenous and endogenous psychoses, and delineated "acute exogenous reaction types." In 1910 he published a paper on the "symptomatic psychoses," which had similar presentations to endogenous psychoses but were caused by independent "noxae," which he referred to as exogenous. These were reactions of inherently healthy brains to damage occurring in the course of life. The latter referred to modes of response by the brain to injury, although confusion was soon engendered by the meaning of *exogenous*.

The psychiatrist Aubrey Lewis (1900–1975) thought that it properly referred to events external to the brain. The terms *endogenous* and *exogenous* had actually been introduced earlier by the physician and philosopher Paul Möbius, the former reflecting disease from within the individual themselves, referencing degeneration, and the latter indicating an impingement on the individual from without. The idea of such "reaction types" was paralleled by the theories of Adolf Meyer in America and Paul Eugen Bleuler's (1857–1939) "psycho-organic syndromes." The term *organic* then became a substitute for *exogenous*, and the latter started to be applied also to the *psychogenic*, culminating in the split between *endogenous* and *psychogenic*. Alternatives such as *cryptogenic* and *idiopathic* then were used for the endogenous disorders, which, if they were not hereditary, lacked causal clarification.

These early discussions largely came from Germany, but the split between *endogenous* and *psychogenic*—the latter then leading to *reactive*—pervaded English and American psychiatry. This terminological Babel has led to much sterile debate and to arguments that still cause confusion and that can interfere with a patient's treatment. Words such as *functional* added to the confusion, especially when attached to neurological disorders. The *DSM* (*DSM IIIR*) and ICD-10 classifications adopted Bonhoeffer's original conceptions, with "Organic Mental Disorders" and "Organic, including symptomatic mental disorders," respectively. *DSM IV* removed the term *organic* on the grounds that it might imply that "non-organic" conditions had no biological foundation. One successful outcome of these deliberations was the birth of organic psychiatry, given its final imprimatur by William Alwyn Lishman's book *Organic Psychiatry*.[27]

The Rediscovery of the Whole

Wernicke, Broca, and others who pursued the parcellation of the human brain into discrete functional units were not without critics, many of whom argued for a Jacksonian perspective.[28] In contrast to a widely held belief that

Gall was an extreme localizer, his "organology" was actually underpinned with a romantic version of the embodied self: a mind–body integration, and a basic unity of animal life as revealed in part by his comparative anatomical studies. He envisaged anatomy as supporting physiology, and his method of anatomical dissection, from the spinal cord upward, revealed to him nature's coherence, with connections and reciprocal influences between anatomical parts. As Edwin Clarke and L. S. Jacyna put it, "Gall maintained that the 'plurality of the organs' on the surface of the brain, each with its own specific moral and mental faculty, functioned together to create the intellect and character." The unity of action of the CNS was central to the tenets of the romantic brain explorers, having already been discussed in the works of Erasmus Darwin, Bell, and others.[29]

In the eighteenth century, the prominent physiologist Albrecht von Haller had set out the view that there was functional equivalence of all parts of the brain (*action commune*), a doctrine supported by Bell and which held sway for half a century. Flourens, who carried out many animal ablation experiments, but also brain stimulation, identified what he called a vital node (*node vital*) in the medulla oblongata, which controlled respiration and was vital for existence itself. These subdivisions of the CNS all influenced each other, acting in reciprocal action by means of their common energy (*action commune*).[30] While the cerebral cortex was the seat of higher intellectual faculties and volition, these were not considered localized. These ideas of metaphysical dualism vied with Gall's materialism.

Advocators of more "holistic" principles of brain function in the nineteenth century may look to Hughlings Jackson's resistance to Broca. These ideas found support from neurologists such as Pierre Marie, one of Charcot's most able juniors, who succeeded to his chair at the Salpêtrière in 1918, and who held that aphasia was a manifestation of diminished intelligence. Constantin von Monakow (1853–1930) put forward the concept of diaschisis. This was linked to the ancient idea of sympathy, body parts having remote effects, first written about by Galen (*consensus partium*—a mutual influence of organs of the body on each other). Von Monakow was impressed by the recovery that was seen after various cerebral lesions, as well as the negative cases of cerebral localization, in which expected losses were not seen after injury to areas that were considered sites of localized function. His explanation was that functional as opposed to structural alterations occurred at sites distant from the primary injury. This led to errors of localization. Although diaschisis was a transient state, it raised questions as to our understanding of brain localization.

Together, von Monakow and Hughlings Jackson shifted the narrow views of localization from a strict anatomical to a physiological localization, and to a perspective of nervous functions as evolved from reflex mechanisms related to structural changes of brain development—referred to by von Monakow as "chronogenetic" localization. This is the principle that "each nervous function is the result of a historical evolution . . . successive processes always involve the activity of several centres or mechanisms . . . early stages of nervous function are fragments of the finished one."[31] No brain activity implying a sequence of processes could be assigned to a focal brain lesion, an idea that has now achieved a revival with modern brain imaging and a shift (see later) to exploration of cerebral circuits and connections.

The concept of cerebral equipotentiality was pursued by experimenters such as the physiologist Friedrich Goltz (1834–1902) and later the psychologist Karl Lashley (1890–1958), whose animal ablation experiments led them to the conclusion that the effects of lesions related not to the localization of the injury but to the amount of tissue removed. One outcome of these debates, summarized by Riese, could be that "there is neither localisation of functions nor localisation of symptoms, which are but disordered functions; the only type of localisation remaining is that of lesions."[32] Lesions are located space; functions are revealed and move in time.

I Remember It Well

Along with attempts to provide an Aristotelian framework to an understanding of the brain and its disorders, as new centers, pathways, and disorders were described by neuroanatomists and clinicians, an independent psychology was emerging, albeit closely allied to medicine. We have noted the importance of the Paris school in this regard, with the huge influence of Janet, and the biographical progression of Freud from a young neurologist to a psychologist. The interest in dreams, twilight states, and such phenomena as somnambulism and the rise of a plethora of "psychological" treatments, ranging from hypnosis and its variants to psychoanalysis, led to the beginnings of what is now called neuropsychology, with as intense an interest in memory and its failings. The first clinical neuropsychologist, pupil of Charcot and therapist of Marcel Proust (1871–1922), was Paul Sollier (1861–1933).[33] Shakespeare's sonnet 30 relates to memory:

When to the sessions of sweet silent thought
I summon up remembrance of things past,
I sigh the lack of many a thing I sought,
And with old woes new wail my dear time's waste.[34]

The words "remembrance of things past" were chosen by C. K. Scott-Moncrieff as the title for his English translation of *À la recherche du temps perdu*. But Proust's penetrating analysis of human memory was not that portrayed in the sonnet, and neither was it that which became of interest to the discipline of neuropsychology. The latter soon became infatuated with and succumbed to paper tests, scales, and classification—the science of psychometrics. Jung's division of extroverts and introverts, Ernst Kretschmer's (1888–1964) studies of the physiques of mental patients (athletic, pyknic, asthenic; *Körperbau und Charakter*, 1921),[35] Alfred Binet's (1857–1911) development of the first practical intelligence test, and the extensive work of Sir Francis Galton (1822–1911) on genius, intelligence, the power of prayer, mental imagery, and other psychological faculties, studying individual differences using ratings and measurements, were all taking psychology away from neurology. Our cerebral achievements became encased in a circinate enclosure from which they hardly have ever escaped.

Work focused only on the taxonomy of memory and its divisions, which seem to continue to proliferate, avoids dealing with memories. To remember is a self-reflective activity, *se souvenir de* in French, *erinnern (ich erinnere mich)* in German. In English we *re*-member and *re*-collect. Things re-membered are associated with places and time; there is a spatiality and an in-habiting.

Plato likened memory to a wax tablet. Aristotle considered memory to be like a picture, an imprint, but he distinguished between mnesis and amnesis, the latter being active and time related in terms of the absent being present: no one can remember the present while it is present. He refuted the ability of animals other than humans to have powers of recollection, and he described what we now refer to as déjà vu experiences. Socrates, who also discussed memory images as analogous to wax impressions (an image which in some people could be shaggy!), distinguished between learning and seeking, the latter implying an active process.

In the nineteenth century, associationist models of memory were popular, especially in England, and cases of fugue states, with loss of autobiographical memory, and examples of what we now refer to as transient global amnesia were reported. Charcot and Janet had both been interested in traumatic amnesias, especially linked to hysteria, and the French psychologist Théodule-Armand Ribot (1839–1916) studied memory disorders in a variety of neurological conditions. Ribot's law (1881) stated that in retrograde amnesia, recent memories are more likely to be lost than remote ones, which is still an important clinical observation.

Sollier was considered to be one of the brightest of Charcot's group, and he wrote widely on medical matters. He was asked by Charcot to review existing theories of memory, which led to two books on the subject. He was especially interested in the neurobiological processes that might underlie memory, as well as the neuroanatomical associations. He anticipated the concept of neural plasticity, suggesting increasing contacts between neurons as a basis for memory and why memory retrieval is quicker with repetition. He implied that there were memory centers, and while not specified, he cited the frontal lobes as an intellectual center regulating learning and retrieval.

Forgotten in studies of memory is that forgetting is an active process. This had been central to the theories of Nietzsche and Freud and was seen by Sollier as a "condition of memory." As expressed by the novelist Milan Kundera, following Proust, memory itself is a form of forgetting. The enigma of the presence of absence still remains. Proust's writings, in contrast to the realistic authors such as Honoré de Balzac (1799–1850) and Zola, take us on a journey into imagination, seeking psychological laws in search of a human essence and his own identity. Through his experiences of involuntary memories (an example of which is given at the start of chap. 11), revealing for him essence covered by time, he portrays the discovery of a metaphysical self—time regained through the involuntary memories, revealing the beauty beneath the surface of the phenomenal world: "Essence . . . the part of which after all our tears have dried can make us weep again." For Proust, Platonic in outlook, the search for essence is found in nature and in art, and time was the greatest artist. His rather dull, vain, intellectually uninteresting characters become for him enlivened by time and his memory to become manifestations and embodiments beyond themselves. This transcendental quiddity, hidden behind physical appearance, is given intense value through metaphor, providing a vast store of memories in his inner world moved from the contingencies of time and habit. The bonds that unite people are psychological, and the very processes of remembering seem proof of embodied time within a creative brain: "Our slightest desire, although striking its own, unique chord, contains within it the fundamental notes on which our whole lives are based."[36]

Proust's father, Adrien, was a physician in Paris, knew Charcot, and published works on neuropsychiatric problems, including neuraesthenia and aphasia. Babinski visited Proust on several occasions, as the latter feared he may develop aphasia, and before he died Babinski affirmed that he had but a short time to live. After Proust's death, he refused the opportunity to examine his brain.

Collapse

The century turned, and the twentieth century would soon witness the total eclipse of civilization and dominance of the Enlightenment agenda. It was more than London Bridge that was falling down, as Eliot wrote:

> A heap of broken images, where the sun beats,
> And the dead tree gives no shelter, the cricket no relief,
> And the dry stone no sound of water. Only
> There is shadow under this red rock,
> (Come in under the shadow of this red rock),
> And I will show you something different from either
> Your shadow at morning striding behind you
> Or your shadow at evening rising to meet you;
> I will show you fear in a handful of dust.[37]

The myth of Prometheus, who for Nietzsche was equated with Dionysus, reminds us that the supreme value of fire, so important in the evolution of our species, is Janus-faced. Energy gives rise not only to creativity but also to destruction.[38]

NOTES

Epigraph: In this book, Roth depicts the decline of the Austro-Hungarian Empire and the events leading up to the First World War, with the assassination of the heir to the throne. The path of four generations of one branch of the Trotta family parodies the decline of the empire and the deterioration of the health of Emperor Franz Joseph. There was a dissolution of all values. The Radetzky March is not a pretty Strauss waltz, but it was composed by the elder Strauss, in honor of an Austrian field marshal.

    1. Sophocles, 1962, lines 977–983. Oedipus then finds out that he has not only married his mother but also killed his father; Jocasta hangs herself, and he blinds himself.

    2. Breger L, 2009, v; Scully S, 1997, 230.

    3. There were, of course, exceptions, including Descartes. For much more on this and a summary, see Cutting J, Toone B, Trimble M, 2013, 13–32.

4. Tauber AI, 2010.

5. Freud took three semesters of philosophy under Brentano at the University of Vienna and considered following his MD with a PhD in that subject. Freud spoke of Nietzsche as a philosopher "whose guesses and intuitions often agree in the most astonishing way with the laborious findings of psychoanalysis." He added that for a long time he avoided reading Nietzsche on that very account, in order to keep his mind free from external influences (see Ellenberger HF, 1970, 277). In 1908 there was a meeting of Freud's inner circle in which the works of Nietzsche were chosen for discussion. Freud stated that "Nietzsche's ideas had not the slightest influence on (his) own work" (Köhler J, 2002, 210). On Freud's 70th birthday Otto Rank (1884–1939) presented Freud with a specially printed edition of Nietzsche's works, causing Freud to complain about the waste of the gold used in the lettering. Rank was expelled from the close inner Freudian circle.

6. Schopenhauer, in Magee B, 1997, 133.

7. Tauber AI, 2010, 154.

8. Wilde O, 1891. The full quote reads, "Beauty is a form of Genius—is higher, indeed, than Genius, as it needs no explanation. It is one of the great facts of the world, like sunlight, or springtime, or the reflection in the dark waters of that silver shell we call the moon. It cannot be questioned. It has divine right of sovereignty. . . . It is only shallow people who do not judge by appearances, the true mystery of the world is the visible not the invisible" (Wilde O, 26, 29).

9. Some examples are Dante Gabriel Rossetti's (1828–1882) portrayal of his lover Elizabeth Siddal as Ophelia and John Waterhouse's (1849–1917) *La Belle Dame sans Merci*.

10. Ellenberger HF, 1970, notes how there was a similarity between patients described by psychiatrists and the characters in novels and plays. He suggests, for example, that some of Janet's patients show similarities with Zola's characters (283). The Vienna Secession movement gave us artists such as Gustav Klimt (1862–1918), Egon Schiele (1890–1918), and Oscar Kokoschka (1886–1980).

11. Nietzsche did not give overriding dominance to the sexual instinct but to aggressive and self-destructive drives. Nietzsche's ideas fit with the so-called defense mechanisms developed by Freud (especially sublimation—a term that appears in Nietzsche's own works), repression (inhibition), and the turning of instincts toward the self. Nietzsche's description of *ressentiment* and false morality anticipated Freud's descriptions of neurotic guilt and of the superego. According to Ellenberger, Freud's *Civilisation and Its Discontents* shows a parallelism with Nietzsche's *On the Genealogy of Morals*. Both give rise to the idea that modern man is afflicted with the illness of civilization, a civilization that demands renunciation of natural instincts. See Ellenberger HF, 1970, 271–278.

12. Nietzsche F, 1886/1998, 58, aphorism 68.

13. Translations as given in Ellenberger HF, 1970, 273, 233.

14. Nietzsche noted the disorder and incoherence of dreams, suggesting that the dream was a reenactment of fragments from our own past and also the past history of mankind. Nietzsche incidentally went on to attribute similar processes to the development of mental illness; see Ellenberger HF, 1970, 273–274. Nietzsche's idea of the Will to Power became reinterpreted by Freud as a death urge or instinct. The observation that psychological illnesses can be caused by a trauma in childhood was suggested by Nietzsche in 1884 when he referred to "the wave effect," which can often manifest itself throughout the whole of one's life, having been triggered by a nervous shock as a child (Köhler J, 2002, 209). Köhler wrote that "almost everything that is today associated with the name of Sigmund Freud can be found in the writings of Friedrich Nietzsche, if only one knows where to look" (209).

15. Silk MS, Stern JP, 1981, 232, 288.

16. Tauber AI, 2010, 164.

17. Steiner G, 2011, 149.

18. Freud S, 1923, 636.

19. The problem was that since the classical Greek philosophers, mankind had been considered a rational being, separated from animals by *logos*. Descartes gave the logical mind complete freedom from the body, and the identity of the self (and soul) was a perquisite, given by God. Kant laid limits to human understanding, although rationality for him was a priori, the self (ego) being transcendental and the source of agency. Recall that for Kant, the external world is given, but we can only ever experience phenomena, not noumena. Georg Wilhem Friedrich Hegel (1770–1831) shifted the view of continental philosophy to being concerned with consciousness, and "forces" were for his philosophy embedded in history, history itself as a significant creator of history. The world was constructed through consciousness; absolute consciousness was ultimate and enfolded everything. The "ego" was universal; a community, a nation, even humanity as a whole could have a collective consciousness, a part of what in German is referred to as *Weltgeist*, universal spirit, the individual being integrated with the whole. But Hegel's scheme was teleological: reason was a final aim and an energizing power. Further, individuality becomes lost as a historical and social construction. Hegel shifted the Kantian balance between inside and outside, and the neo-Hegelian Karl Marx (1818–1883) shifted the balance further, "setting Hegel on his feet" by insisting on the reality of external social and economic structures and the way they created the individual. Nietzsche's philosophy was concerned with forces, humans being animated by creative and destructive instincts, and this transformative energy being linked to his (so often misunderstood) concept of Will to Power. His philosophy was not concerned with being, but with becoming, and the forces of the unconscious mind were integral not only to the evolution of the human psyche but also to creativity. His ontology questioned where morals come from, especially if, as he so boldly pronounced, "God is dead," and he proposed a philosophy based on a reevaluation of all values.

20. Breuer J, Freud S, 1974; Breger L, 2009, 34. While Freud used the same term in his write-up of the second case history, that of Frau Emmy von N, it is unclear whether the attribution of the word in the case of Anna O was referring to Freud or to Breuer; Breger L, 2009, 100.

21. Gay P, 1988, 563. Among other problems he had was the Americans' insistence against lay analysts, since several members of his inner group were not medically qualified, and neither was his daughter Anna Freud (1895–1982), herself an analyst and founder of psychoanalytic child psychology. Another quote about Americans: "This race is destined to extinction. They can no longer open their mouths to speak; soon they won't be able to so do to eat" (Gay P, 567).

22. For a good critique of the biopsychosocial model, see Ghaemi N, 2012. Ghaemi states, "I have doubts about the biopsychosocial model. If not untrue, it has at least far outrun the purposes it originally served. Psychiatry is currently eclectic, verging on anarchic. For all the advances in neurobiology and psychopharmacology, the field has no overarching conceptual structure. Or, perhaps better said, what passes for a conceptual scheme for the field—the biopsychosocial model—rose from the ashes of psychoanalysis and is dying on the shoals of neurobiology" (ix).

23. Wagner-Jauregg was a member of this committee, and he resigned from it. It was argued in Kauders's case that electrical treatment did not cure but caused harm, that holding people in hospitals in solitary confinement was cruel, and that the whole system was one of torture designed to compel the patients back to the war front as soon as possible. The full story is told in Eissler KR, 1986. Wagner-Jauregg himself was somewhat of an outsider. Not

Viennese but Tyrolean, his German was difficult to understand; he had been a classmate of Freud.

24. Published in the *New Republic* on June 17, 1936.

25. Mott's book *War Neuroses and Shell Shock* (1919) was important in outlining his theories.

26. There were those such as Yealland who believed that acknowledging the need for organic treatments did the patients justice, on the grounds that if only a psychological explanation was given, the patient would feel dishonored and weak and possibly be perceived as a malingerer. Other physicians considered such approaches alien to an understanding of the psychological underpinning of the symptoms and damaging to the prospect of eventual cure.

27. For more on the endogenous/exogenous dichotomy, see Lewis A, 1979, 173–178. Lewis's accompanying paper on the various meanings of the word *psychogenic* is also a thoughtful reflection on a word that continues to baffle, and which he referred to as a "confused but speciously attractive and convenient concept" (185–191). Alwyn Lishman worked at the National Hospital for Nervous Diseases for a short time beginning in 1966. His book, now in its fourth edition, is considered to be the standing reference text on the subject.

28. There are many texts that discuss this, but highly recommended is that by Riese W, 1959.

29. Clarke E, Jacyna LS, 1987, 234.

30. Clarke E, Jacyna LS, 252.

31. Riese W, 1950, 92.

32. Riese W, 148.

33. Bogousslavsky J, Walusinski, O, 2011.

34. Shakespeare W, 1998, 30, lines 1–4.

35. The English translation, *Physique and Character*, was published in 1925.

36. Proust M, 1919/2002, 222. 1925/2002, 591.

37. Eliot TS, 1940, 28, lines 25–30.

38. Nietzsche F, 1872/1993, 51.

REFERENCES

Bogousslavsky J, Walusinski O. Paul Sollier: The first clinical neuropsychologist. In: Bogousslavsky J, ed. *Following Charcot: A Forgotten History of Neurology and Psychiatry. Frontiers of Neurology and Neuroscience*, vol. 29. Karger, Basel; 2011.

Breger L. *The Dream of Undying Fame: How Freud Betrayed His Mentor and Invented Psychoanalysis*. Basic Books, New York; 2009.

Breuer J, Freud S. *Studies on Hysteria*, vol. 3. Trans. Strachey J, Strachey A. Pelican Books, London; 1974.

Clarke E, Jacyna LS. *Nineteenth-Century Origins of Neuroscientific Concepts*. University of California Press, Los Angeles; 1987.

Cutting J, Toone B, Trimble M. The merits of a phenomenological analysis of dreams. *Appraisal* 2013;9:13–32.

Eissler KR. *Freud as an Expert Witness: The Discussion of War Neuroses between Freud and Wagner-Jauregg*. Trans. Trollope C. International Universities Press, Madison, CT; 1986.

Eliot TS. *The Waste Land and Other Poems*. Faber and Faber, London; 1940.

Ellenberger HF. *The Discovery of the Unconscious: The History and Evolution of Dynamic Psychiatry*. Basic Books, New York; 1970.

Freud S. *The Ego and the Id.* In: *The Standard Edition of the Complete Psychological Works of Sigmund Freud, Volume 19 (1923–1925): The Ego and the Id and Other Works.* Translated Strachey J. Hogarth Press, London; 1923.

Gay P. *Freud: A Life for Our Time.* J. M. Dent and Sons, London; 1988.

Ghaemi N. *The Rise and Fall of the Biopsychosocial Model: Reconciling Art and Science in Psychiatry.* Johns Hopkins University Press, Baltimore; 2012.

Köhler J. *Zarathustra's Secret: The Interior Life of Friedrich Nietzsche.* Trans. Taylor R. Yale University Press, London; 2002.

Lewis A. *The Later Papers of Sir Aubrey Lewis.* Oxford Medical Publications, Oxford University Press, Oxford; 1979.

Magee B. *The Philosophy of Schopenhauer.* Oxford University Press, Oxford; 1997.

Nietzsche F. *The Birth of Tragedy.* Trans. Whiteside S. Penguin Classics, London; 1872/1993.

Nietzsche F. *Beyond Good and Evil.* Trans. Faber M. Oxford World Classics, Oxford; 1886/1998.

Proust M. *In the Shadow of Young Girls in Flower.* Trans. Grieve J. Penguin Books, London; 1919/2002.

Proust M. *The Fugitive.* Trans. Collier P. Penguin Books, London; 1925/2002.

Riese W. *Principles of Neurology: In the Light of History and Their Present Use.* Smith Ely Jelliffe Trust, New York; 1950.

Riese W. *A History of Neurology.* MD Publications, New York; 1959.

Roth J. *The Radetzky March.* Trans. Joachim Neugroschel. Overlook Press, New York; 1932/2002.

Scully S. Freud's antiquities: A view from the couch. *Arion* 1997;5:222–233.

Shakespeare W. *Shakespeare's Sonnets.* Ed. Duncan Jones K. The Arden Shakespeare, London; 1998.

Silk MS, Stern JP. *Nietzsche on Tragedy.* Cambridge University Press, Cambridge; 1981.

Sophocles. *Oedipus the King.* Trans. Kitto HDF. Oxford World Classics, Oxford; 1962.

Steiner G. *The Poetry of Thought: From Hellenism to Celan.* New Directions Books, New York; 2011.

Tauber AI. *Freud, the Reluctant Philosopher.* Princeton University Press, Princeton, NJ; 2010.

Wilde O. *Picture of Dorian Gray.* Simpkin, Marshall, Hamilton, Kent, London; 1891.

CHAPTER TEN

The Turn of the Screw

Now, to pry into roots, to finger slime,
To stare, big-eyed Narcissus, into some spring
Is beneath all adult dignity. I rhyme
To see myself, to set the darkness echoing.

Seamus Heaney, "Personal Helicon," 1966, *Death of a Naturalist*

"For there again, against the glass, was the hideous author of our woe—the white face of damnation. I felt a sick swim. . . . I shrieked, as I tried to press him against me. . . . But he had already jerked straight round, stared, glared again, and seen but the quiet day. . . . He uttered the cry of a creature hurled over an abyss, and the grasp with which I recovered him might have been that of catching him in his fall. . . . I caught him, yes, I held him—it may be imagined with what a passion; but at the end of a minute I began to feel what it truly was that I held. We were alone with the quiet day, and his little heart, dispossessed, had stopped."

The Turn of the Screw, by Henry James (1843–1916), is a story about a governess who has been sent to a country house to look after two small, "pure" children. She has several "turns," and in some of them she has hallucinatory experiences, visual apparitions of a sinister nature accompanied by a strange sensation of dread and death, associated with feelings of familiarity. These dreamlike states portray not only déjà vu reminiscences but also a strange loss of temporal awareness. In one of her attacks she describes how she fell to the ground, only to regain awareness some time later in the day. At the climax of the story, the governess is alone with the boy, Miles; she has another turn, but in order to protect the boy from the specter "in a blind moment," with a tremor of the hands and falling, she gets hold of him, and he dies in her arms.

The James Family

Henry James was the brother of William James (1842–1910). The latter qualified as an MD at Harvard and was influenced by the transcendentalist Ralph
Waldo Emerson (1803–1882). Like Hughlings Jackson, William James had a
significant interest in psychology and philosophy. His *Principles of Psychology*
(1890) opened up ideas of the "active mind" to a new perspective, but one
quite in line with much else that was happening in the fin de siècle Western
world. Although isms were proliferating, boundaries were breaking. In
the visual arts, the first impressionist exhibition had been held in 1874,
named after Claude Monet's *Impression, soleil levant* (Impression, sunrise);
other scenes he painted revealed the effects of light and time on perception
(e.g., Rouen Cathedral). In 1910, the English artist Roger Fry (1866–1934)
held an exhibition in London of modern French painters called *Manet and
the Post-Impressionists* (including Vincent van Gogh [1853–1890] and Paul
Gauguin [1848–1903]). Paul Cézanne (1839–1906) ingrained spheres, cones,
and cylindrical shapes to nature, shifting perceptual and perspectival planes.
The cubism of Georges Braque (1882–1963) and Pablo Picasso (1881–1973)
followed, while elsewhere in 1917, Marcel Duchamp (1887–1968) exhibited an
upside-down urinal (*Fountain*), challenging the need of the artistic creative
process itself. Surrealism originated in Paris, with the first surrealist manifesto written by André Breton (1896–1966) and released to the public in
1924. This was a movement concerned with such phenomena as automatic
writing, dream résumés, and thought that was devoid of reason and aesthetic or moral implications. The unconscious, especially sexual forces were
brought to the fore, as in the work of Salvador Dalí (1904–1989). René Mag-
ritte (1898–1967), influenced by Giorgio de Chirico (1888–1978), turned art
and philosophy inside out with his astonishing and disturbing juxtaposi-
tions of images, verbal and visual. His *The Treachery of Images*, showing a
pipe with the caption *"Ceci n'est pas une pipe"* (This is not a pipe), with its
visual, linguistic, and philosophical ambiguities, not only undermined empirical philosophy but challenged the nature of art itself.

Surrealism in art was one outgrowth of the medical and psychological
theories of Charcot, Freud, and Janet. Breton dropped out of medical school
and learned about automatic writing from Janet. These artists were fascinated by psychic automatisms and endeavored in their art to suspend the
conscious mind and liberate the unconscious. Luis Buñuel and Salvador
Dalí, in films such as *Un Chien Andalou*, presented dreamlike allusions with

unconnected events and no temporal sequence, mixing their aesthetics with Freudian psychology. They wanted a beauty that was "convulsive." Freud seemed remarkably unimpressed, for he believed that the latent contents of the dream were discoverable and had meaning, which was quite contrary to the surrealist manifesto.

William James is perhaps most famous for his discussion of emotions and their representation in the body. Acknowledging the earlier work of Bell and Erasmus Darwin, who emphasized the visceral factors involved in the experience of emotion, he opined that a disembodied human emotion was not possible. He referred to the "cutaneous shiver which like a sudden wave flows over us" and "the heart-swelling and the lacrymal effusion that unexpectedly catch us at intervals." This could be evoked by poetry and drama. He reminded us that "in a rage we 'work ourselves up' to a climax by repeated outbreaks of expression."[1] He considered that our emotions were "nothing but the feeling of the reflex bodily effects . . . *the bodily changes follow directly the* PERCEPTION *of the exciting fact, and that our feeling of the same changes as they occur* IS *the emotion.*"[2] We feel sorry because we cry, angry because we hit out, and afraid because we tremble. The central representations of emotion reside therefore in the cortical receptive areas of our somatic and proprioceptive afferents. Similar ideas had been expressed by the Danish physician Carl Lange (1834–1900), and the hypothesis is thus referred to as the James–Lange theory of emotion. It has been well revived in more recent times, but in a much-modified form, by Damasio.

But William James's greater influence should relate to his theories of consciousness. He argued that the tenets of associationism simply failed to account for our experiences, and he emphasized the goal-directed nature of our activities. The mind was composed of ideas that could not be generated by purely passive influences. He was after a physiological psychology, without a Cartesian homunculus or the hidden mental processes of Kantian categories. There was no transcendental self, as Kant believed. Our inner life is a stream of consciousness, not a linking together of disconnected ideas. Thoughts have a "feeling" of belonging together with other thoughts; we have a sense of continuity through time, experience being unified, not discontinuous. While associationism relegated thoughts to being composed of simple, unchanging parts, James tells us that thoughts are creative and ideas are complex.

He adopted a stance of psychophysical parallelism; there was no place in the mind or brain where thoughts are stored. His developed philosophy was one of functionalism, which implied that mental life was organic, active, and pragmatic, but he made us aware of the fourth dimension that permeated thoughts.[3]

He enumerated the four characteristics of consciousness as follows:

1. Every state tends to be part of a personal consciousness.
2. Within each personal consciousness states are always changing.
3. Each personal consciousness is sensibly continuous.
4. It is interested in some parts of its object, to the exclusion of others, and welcomes or rejects—*chooses* from among them, in a word—all the while.[4]

Further, "all our thinking has a TOPIC or SUBJECT." Words are not essential, and while the feelings of relation are everything, some trains of thought are tinged with emotion, others not. "We notice only those sensations which are signs to us of *things* which happen practically or aesthetically to interest us . . . the mind selects."[5]

Henry James remained secretive about the nature of the governess's visions in his story, not wanting to spoil the mystery he had created. His father, Henry James Sr., had an episode in May 1844, in which he was suddenly seized with terror, which came to him "in a lightning flash." In this he saw a "damnèd shape squatting invisible to me within the precincts of the room, and radiating out from his foetid personality influences fatal to life." Perhaps this was a clue. However, both brothers spent a lot of time in London, and their sister, Alice, was a chronic invalid, probably suffering from what today would be referred to as a somatoform disorder.[6]

Growing up "like a rare, fragile tropical plant," in a male-dominated household, Alice had her first mental breakdown at age 19; her multiple symptoms included "violent turns of hysteria."[7] She was well acquainted with brother William's probings into abnormal states of mind and, through his writing, with the theories of Janet. She consulted several doctors in London, at one time residing quite close to the National Hospital for the Paralysed and Epileptic. It is hard to believe that William or Henry failed to contact any of the famous physicians in attendance there. The neurologist James Purdon Martin (1893–1994) doubted little that Henry James would have been acquainted with the writings of his brother and the latter's knowledge of Hughlings Jackson's ideas, which are quoted in William's textbook. Purdon Martin even speculated that Henry James visited Hughlings Jackson at his home in Manchester Square: "in so far as the basis of the governess's behaviour was a secret, I think it is most likely that Dr Jackson was a party to it." Temporal lobe epilepsy, as it became referred to, with its subtle but sometimes dangerous manifestations perhaps reflected in the story, leading to the death of Miles by strangulation, was now in the literature. The link to a "turn" seems obvious, and the implication is made clear in some productions of the opera of the same name by Benjamin Britten (1913–1976).[8]

Return to Germany

On Sunday, June 13, 1886, King Ludwig II of Bavaria (1845–1886) and his doctor, Bernhard von Gudden (1824–1886), were found drowned in Lake Starnberg, not a few paces away from one of the king's castles. The king, having depleted the financial resources of the royal coffers on building his famous "fairy-tale" castles, had been certified as unfit to govern and was confined to Berg Castle. Together with von Gudden, he left the castle to go for an evening walk. The circumstances surrounding their deaths will forever remain a mystery, but von Gudden was there because he was one of Germany's leading psychiatrists. He had provided the document of deposition asserting that the king was insane, formally placing him in custody.[9] He founded a group of "neuromorphologists" in Munich, which included Bleuler, Auguste Forel (1848–1931), Ganser, Kraepelin, and Franz Nissl (1860–1919). He was a skilled neuroanatomist, and there are several brain structures named after him.

The search for a somatic basis of mental disorders was continuing apace in Germany, especially in Munich, with the school developed by Kraepelin, serving as professor of psychiatry there, as well as head of the University Clinic. The roll call included Alois Alzheimer (1864–1915), Constantin von Economo (1876–1931), Robert Gaupp (1870–1953), Frederic Lewy (1885–1950), Brodmann, and Kleist, the last of whom in turn working with others such as Otto Detlev Creutzfeldt (1927–1992) and Kurt Schneider. Their contributions were considerable and cannot be discussed in any detail here, but the very list of these names catches the breath of anyone interested in neuropsychiatry, since at these times there were intellectual and financial forces pulling apart what became psychiatry and neurology. These men were nearly all psychopathologists and neuropathologists, explorers of the brain and mind, and should have inspired a continuing line of the clinical neurosciences, different from neurology and better referred to as neuropsychiatry in its truest sense.

Kraepelin insisted on systematic collection of facts, delineating disease entities, and following the progress of patients over time. He encouraged work in the basic sciences, leading to the development of new stains for examining the nervous system (Nissl), to pathological descriptions of diseases (Alzheimer), and to studies in experimental psychology. He had been a student of Wilhelm Wundt (1832–1920), who at one point was an assistant to von Helmoltz, and who established psychology as an independent discipline. Wundt was the first person to refer to himself as a psychologist, and his *Principles of Physiological Psychology* (1874) was the first textbook of what we today may refer to as psychology.

Kraepelin's *Lehrbuch der Psychiatrie* went to nine editions, and the ideas expressed, which evolved with his gathering experience, remain so influential today. In the sixth edition, his *dementia praecox*, which became Bleuler's schizophrenia, was distinguished clearly from manic-depressive psychosis (between deteriorating and nondeteriorating periodic psychoses). This was the outcome of precise psychopathological descriptions observed over time. Psychiatric disorders could be allocated separate categories, for which later discoveries should lead to underlying pathological causes; insanity was not to these authors a single disease entity. The alternative, the idea that there was just one psychosis, an *Einheitspsychose* (unitary psychosis), originated with Ernst Albrecht von Zeller (1804–1877) and was advocated by Griesinger. The position was that different *forms* of illness represented different *stages* of one morbid process. Karl Ludwig Kahlbaum (1828–1899) challenged such ideas, as they signaled the end of psychopathology as a discipline. His 1863 publication *Die Gruppierung der psychischen Krankheiten* (The classification of psychiatric diseases) delineated four distinct types of mental illness (vesania)—namely, vesania acuta, vesania typica, vesania progressiva, and vesania catatonica—and was influential for Kraepelin's ideas and his taxonomy.

Meanwhile in America

The American Neurological Association was founded in 1875, with the *Journal of Nervous and Mental Disease* (1874) as its official journal. The debates in Europe over the causes of traumatic neurosis, the use of hypnosis, and the ideas of Freud permeated American neurology, as many neurologists were seeking some way to help the many patients with neurosis who came to their clinics. They were aloof from the asylum psychiatrists and were seeking a scientific basis to underpin any therapies.

Beard had given the term *neurasthenia* to a disorder of exhaustion of the nervous system, linked to the pressures of contemporary life. He called it an

American neurosis, while William James, who was diagnosed as having the condition, used the moniker *Americanitis*. Beard claimed that he had a cure, in part using electrotherapy. To explain the disorder, he used reflex theory and the nature of electrical energy within the brain, giving the diagnosis not only respectability but also an apparent scientific basis.[10]

Weir Mitchell was a neurologist who, along with William Alexander Hammond (1828–1900), had essentially inaugurated neurology as a discipline in America. Their observations and management of peripheral and central nervous injuries in the American Civil War were outstanding, and they also took an interest in trancelike states and neurasthenia. But, in opposition to Beard, Weir Mitchell invented the "rest cure" (complete rest, plenty of good food, and daily bodily massages). He published several papers on poisonous snakes, and ideas about their venom occupied his mind incessantly. He had a bad hand tremor but wrote extensively, and his books *Wear and Tear* and *Fat and Blood* brought him fame and wealth. When he was in Paris, he consulted Charcot for an opinion about his own "run-down" state of health, but he did not give his name. Charcot asked him where he was from, and when he heard that he was from Pennsylvania, Charcot told him that he needed to go home and see Dr. Weir Mitchell, offering to give him a letter of introduction, since that doctor knew more about "run-down" conditions than anyone else!

Weir Mitchell was hardly an academic, and when he was appointed as a professor at the University of Pennsylvania, he resigned within minutes of the interview at which he was offered the position. However, he was considered a great authority and teacher of neurology, and he also wrote fictional books and poetry. During his last days, his wandering talk was all about mutilation and bullets. One obituarist noted that "it is when we read his tales and poems, no matter what their subject, all come from a spirit over which had passed the great vision; every drop of ink is tinctured with the bloods of the Civil War."[11]

Pierre Janet visited Harvard in 1906 and lectured on the major symptoms of hysteria. While most psychiatrists working in hospitals had little or no interest in such psychological approaches to causality and treatment, neurologists perhaps paradoxically did. Some neurologists, such as those in Boston, were more enthusiastic than others (not so in Philadelphia). They included Putnam, who became the leading American proponent of psychoanalysis, and Morton Prince (1854–1929). The latter had visited Charcot and used hypnosis, especially to study the mind and its fragility. He wrote widely on dissociation, and his book *Dissociation of a Personality*, a study of a patient with multiple personality disorder, reverberates clinically and controversially to this day.[12]

Perhaps not surprisingly, many neurologists were less than enthusiastic about the incursion of psychological influences into their growing somatically based discipline, with some considering the psychologically orientated physicians—with a chargeable asset and a quasi-scientific basis for their treatments—as a threat to their income. The gulf between neurology and psychiatry inevitably widened, as neurologists more and more were concerned only with organic neurological disorders. However, at this crucial time at the interface between neurology and psychiatry in America, there were several who rightly should be called neuropsychiatrists and deserve special mention.

Twentieth-Century Neuropsychiatrists

Smith Ely Jelliffe (1866–1945) and William White's (1870–1937) *Diseases of the Nervous System: A Text-Book of Neurology and Psychiatry* (1915) was a compendium of up-to-date information in neuropsychiatry. Jelliffe was professor of psychiatry at Fordham University in New York, and White a professor of nervous and mental diseases at Georgetown University in Washington, DC. They united what they referred to as "sensorimotor neurology" with knowledge of the "historically oldest portion of the nervous system, the sympathetic and autonomic (vegetative) and . . . the increase in our knowledge of the mechanisms that operate at the psychic or mental levels." The individual was acknowledged as a biological unit, with broadly defined goals, the nervous system being a part of the larger whole: "Man is not only a metabolic apparatus . . . nor do his sensorimotor functions make him a feeling, moving animal . . . nor yet is he exclusively a psychical machine . . . he is all three, and a neurology of today that fails to interpret nervous disturbances in terms of all three of these levels, takes too narrow a view of the function of that master spirit in evolution, the nervous system."[13]

While Freudian ideas are embedded in their text relating to the psychoneuroses and psychoses and the technique of psychoanalysis outlined, this textbook, running to over 1,100 pages, must be the last comprehensive book of neuropsychiatry written to date.

Jelliffe, a devoted botanist, often traveled in Europe and was much impressed by European culture, learned German, knew Oppenheim, studied with Kraepelin, heard Charcot and Babinski lecture, met Jung and Bernheim, visited the National Hospital for the Paralysed and Epileptic, and had correspondences with Freud. At first convinced by the European concept of degeneration and the ideas of Lombroso, his experiences gave him much contact with the developing psychotherapeutic movement in Europe and America. He was converted to psychoanalysis by Abraham Brill (1874–1948), America's first psychoanalyst. In his practice Jelliffe was a psychoanalyst, but he cautioned against the excesses of Freudianism. He wrote to White in 1909, "This whole Freud business is done to death. The lamp posts of Vienna will cast forth sexual rays pretty soon. They keep writing new stuff. The poor unborn children cannot be told fairy tales any more because of their sexual significance. I suspect that William Tell's apples must have been a pair of testicles and as for George Washington's cherry tree—well . . ."[14]

Jelliffe first met White at Binghamton State Hospital in 1896, and the book they cowrote went through many editions. Most of the neurology was written by Jelliffe, who continually updated the introductory chapter to expound his biological perspective. The emphasis on evolution from the simple to the complex and structural anatomical hierarchies that parallel evolutionary development all hinted back to both Freud and Hughlings Jackson. He added chapters on endocrine influences and gave special importance to the reproductive instinct from both conscious and unconscious psychological and symbolic perspectives.

Jelliffe's strong advocacy of psychoanalysis led to some dissatisfaction with the content of the *Journal of Nervous and Mental Disease*, which Jelliffe had been involved with as associate editor since 1899, acting as editor in chief from 1902 to 1944. For a long time it was the only neurological journal in America. There was the *American Journal of Insanity* for asylum physicians, later to become the *American Journal of Psychiatry* (1921), but in 1919 the *Archives of Neurology and Psychiatry* was first published, sponsored by the American Medical Association.[15] This rivaled Jelliffe's enterprise and developed a more secure neurological content as the split between neurology and psychiatry continued. Applauded by Freud as a man of the medicine of the future, and heralded by some as a prophet, Jelliffe was a cultured polymath whose attempts to develop an identifiable balance between neurology and psychiatry, today's neuropsychiatry, remained as a candle in the wind which refused to be blown out.[16]

But Jelliffe was not the last or least of the neuropsychiatrists of the early part of the twentieth century. In America more and more university departments of psychiatry were established, and very influential was that at the Phipps Clinic in Baltimore, headed by Adolf Meyer. He was Swiss, studied medicine and neurology in Zurich, and immigrated to America in 1893. He worked in the pathology department of the Illinois Eastern Hospital for the Insane and then moved to the Massachusetts Insane Asylum in Worcester (1895–1910). He went to the New York State Psychiatric Institute as chief pathologist, before going to Cornell University Medical College as professor of psychiatry. His wife, Mary Potter Brooks, helped with his evaluation of patients, and since he came from a European perspective, he was surprised at the lack of attention given by psychiatrists to history taking. His wife would thus visit patients to explore their backgrounds, essentially the first American social worker.

Meyer's background was in pathology, and he entered psychiatry through the autopsy room. His thorough acquaintance with European ideas in neurology and psychiatry and his interest in patients' social surroundings and personalities led him to his "psychobiological" approach to mental illness. This involved bringing the person and known biological facts together.

Meyer had spent a brief time with Charcot in Paris, and in London he watched Gowers documenting histories from patients in shorthand (so as not to miss anything). He was influenced by the ideas of von Gudden and von Monakow, the scientific and evolutionary approaches of Thomas Huxley (1825–1895), and the writings of Hughlings Jackson, especially his principles of evolution and dissolution. Meyer was disappointed that the last had not more firmly linked his interest in psychology to neurology; there was no Cartesian dualism for Meyer, only integrated mental activity, which he termed *ergasia*. Meyer liked the sober conservatism of British thought, in contrast to the then overelaborate Germanic *Naturphilosophie*, and he helped to introduce the works of Kraepelin to America.

Clinically, he introduced the life chart, a clinical aid that would determine the life history and experiences of a patient in a graphical manner. Space is given for each year of life, and a weight curve for the most comparable part of the principal organs is on the chart, for example, for nervous and mental conditions the weight curve of the brain. The periods when various disorders occur are marked, including situations and reactions of patients, giving a "mental record." Also included are life events and important environmental influences. In such a way a graphic picture of what is important for understanding the individual is given as an integrated whole (see fig. 10.1).[17]

By all accounts, Meyer's wide approach and his poor powers of expression in English, in addition to his lack of enthusiasm for psychoanalysis, led him to being poorly understood and undervalued by American psychiatry.[18] He

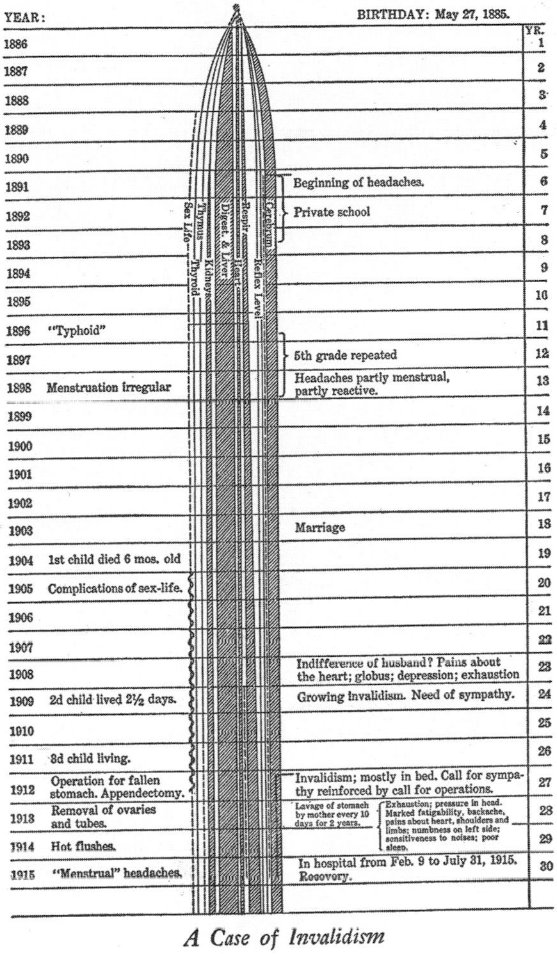

YEAR: BIRTHDAY: May 27, 1885.

YEAR			YR.
1886			1
1887			2
1888			3
1889			4
1890			5
1891		Beginning of headaches.	6
1892		Private school	7
1893			8
1894			9
1895			10
1896	"Typhoid"		11
1897		5th grade repeated	12
1898	Menstruation irregular	Headaches partly menstrual, partly reactive.	13
1899			14
1900			15
1901			16
1902			17
1903		Marriage	18
1904	1st child died 6 mos. old		19
1905	Complications of sex-life.		20
1906			21
1907			22
1908		Indifference of husband? Pains about the heart; globus; depression; exhaustion	23
1909	2d child lived 2½ days.	Growing invalidism. Need of sympathy.	24
1910			25
1911	3d child living.		26
1912	Operation for fallen stomach. Appendectomy.	Invalidism; mostly in bed. Call for sympathy reinforced by call for operations.	27
1913	Removal of ovaries and tubes.	Lavage of stomach by mother every 10 days for 2 years. / Exhaustion; pressure in head. Marked fatigability, backache, pains about heart, shoulders and limbs; numbness on left side; sensitiveness to noises; poor sleep.	28
1914	Hot flushes.		29
1915	"Menstrual" headaches.	In hospital from Feb. 9 to July 31, 1915. Recovery.	30

Vertical labels: Sex Life — Thymus — Thyroid — Kidneys — Digest. & Liver — Heart — Respir. — Metabol. — Reflex Level

A Case of Invalidism

Figure 10.1. A life chart as used by Meyer, for a case of invalidism. From Lief A, 1948, 421

likened traditional psychiatry to a geology that tries to explain the formation of the Alps by studying only the regions above the snow line, yet his psychobiology was difficult to grasp. Meyer referred to himself as a neuropsychiatrist and adopted what he called a commonsense approach to the individual. Organization of structure of the CNS needed an organization of function, the latter meaning conscious activity, bringing a psychobiological organization to the activity of a cerebrally integrated organism. He applauded the textbook of Jelliffe and White and appealed as follows: "We want neuropsychiatrists—not merely neurologists and not merely psychologists, but primarily physicians able to study the entire organism and its functions and behaviour and more especially the share of the nervous system and of the problems of adaptation."[19]

Body Image and the Image of the Body

Wagner-Jauregg brought several very talented people to his department in Vienna. Paul Schilder (1886–1940) was well acquainted with the cultural and psychoanalytic movement of that city and was himself an analyst. But he was also immensely interested in philosophy, childhood development, and language and wanted to write a textbook linking phenomenology, psychoanalysis, and neurology.

Schilder went to America in 1929 and became an assistant at the Bellevue Hospital. He became a member of the Psychoanalytic Society of New York, but he was later expelled, as the members did not consider him a real analyst (coming from Vienna!). It was in Baltimore, with Meyer at the Phipps Clinic, that he compiled his book *Brain and Personality*, which summarized lectures given by him when there. His most significant text was his book on the body image, *The Image and Appearance of the Human Body* (1935), elaborating the concept of a *Körperschema*, a body scheme that orders sensations and motor expressions, schemes in which perception and action, impression and expression form a unit.

Schilder's philosophical and psychological background was that of phenomenology, seeking a methodological foundation for all realms of experience and being, but for him with a scientific and integrated biological framework. More on phenomenology will be discussed later, such as how through Karl Jaspers (1883–1969) it became important for later twentieth-century neuropsychiatry. Suffice it for now to note that it emerged from the philosophy of Edmund Husserl (1859–1938) as a field of philosophical inquiry putting individual consciousness at the center of human experience.

Here is a summary of Schilder's perspective: "The basic notion, originally of Brentano and Husserl, was that the human organism was driven (intended towards, the 'intentionalism' of the phenomenologist), towards objects. These 'drive intendings' arise from somatic sources (sense data), becoming structuralised into the psychological apparatus. Both physiological changes, as in brain injuries, and psychological changes, as in the neuroses, can basically change behaviour. Second there is a *sphere*, an 'unformed background of experience' which is similar to the unconscious, from which object-directed 'intentions' arise to emerge as 'reality-relevant' thought and actions. This concept represented the primary process of Freud in his 'unconscious-conscious dichotomy.'"[20] Further, Schilder's theories embraced a synthetic (active) psyche and the body-image concept.

The Body Image in Neuropsychiatry

Schilder had taken his term "body scheme" from the neurologist Henry Head. He and Gordon Holmes (1876–1965) had written on this subject, especially in relationship to the parietal lobes. For Schilder, the body image linked motility and perception, since "perception of one's body image is followed by reaching out to other bodies, eventually building up a socialised picture of the outside world"; in the course of development, the child's mental processes are constructive and linked to motility. His book explores the nature of creative processes, and later he wrote about the image of the body and its structural breakdown in the works of such artists as Picasso, Braque, and van Gogh.[21] He carried out experiments relating the perception of sense data to movement tendencies, drawing particular attention to the role of vestibular influences on the body image. A motor response was common to all senses, and, following Coleridge, Hughlings Jackson, and Freud, he considered that thought might be considered in terms of movements rather than words, motor sequences changing our inner attitudes.

Schilder defined the body image as "the picture of our own body which we form in our mind, that is to say the way in which the body appears to ourselves."[22] The midline of the body, the head (as the localization of the ego), the vestibular system, and the effects of gravity on it were of considerable importance. This moved away from the ideas of the "scheme" of Head and toward a more relevant neurobiological understanding of the body image and its importance for neuropsychiatry.

The first part of the book covers neurological literature, including conditions such as apraxia, phantom limbs, autotopagnosia (loss of knowledge of the location in space of parts of the body), allochiria (in which a patient responds to a sensory stimulus on one side of the body as if it had been presented on the other), and anosognosia (lack of awareness of disability, secondary to neurological disorder). Head had emphasized visual, tactile, and proprioceptive sensations leading to our ability to locate the position of the body and its parts in space, heavily dependent on the parietal cortex. In parts 2 and 3 of Schilder's book he draws on much wider sources, discussing the importance of the libidinous (psychoanalytic, but discussing such syndromes as depersonalization and hysteria) and social aspects (importance of other human beings, and the social body image) of body-image construction.

Since the work of Head and Schilder, few have ventured much discussion of the body image, in spite of its relevance for neuropsychiatry. At one end of the spectrum are the more neurological disorders, some of which are noted above, yet within the panoply are such conditions as anorexia nervosa, transgender syndromes, body-image dysphoria, and out-of-body experiences.

Macdonald Critchley was appointed to the staff of the National Hospital for the Paralysed and Epileptic in 1928 and was a renowned neurologist who wrote extensively about states and conditions on the fringes of the central discipline. He was critical of Head and Holmes because "they never for a moment realised that they had opened up a Pandora's box which let loose a spate of metaphysics, much of it sheer verbiage." By contrast, he considered Schilder's book as "a work which just falls short of being one of the great monographs of neurology."[23] Critchley used the term "corporeal awareness" and emphasized in his work many of the features in Schilder's scheme. Critchley emphasized the importance of the face and reflection, always incomplete and stylized, and information on the inspection of the bodily features of others. The image is not identified with a homunculus, and it expands to include our accoutrements, inanimate objects with which we identify and by which we are identified (e.g., a doctor's stethoscope, a blind man's white stick). Our own image never keeps in touch with the passage of time and aging, and any understanding of the neuropsychiatry has to take into account heutoscopy, hallucinatory self-images, out-of-body experiences, states of denial of illness and lack of concern for paralysis (anosognosia through to anosodiaphoria), and misoplegia, in which a paralysis becomes hated or is treated like a puppet. Although he reaches no clear conclusions on some points, he suggests that the nondominant cortex and the patient's personality require consideration in the various disturbances of corporeal awareness. In any event, the corporeal image is not tied to anatomical boundaries and is in part related to the parietal cortex, but disturbances such as macro- and microsomatognosia, which occur in epilepsy and migraine, and bizarre body-image abnormalities reported in schizophrenia require a broad perspective on the subject. Penfield's homunculus does not resemble Michelangelo's David (see fig. 10.2)—there is more to the image of the body than meets the eye.[24]

Figure 10.2. The homunculus derived from cortical stimulation (*left*), compared to the ideal male image, Michelangelo's David (*right*). A comparison of the body image in the brain with the body image as viewed.

Schilder is remembered eponymously by a disorder of diffuse demyelination in a subcortical distribution which rarely penetrated the cortex (encephalitis periaxalis diffusa—Schilder's disease), although he also described adrenoleukodystrophy. He died in a car accident at the young age of 54.

The Disorder That Can Scarcely Be Forgotten

Another of Wagner-Jauregg's protégées was Constantin von Economo, a Romanian neuropsychiatrist of Greek origin, wealthy, highly educated, who in the First World War was a pilot and director of Lufthansa. A musician and champion bridge player, he was also a brilliant neuroanatomist; he published one of the most important books of neuropsychiatry ever written.

He extended studies of cortical architecture, enumerating twice as many areas as Brodmann; he identified seven lobes of the brain, which included a limbic lobe and an insula lobe, and identified what have become referred to as von Economo neurons, with a large spindle-shaped soma, a single axon, and a single dendrite. In primates these are found only in the brains of humans and great apes, notably in the anterior cingulate and dorsolateral prefrontal cortex.[25]

Toward the end of 1916, von Economo reported on a number of patients who presented with an unusual variety of symptoms that followed an influenza-like prodrome. Some had marked lethargy and disturbance of eye movements and at postmortem had inflammatory changes almost exclusively confined to the gray matter of the midbrain. He referred to this disorder as encephalitis lethargica (von Economo's disease), and it soon became recognized as an encephalitis secondary to the influenza pandemics that spread across Europe in the first few years of the twentieth century. Survivors had a variety of clinical pictures, but especially noted were lethargy, sleep disorders, seborrhea, anxiety, obsessive compulsive disorders, and psychoses. Motor disorders were prominent, including dyskinesias and dystonias, blepharospasm, abnormal breathing, and a later-developing Parkinson-like syndrome. The importance of the disorder, with a combination of neurological signs and psychopathological symptoms due to pathology in the midbrain, especially in the basal ganglia and related structures, was recognized by von Economo:

> The dialectic combinations and the psychological constructions of many ideologists will collapse like a pack of cards if they do not in future take into account these new basic facts. Every psychiatrist who wishes to probe into the phenomena of disturbed motility and changes of character, the psychological mechanism of mental inaccessibility, of the neuroses, etc., must be thoroughly acquainted with the experience gathered from encephalitis lethargica. Every psychologist who in the future attempts to deal with psychological phenomena such as will, temperament, and fundamentals of character, such as self-consciousness, the ego, etc., and is not well acquainted with the appropriate observations on encephalitic patients, and does not read the descriptions of the psychological causes in the many original papers recording the severe mental symptoms, will build on sand.[26]

Some of the other great talents to work with Wagner-Jauregg were Joseph Gerstmann (1887–1969; eponymous syndrome with key signs of agraphia, acalculia, finger agnosia, and right/left disorientation), Heinz Hartmann (1894–1970; Freudian psychoanalyst, particularly interested in ego psychology[27]), Otto Kauders (1893–1949; continued to study malarial therapy and encephalitis lethargica), and Erwin Stengel (1902–1973; described the condition of pain asymbolia, and translated Freud's book on aphasia). Stengel emigrated to England and was professor of psychiatry in Sheffield.[28]

Integrative Action of the Nervous System

Neuropsychiatrists have always been interested in what may be referred to as the integrated action of the whole human organism, as exemplified in this chapter by the likes of Jelliffe and Meyer, following a tradition from much earlier times. Today, however, the term *integrated action* is most associated with Sherrington. The theme has been a constant echo from the Greek physicians' theories of balance and harmony between the bodily constituents of man (such as the four humors) through to the conceptions of the romantic poets and neuroscientists discussed in chapter 5.

As noted earlier, Sherrington's hero was the sixteenth-century physician Fernel, especially the latter's interest in nature as the underlying cause of the indivisible unity of things, a vegetative life principle that determined the continuity of all life—form was central, and man was the greatest work in nature. Fernel's writings appealed to Sherrington, as Fernel was a physician and a skilled anatomist. Even if he considered the mind as incorporeal, he viewed the action of the mind in physiological terms, even though the integration of the body was an achievement of the soul. Sherrington considered him the greatest physician of his century.

Although Sherrington is best known for his book *The Integrative Action of the Nervous System* (1906, with several reprints and additions, the last one being in 1947), his most endearing book (in my opinion), *Man on His Nature*, is based on his Gifford lectures and is quite devoted to his hero Fernel. Sherrington considered that it is the physician, not the philosopher, who more intimately studies nature, via study of the human body. His discussion of integrative action begins with the cell, as evidence of "organization . . . an eddy in the stream of energy, maintaining itself as a self- balanced unity," continuing "whether atom, molecule, . . . virus or cell or plant or animal compound of cells, each is a system of motion in commerce with its surround, and there is dynamic reaction between it and the surround." Nature and evolution deal with the individual, body and mind as a unity, and the integrative function of the nervous system implied the integration of all the bodily organs such that "from such integrated organs the organism itself is in its turn integrated . . . integration by the nervous system is *Sui generis.*"

For Sherrington, the urge to live formulates the beginning of the mind, the drive to live and increase, which he referred to as "zest." This was a biological fact. Sherrington's take on this was that "the motor act is set as between a future and a past, it gets temporal relations; also it gets space relations." He refers to the cortex as the "roof-brain," and while he recognized that there was localization revealed by centers, this view "gravely mistook the scale of collaboration of action within the brain."[29]

Sherrington's investigations led him to study reflex action in particular, but in his foreword to the 1947 edition of *The Integrative Action of the Nervous System* he is critical of Descartes's vision of man as an automaton, a stimulus response machine, happily disconnected from the mind and hence the soul. Descartes, not a physician, had overemphasized the importance of the reflex, since although studying reflex action was a useful experimental model for exploration of the nervous system, it cannot allow exploration of instincts, urges, and drives. He was also critical of the work of Ivan Pavlov (1849–1936), and he commented that Descartes could never have kept a pet.

Sherrington's work led to many advances in neuroscience. In 1906, Ramón y Cajal won a Nobel Prize for his superb histological work on the brain using a modified Golgi silver stain. Then von Waldeyer-Hartz introduced the term *neuron* to refer to independent cells in the brain without fusion of terminal arborizations. This sparked a huge controversy, in that brain cells had previously been considered as part of a reticular net, a view supported by Camillo Golgi (1843–1926), who shared the Nobel Prize with Cajal. Their respective speeches clashed. Golgi emphasized his support for the reticular theory, attacking the neuron theory supported by his fellow prizewinner. Cajal soon won the argument and the neuron theory became accepted, but if neurons were independent, how did they communicate, and what was the direction of the flow of information?[30] Sherrington took this argument further and gave us the word *synapse.*

Sherrington's work on reflexes followed the path taken by Fernel, Descartes, Willis, Whytt, and Hall, the last giving us the term *arc.* Fernel had argued that the outside world interacted with internal body conditions to determine the nature of a motor act, in contrast to Aristotle's view that motor acts originate only from the mind, a place of "forms," motion actualizing a potentiality. But it is generally held that Descartes was the origin of our modern ideas of reflex action (see fig. 3.3), and Whytt, who was concerned with involuntary motions, linked these to the idea of sympathy. In contrast to Willis, he was concerned with the viscera and organs of the body, and he discussed pathological sympathy, which may lead to symptoms such as hysteria via the uterus in females. As noted, these were only starting points for Sherrington, who discovered that inhibition, as well as excitation, was part of the reflex, which graded responses and allowed adaptation, which was an underlying basis for homeostasis.

Although Sherrington espoused a kind of dualism, mind not being localizable, he was clear that the finite mind was embodied, interlinked with the energy system of the body. The romantic overtones of his views of the nervous system cannot be ignored, and in keeping with an earlier generation discussed in chapter 5, he was also a poet. His collection *The Assaying of Brabantius and Other Verse* was published in 1925, receiving much positive acclaim from the critics. He was awarded a Nobel Prize in 1932.[31]

The Body Electric

Sherrington shared the Nobel Prize in 1932 with Edgar Adrian (1889–1977). The latter had noted papers published by the German psychiatrist Hans Berger (1873–1941). Berger had succeeded Binswanger as physician in chief at Jena and was a collaborator with Brodmann. He was interested in discovering more about the physiology of mentation and found that it was possible to record electrical potentials from the brain from the surface of the skull. Initially, he looked for electrical activity in the exposed cortex of dogs, but his early equipment gave him insecure results. One method he used on humans was to insert silver wires under the scalp, then bind silver foil to the head with a rubber bandage, with leads from the front and the back of the head connected to a galvanometer. He came across a patient who, as a result of an accident, had lost a part of his skull but survived, and this provided the opportunity to carry out epidural recordings.

On October 14, 1927, with a double coil galvanometer, he recorded brain waves from his 15-year-old son; realizing that the recordings were the same as the epidural recordings, he cried "Eureka!" He kept his investigations quite secret, until in 1929 he presented his results, in German, as *Über das Elektrenkephalogramm des Menschen* (On the electroencephalogram of man), which initially attracted little attention. Unfortunately, he was ignorant of the technical basis of the mechanics, physics, and electricity of the methods he was using, which made it difficult for him to advance his methodology (see fig. 10.3).

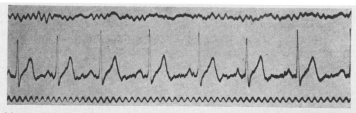

Fig. 14. 36 year old bald-headed man. Double-coil galvanometer. Condenser inserted. Record from forehead and occiput with lead foil electrodes. Resistance = 140 Ohms. Electrocardiogram with lead foils from both arms. At the top: the curve recorded from the scalp; in the middle: the electrocardiogram; at the bottom: time in 1/10ths sec.

Figure 10.3. An early EEG from Berger. Berger commented, "One nevertheless recognises in it the characteristic larger and smaller waves of the current oscillations which are sufficiently well known to us from the epidural recordings." From Gloor P, 1969, 59

Adrian was impressed and, like Berger, had been aware of the earlier work of Richard Caton (1842–1926), who had used a galvanometer to record electrical potentials from the brains of animals. Berger had recorded slowish waves of around 10 cycles per second (cps) while people had their eyes closed, which increased in frequency at lower voltage when the eyes were opened. Adrian, in collaboration with the neurologist Brian Matthews (1920–2001) at Cambridge, using an oscillograph and scalp electrodes, began making recordings of the potentials from the human brain. It is rumored that Berger too was nominated for a Nobel Prize but was prevented from accepting it by the Nazi regime, which also removed him from his university post. He committed suicide in his hospital in 1941.

Neuropsychiatry Eclipsed

The number of talented Jewish neuropsychiatrists who were forced to leave Austria and Germany in the middle of the twentieth century was a considerable loss to the Germans and Austrians and a considerable gain to America and the United Kingdom. Yet their place in mainstream psychiatry evaporated, especially as chairs of psychiatry in America were populated by psychoanalysts, who trod the path of the brainless mind quite mindlessly. But they were a part of the interregnum between the decline of neuropsychiatry at the end of the nineteenth century and its rejuvenation in the second half of the twentieth century, as will be the theme of the next chapter.

In later editions of Jelliffe and White's textbook, they write that "the nervous system is the organiser of all experience . . . and is the coordinator of that experience. The human being is here regarded in the light of an open energy system, by which energy is captured, transformed and delivered . . . the energy comes from the cosmos." They express similar sentiments to Sherrington's, with an emphasis on the active energetic nature of the nervous system.[32] But with the advances in neuroscience and the evasion by psychiatry of matters to do with the brain, the wisdom of the few who were still yearning for a unification of mind, brain, and psychopathology based on some unified framework was barely heard.

Hunter noted in 1973 that "the concept of psychosis or schizophrenia is a historical accident. The abnormal mental state is not the illness, nor even its essence or determinant, but an epiphenomenon. Had the epidemic of encephalitis broken out only 10 years earlier, or had its manifestations in endemic form been recognised for what they were, psychiatry would look very different today." He noted that the majority of psychotic patients displayed motor disorders, and in his view schizophrenia was consequent on a viral etiology, a view that is still pursued by some researchers. The last, ironic words in the text by von Economo were, "Encephalitis lethargica can scarcely again be forgotten."[33]

NOTES

Headnote: From James H, 1956, 408.

1. From "What Is an Emotion?," in James W, 2011, 55, 57.

2. James W, 54, 49; italics and capitals in the original.

3. The transcendentalist movement originated on the East Coast of America and was based on romantic ideals of self-reliance and individuality, rejecting the growing emphasis on intellectualism and empiricism and emphasizing inner spiritual values. "Transcendental self" implied that the self could only be known by inference, but which concept was necessary for a unified experience of self-consciousness. For Kant, there is one self-consciousness, an a priori subjective unity.

4. From "The Stream of Consciousness," in James W, 2011, 68.

5. James W, 81, 85; capitals and italics in the original.

6. Strouse J, 1980, 14.

7. For a good account of the life of Alice James and her symptoms and treatments, see Strouse J, 1980; quotes from 55, 118. See also Dillon B, 2009, 131.

8. Purdon Martin J, 1973. Purdon Martin was a neurologist at the National Hospital for Nervous Diseases, with a particular interest in disorders of the basal ganglia. In Britten's opera Miles dies in the arms of the governess, and in Jonathan Millers's production she suffocated him while in a seizure. Miller stated that Henry James had actually consulted with Hughlings Jackson (see Bassett K, 2012, 227). It is known that Alice was treated by Sir Andrew Clark (1826–1893), who was on the staff at the London Hospital, where Hughlings Jackson was also on the staff, from 1863 to 1894. Clark was very distinguished. He was made a baronet in 1883 and elected an F.R.S. in 1885. He had a large private practice and was physician

to Edward VII. Alice, Henry, and William all suffered from neurotic symptoms, especially headaches, back problems, and stomach ailments. Another brother, Wilky, was an alcoholic.

9. King Ludwig II was best known for his patronage of the composer Richard Wagner and for his love of the myths of Lohengrin. His most famous castle is that of Neuschwanstein, built in the style of the legendary German knights. His death is shrouded in mystery. Possibilities include that there was a plot to kill him, it was suicide, or he had a fatal fight with von Gudden while they were out walking. Ludwig's other lasting legacy was funding the Festspielhaus in Bayreuth for the performance of Wagner's operas. Von Gudden made his diagnosis without having examined the king, based only on the reports and testimonies of others.

10. The disorder did not remain an American neurosis, and *neurasthenia* became a worldwide appellation. It was fashionable with the worn-out middle-class businessman; his symptoms were not "in the mind" but were a well-recognized disorder with an established scientific basis. It was fashionable to carry such an appellation. The name still lingers on in some places, but it has been refashioned in the late twentieth century as myalgic encephalomyelitis, fibromyalgia, chronic fatigue syndrome, and other related conditions.

11. See Freeman W, 1968, 153–159.

12. Edward Brown draws attention to the link between Boston and the transcendentalist philosophers, as well as to William James, which may have been the reason why these ideas found acceptance there (Brown E, 2008). There are continuing debates as to the validity of multiple personality disorder as an entity. The history and arguments are well rehearsed in Merskey H, 1995, who concludes that "on both empirical and logical grounds psychiatry would be better off without this diagnosis" (308).

13. Jelliffe SE, White WA, 1929, vi–vii.

14. Burnham JC, 1983, 70.

15. The American Neurological Association was founded in 1875; Weir Mitchell was the first nominated president, but he declined the office. The Association of Medical Superintendents of American Institutions for the Insane was founded in 1844, which became the American Medico-Psychological Association in 1894 and then the American Psychiatric Association in 1921.

16. The *Archives of Neurology and Psychiatry* became the *Annals of Neurology* in 1975.

17. Lief A, 1948, 418–422.

18. He wrote, "I am a bit suspicious of those who have been psychoanalytically trained chiefly because they needed to be psychoanalysed on account of their own maladjustment or who had been aroused by the lure of the revelations." Freeman W, 1968, 127.

19. Lief A 1948, 574. The quote is from a presidential address that Meyer gave to the American Neurological Association in May 1922 titled "Interrelations of the Domain of Neuropsychiatry."

20. See Bromberg W, 1985, 83–84. The quote is from a summary of another author, David Rapaport (1911–1960), who translated some of Schilder's works and was himself trying to integrate psychoanalysis, psychology, and the theories of child cognitive development in the works of Jean Piaget (1896–1980).

21. Bromberg W, 84.

22. Schilder P, 1935, 11.

23. Critchley M, 1979, 92. He noted how the terminology blossomed and how *body image, body schema, corporeal scheme,* and *image de soi* became used interchangeably. His preferred term was *corporeal awareness.*

24. For a fuller discussion of the neuropsychiatric perspectives, see Smythies J, 1953; Trimble MR, 1988. Smythies makes clear distinctions among (1) *the body image in the brain*—coincident in space with neuronal activity in the brain; (2) *the body image*—visual,

mental, or memory of a human body; (3) *the perceived body*—the collection of somatic sensations present at any one instant in consciousness; (4) *the physical body*; (5) *the body concept*—a collection of beliefs and knowledge we have about our bodies; and (6) *body scheme*—the subconscious entity described by Head and Holmes. The perceived body is distinct from the physical body, as with a phantom limb. The importance of the whole area has profound philosophical implications, especially with regard to the mind–body–consciousness conundrums. Smythies, as a neurophilosopher-neuropsychiatrist, points out that if the philosophical theory of psychoneural identity is true, the *perceived body* must be coincident in space with the *body image in the brain*; clinical experience rules out such naïve realism.

25. Von Economo neurons or variants are found in other species, such as some cetaceans, but their presence in the human brain may have evolutionary significance and at present is an active area of research interest especially in dementia.

26. Von Economo C, 1931, 167. The disorder almost disappeared, and there was always some doubt as to what the causes of the pathology were. Sporadic cases still appear, and it may be that it was an autoimmune illness provoked by the influenza virus, as opposed to the virus itself. See Dale RC, et al., 2004.

27. Ego psychology emphasizes the ego as adapting the individual to their environment, essentially a conflict-free ego, and not viewed as a mediator between the demands of the instincts and the superego.

28. Gerstmann considered the finger agnosia to be a body-image disturbance. The syndrome occurs as a consequence of left-sided lesions to the parieto-occipital region. Hartmann was analyzed by Freud, and when he had to leave Austria, he went to the Johns Hopkins Hospital in Baltimore.

29. Sherrington C, 1940, 79, 85, 209, 208, 193, 205, 221, 221 (in order of quotation).

30. It is the case that Cajal also won the day, as Golgi is reported to have been quite abrasive toward Cajal, which, while leaving the latter feeling badly bruised, caused ill feeling among the audience of the day.

31. Other neurologists who have published poetry include Henry Head, Walter Russell Brain (1895–1966), and the neuropsychiatrist Eliot Slater (1904–1983).

32. Jelliffe SE, White WA, 1929, 17.

33. Hunter R, 1973, 364; von Economo C, 1931, 167.

REFERENCES

Bassett K. *In Two Minds: A Biography of Jonathan Miller.* Oberon, London; 2012.

Bromberg W. The Schilderian world of phenomenology and psychiatry. In: Shaskan DA, Roller WL, eds. *Paul Schilder: Mind Explorer.* Human Sciences Press, New York; 1985.

Brown E. Neurology's influence on American psychiatry: 1865–1915. In: Wallace ER, Gach J, eds. *History of Psychiatry and Medical Psychology.* Springer, New York; 2008.

Burnham JC. *Jelliffe: American Psychoanalyst and Physician and His Correspondence with Sigmund Freud and C. G. Jung.* University of Chicago Press, Chicago; 1983.

Critchley M. *The Divine Banquet of the Brain.* Raven Press, New York; 1979.

Dale, RC, Church AJ, Surtees RAH, et al. Encephalitis lethargica syndrome: 20 new cases and evidence of basal ganglia autoimmunity. *Brain* 2004;127:21–33.

Dillon B. *Tormented Hope: Nine Hypochondriac Lives.* Penguin Books, London; 2009.

Freeman W. *The Psychiatrist: Personalities and Patterns.* Grune and Stratton, New York; 1968.

Gloor P. *Hans Berger on the Encephalogram of Man: The Fourteen Original Reports on the Human Electroencephalogram.* Trans. Gloor P. Elsevier, Amsterdam; 1969.

Heaney S. *Death of a Naturalist.* Faber and Faber, London; 1966.

Hunter R. Psychiatry and neurology, psychosyndrome or brain disease. *Proceedings of the Royal Society of Medicine* 1973;66:359–364.

James H. *The Turn of the Screw / The Aspern Papers / and Other Stories.* Collins, London; 1956.

James W. *The Essential William James.* Ed. Shook JR. Prometheus Books, New York; 2011.

Jelliffe SE, White, WA. *Diseases of the Nervous System: A Text-Book of Neurology and Psychiatry.* 2nd ed. H. K. Lewis, London; 1929.

Lief A. *The Commonsense Psychiatry of Dr. Adolf Meyer.* McGraw-Hill, New York; 1948.

Merskey H. *The Analysis of Hysteria: Understanding Conversion and Dissociation.* Gaskell, London; 1995.

Purdon Martin J. Neurology in fiction: The turn of the screw. *British Medical Journal* 1973; 4:717–721.

Schilder P. *The Image and Appearance of the Human Body.* Kegan, Paul, Trench, Trubner, London; 1935.

Sherrington C. *Man on His Nature.* Cambridge University Press, Cambridge; 1940.

Smythies J. The experience and description of the human body. *Brain* 1953;76:132–145.

Strouse J. *Alice James: A Biography.* New York Review of Books, New York; 1980.

Trimble MR. Body image and the temporal lobes. *British Journal of Psychiatry* 1988;153(Suppl. 2):12–14.

von Economo, C. *Encephalitis Lethargica.* Trans. Newman KO. Oxford University Press, Oxford; 1931.

CHAPTER ELEVEN

The In-Between Years

All's well that begins well and has no ending.

From the libretto of the opera *Victory over the Sun*

"Oppressed by the gloomy day and the prospect of a sad future, I carried to my lips a spoonful of the tea in which I had let soften a piece of madeleine. But at the very instant when the mouthful of tea mixed with cake-crumbs touched my palate, I quivered, attentive to the extraordinary thing that was happening to me. A delicious pleasure had invaded me, isolated me, without my having any notions as to its cause. It had immediately made the vicissitudes of life unimportant to me . . . acting in the same way that love acts, by filling me with a precious essence."

The narrator of this famous "madeleine" moment, Marcel, tries to understand what has happened to him and what this experience could mean. He searches the very depth of his consciousness and concludes that there is a visual memory attached to the taste. The memory appeared suddenly, which was a childhood reminiscence of having been given a madeleine dipped in tea by his aunt at Combray, his childhood home village. He muses further that "when nothing subsists of an old past, after the death of people, after the destruction of things, alone, frailer but more enduring, more immaterial, more persistent, more faithful, smell and taste remain for a long time, like souls, remembering, waiting, hoping, on the ruin of all the rest, bearing without giving way, on their almost impalpable droplet, the immense edifice of memory." The whole of Combray and its associations emerged, from this sensory moment.[1]

Out of Focus

As the twentieth century moved on, interest in narrated time instead of actual time, and in interior monologues as opposed to description and depiction of external things and events, dominated the intellectual and cultural *Zeitgeist*. There was a breakdown of the very idea of a self as an entity, the ego (being) being overtaken by becoming, the subject's subject being pulled apart by forces of nature, the unconscious, and time itself. The Italian poet Filippo Marinetti (1876–1944) published the *Futurist Manifesto*, rejecting the art of the past, and ushering in an art devoted to the machine, technology, and speed. The artist Kazimir Malevich (1879–1935) and the antirationalist Russian futurists took this further, calling for a dissolution of language, wanting a language of sounds without meaning. In their opera *Victory over the Sun*, Apollo was extinguished, and time itself was abolished. If words had no meanings, then pictorial representations could be abandoned, leading to the alogical paintings of Malevich, in which colors and forms had no links to things in the physical world, no perspective, and no logic. They wanted to liberate the unconscious, freeing us from all preconceptions. It was a short step to removing even color from paintings. Malevich's celebrated *Black Square* (1915), considered an iconic twentieth-century artwork, and consisting only of black bordered by a white frame, rejected 700 years of traditional painting. Since from the Renaissance we have looked at the world knowing where to look, suddenly there was no focus.[2]

Following on from the stream of consciousness psychology of William James, James Joyce (1882–1941), especially in his *Ulysses*, a book with no overall plot, in which there is much interior monologue and verbal artifact, explored the psyche of its protagonist, Leopold Bloom, over the course of one day in Dublin. This vied with Virginia Woolf (1882–1941) and William Faulkner (1897–1962) developing, in a Proustian manner, new ways of exploring the mind and our emotions. In music, the great romantic sounds of Giuseppe Verdi (1813–1901) and Giacomo Puccini (1858–1924) gave way to the style of sustained chromaticism, developed by Wagner, whose operas matured from the dramas of action, such as his *Rienzi*, to *Tristan and Isolde* and his *The Ring of the Nibelung*, in which the music depicts the lived emotional strata of the protagonists. In the latter, musical phrases and leitmotifs constantly intermingle and undergo transformations through which the present is entwined with the presentiment of the past and future, a *déjà acuté* presenting ideas as yet to be fully developed, in which time and space are unified.[3] Maurice Ravel's (1875–1937) *La Valse* gave us a whirl without end, and the expressive primitive content of Igor Stravinsky's (1882–1971) *Rite of Spring* sprung asymmetrical accentuated rhythms of creative energy. With the serial 12-tone atonality of Arnold Schoenberg (1874–1951), all 12 pitches of the octave being treated as equal, music lost its tonal gravity. In this intellectual surround, neuroscience too lost track of the self and interest in studying consciousness. Psychiatry, now well divorced from neurology, especially through the great influence of Freud and the post-Freudians, saw a split from psychology, with their development of Skinnerian behavioralism, in which any intervening ego, hovering somewhere between the stimulus and response, was simply considered as a ghost in the machine. Pavlov's dogs and Skinner's pigeons were the only models needed to understand human behavior.

The Enchanted Loom

Adrian and Matthews (see chap. 10) gave the first convincing demonstration of the "Berger rhythm" in Cambridge in 1934. The "wobbly line," discovered by Berger, of about 10 cps was confirmed, but it was shown to come from the visual association areas, not from the whole brain as Berger had suggested. The EEG soon was used to study neurological and psychiatric disorders, epilepsy being an obvious choice. The number of electrodes applied increased, and developing technology allowed the moment-by-moment changes of the potentials of the brain to be displayed. William Grey Walter (1910–1977), at the Burden Neurological Institute in Bristol, referred to this as a *speculum speculorum*, a mirror for the brain, providing a code waiting to be cracked.[4]

Initially, it was hoped that the EEG would provide some greater insight into the mind and mental illness, ideas smartly rejected by those who were unhappy about any attempts to attribute brain properties to the mind. At the Maida Vale Hospital in London, four main types of rhythm were recorded: delta (0.5–3.5 cps), theta (4–7 cps), alpha (8–15 cps), and higher frequencies (14–30 cps). In the Second World War, the use of the EEG substantially increased, such that the two laboratories at the beginning had expanded to 50 by the end. Automatic decoding of the rhythms by a kind of Fourier transformation gave frequencies and amplitudes over selected time epochs. Rhythmic stimulation techniques such as flickering power-lights in the eyes allowed for the recording of evoked potentials, and the monitoring of epileptic seizures and interseizure epochs gave a renewed interest in epilepsy and its classification.

The alpha rhythm (constant in an individual after maturity, but differing between individuals), which appears when the eyes are closed, attracted much interest. It was thought perhaps to interlink with attention and some aspects of personality, but these were false leads. But there were many interesting early observations, such as the rhythms of sleep and the pattern of development of EEG activity through the ages of maturation. Associations between psychopathic behavior and theta rhythms, and observations of wave irregularities in disorders such as schizophrenia, even if hardly specific, were early clues as to the disorganized brain that underpins that disorder. A compilation of knowledge on the EEG published in 1950 contained a comprehensive review of methods and findings. There was a chapter on psychiatry written by Sir Dennis Hill (1913–1982), one of the pioneers of the use of the EEG, then at the Institute of Psychiatry and Maudsley Hospital, London. In addition to the above-mentioned topics, he discussed genetic studies, links to intelligence, and the rather specific changes in the organic psychoses, identification of the latter being a valuable clinical use of the EEG.[5]

There were other pioneers of the EEG in the United Kingdom, including William Cobb and Denis Williams (1908–1990) at the National Hospital for Nervous Diseases. The latter described to the author an important consequence for him of the clinical use of the machine. He told how so many patients came to the neurological clinics, and that he had little time to talk with them. However, since in those days there were no EEG technicians, Williams had to put the scalp electrodes on his patients himself. This gave him more time to converse with them and find out about their symptoms and personalities.[6]

The more significant contributions of the EEG for neuropsychiatry, however, developed in America. Stanley Cobb (1887–1968), George Engel (1913–1999), and Fred (1903–1992) and Erna Gibbs (1904–1987) were all important.

Cobb taught neuropathology at Harvard Medical School, established a neurological unit at Boston City Hospital, and in 1934 went to the Massachusetts General Hospital. There he succeeded Tracy Putnam (1894–1975), the latter, along with H. Houston Merritt (1902–1979), having discovered dilantin (phenytoin), a significant advance in the management of epilepsy. Cobb developed a psychiatric unit within Massachusetts General, the first of its type in America, with 12 beds on the same floor as neurology and neurosurgery. Cobb's book *Foundations of Neuropsychiatry* (1936) went to four editions, each one adding new information about anatomy and physiology and continually emphasizing the importance of maintaining integration between neurology and psychiatry: "New points of view are continually emerging and disturbing the neurologist and the psychiatrist who had too soon settled themselves into orthodoxy." He considered the divisions between the

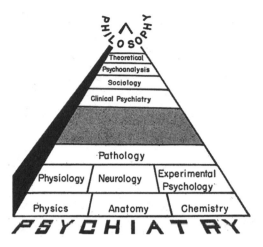

Figure 11.1. Cobb's scheme of the integration of neurology and psychiatry. From Cobb S, 1944, 3

departments purely arbitrary, with no biological significance. He set out his scheme for integration as a pyramid, in which neurology is tightly embedded within psychiatry, as is philosophy (see fig. 11.1). But there is a void in the middle, that area that lies between building blocks with more secure scientific foundations and clinical psychiatry and sociology. He looked forward to many years of future research that would fill out this void, "psychiatry being the oldest of the medical arts but the youngest of medical sciences."[7]

Cobb noted that the motor cortex was really sensorimotor. He admired Sherrington's ideas of integration as the basis to eventually explain higher mental functions, opining that "widespread activity of association paths is one of the essential phenomena of 'mind.' . . . The parts (of the brain) have more or less individual functions, when put together something new is built. This new coordinated organism loses something when any part is injured." Dichotomies between "functional" and "organic" he considered simply misleading since the line between the "physical" and the "mental" was entirely arbitrary.[8]

Cobb had a notable stammer and sought treatments for himself, including hypnosis and psychoanalysis. In a lecture he once gave in New York, he said, "I presume that you have noted my speech defect—I am not referring to my Harvard accent."[9] He has been considered by some to be the founder of biological psychiatry in America, but he was surely a neuropsychiatrist *par eminence.*

Entering the Brain

It was in Cobb's lab that William Lennox (1884–1960) and the Gibbses made their first observations of epilepsy. The Montreal Neurological Institute was opened in 1934, and Wilder Penfield (1891–1976) was making headway into exploring the psyche using cortical stimulation of the brain surface. In 1936, Penfield operated on a 14-year-old girl with a right temporo-occipital lesion. Her seizures were preceded by an aura, in which she saw a scene of a man following her as she walked in a meadow; he told her that he had snakes in a bag he was carrying, which frightened her. This was reportedly a true experience, confirmed by her brothers, who were with her at the time. Under local anesthesia, Penfield placed the stimulator on the temporal cortex, and the girl said that she was frightened and that she saw someone coming toward her. Stimulation a little anteriorly evoked hallucinations of the voices of her brothers and mother. These Penfield equated with the "dreamy states" of Hughlings Jackson, and in subsequent studies he noted that such experiences tended to be elicited by stimulation at sites posterior to the central (Rolandic) fissure and below the Sylvian fissure. In other patients he recorded episodes of forced thinking that seemed to have a frontal location.[10]

Penfield's results remain interesting to this day. In a review of his cases in 1963, he reported that out of 1,288 craniotomies on 1,132 patients being treated for focal seizures, the temporal lobe was stimulated in 520. Only 40 reported these experiential events, and only the temporal cortex gave rise to them. Further, they all derived from the nondominant side and were usually accompanied by emotions, often fear and seldom pleasurable. The divide between neurology and psychiatry, as Fred Gibbs later wryly stated, was now set by the Sylvian fissure.

Penfield thought that cerebral vascular constriction was important at the onset of a seizure. Fred Gibbs suggested to Penfield that he should take an interest in the EEG, which he did with Herbert Jasper (1906–1999). This association proved so fundamental in understanding epilepsy and its symptoms that their works are still often quoted today, but these are beyond the scope of this book to discuss any further.[11] Penfield wrote two novels: *The Torch*, on Hippocrates, and *No Other Gods*, a story of Abram and Sara set in Mesopotamia, at a time before their journey to found Israel. A deeply religious man, he spent much time on trying to understand links between the mind and the brain. *Mystery of the Mind: A Critical Study of Consciousness and the Human Brain* was published in 1978.

Temporal Discoveries

Adrian's findings were replicated in 1934, around the same time that the Gibbses began using the EEG. Fred Gibbs became involved in a considerable controversy regarding the defense of Jack Ruby, who assassinated Lee Harvey Oswald. Ruby had an EEG examination, which was reported to show a "psycho-motor variant." A defense was admitted that Ruby was unconscious at the time of the shooting, and the latter was consequent on epileptic muscular activity in his hand pulling the trigger. Gibbs attended the trial, and the judge asked him whether or not Ruby could tell right from wrong. Gibbs replied that the EEG could not answer that question, and he was summarily dismissed. Somewhat reminiscent of the use of brain imaging in criminal trials today, there was skepticism as to any suggestion of links between mind and brain, or brain abnormalities and behavior, the law being totally and apparently unashamedly Cartesian in its structure. But also lurking was a lingering prejudice against neuropsychiatry. Gibbs's presence at the trial caused outrage among epilepsy groups, fearing any contamination of epilepsy with psychiatry.[12]

Gibbs went to the University of Illinois as an associate professor of psychiatry, where there were many patients with severe psychotic problems. Some could be quite violent. He carried out many EEGs on these patients and considered that "they were all temporal lobe epileptics from our point of view. That is why I identified very clearly an association between temporal lobe disorders and these states. It was a very clear overlap."[13]

Gibbs noted that anterior temporal lobe foci were associated with the highest frequency of psychiatric disorder, especially severe personality disorders and psychoses. His results are shown in figure 11.2. He summed up his conclusions as follows: "The patient's emotional reactions to his seizures, his family and to his social situation are less important determinants of psychiatric disorder than the site and type of the epileptic discharge." The foundation stone for the beginnings of the modern era of neuropsychiatry was by now well and truly laid.[14]

Engel and Romano

Engel was influenced by the teachings of Meyer at the Johns Hopkins Hospital, and early on he did work on metabolic–EEG relationships at Mount Sinai, New York. The head of the department of psychiatry there was Lawrence Kubie (1896–1973), a psychoanalyst with an appointment in the neurology

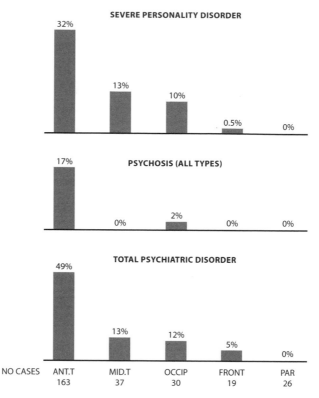

Figure 11.2. The incidence of psychopathology in patients with various sites of epileptic abnormality. From Gibbs FA, 1951, 525

department. He was much interested in the creative processes driven by a "preconscious," a way station between the unconscious and the conscious, which is allegorical, symbolic, and nonverbal. Kubie was responsible for helping to place many Jewish émigrés from Germany in posts that increased the powerbase of psychiatry substantially, and his psychiatry program was considered a blueprint for psychiatry in a general hospital.[15]

In 1941 Engel went to Boston, where he first met John Romano (1908–1994): "I saw the attending, John Romano . . . pull up a chair, and sit down with the patient and, in effect, invite him to tell his own story . . . the drama of Romano's 'pulling up the chair' and listening to the patient, as he was accustomed to do on psychiatric rounds, changed my life forever. My entire career since can be traced to that happy concordance of vision and action."[16]

Romano had trained in neurology at Boston City Hospital, in part supported by a Sigmund Freud scholarship. In 1939 he went to the Peter Brent Brigham Hospital, where he lectured in neuroanatomy and developed his ideas of teaching methods. When Engel arrived in Boston, given a mutual interest in delirium with Romano, it is hardly surprising that they developed a close collaboration. Then in 1942, Romano went to Cincinnati as chairman of the Department of Psychiatry. Charles Aring (1904–1998) was the chairman of neurology, and Engel joined him there. Aring had done neurophysiological research but was also very interested in psychosomatic medicine, and in his department he emphasized the interrelationships between neurology and psychiatry.

In 1946, Romano became chairman of psychiatry at the University of Rochester, and Engel followed him there. They collaborated closely for the next 20 years. Engel had dual appointments in the Departments of Psychiatry and Medicine and was responsible for establishing a medical psychiatric liaison service, staffed largely by internists. He thus was involved in the incorporation of psychiatric training into the medical school curriculum, and he also began his own training in psychoanalysis. A prominent member of the American Psychosomatic Society and editor of *Psychosomatic Medicine*, he began to develop ideas that later became embraced under the rubric *biopsychosocial*, today still a popular and fashionable moniker.

They pursued their work with the EEG, examining the impact of psychological processes on physiological measurements, and vice versa. Their most significant findings in this area related to the close correlation between the degree of slowing on the EEG and the state of delirium. They reported that improvements or deteriorations of the mental state could be followed by changes in the EEG profile. This important observation receives so little interest today, yet it is clinically very important in, for example, distinguishing states of delirium with minimal clouding of consciousness and hallucinations from schizophrenia, or the clouded mental states of a hysterical pseudodementia from organic dementias. In reality, psychiatrists soon lost all interest in the EEG, while in neurology the concept that it could be helpful in the differential diagnosis of the causes of an abnormal mental state was and is rarely heard, except from those who practice neuropsychiatry. Perhaps Engel's most significant contributions, apart from the seminal work on delirium, are his descriptions of the pain-prone patient (1959) and his book *Psychological Development in Health and Disease* (1964).[17]

Engel and Romano were both instrumental in teaching a neuropsychiatric approach to students, although sadly their collaboration began to deteriorate somewhat as Romano became more and more concerned by Engel's research based largely on psychoanalytic theory. Psychiatry there, but generally in America, was losing the few roots it had in clinical medicine, and the emphasis on the "biopsychosocial" approach in teaching waned.[18]

Romano became aphasic after a stroke, but at his memorial service Engel paid tribute to all that Romano had done for him. Engel had an identical twin, and in 1964, almost a year to the day after his twin brother had died from a coronary, he had a myocardial infarction. He too had long-standing heart disease and never worked on the days around the time of his brother's death, remaining at home.[19]

The Importance of Sugar

Meanwhile, the development of somatic therapies for mental disorders was moving ahead. Manfred Sakel (1900–1957) studied medicine in Vienna and was well acquainted with physical therapies for psychiatric illness. Frederick Banting (1891–1941) and Charles Best (1899–1978) had discovered insulin in 1922, which was tried as a sedative and appetite stimulant, and Sakel started to use it in the withdrawal treatment of addicts. However, control over the hypoglycemia was difficult, and some patients had seizures, or even became comatose. He noted improvement of mental symptoms, especially with coma, and so started to treat schizophrenia with insulin, reporting his successes in 1933 to the Viennese Medical Society. He then left Europe and went to New York, where he remained until he died.

Behind his use of insulin in schizophrenia lay the idea that a noxious agent had altered the metabolism of the nerve cells, reducing their energy, and that insulin could induce a hibernation of these cells, conserving energy and reinforcing them. He reported full remission in up to 70 percent of patients and a further 18 percent with social remissions. This was a remarkable rate of achievement, considering the few available therapies and that the disorder was largely regarded as incurable. Insulin coma therapy brought Sakel much renown, became widely used, and might be regarded as one of the most important discoveries made in neuropsychiatry. It had its dangers, including irreversible coma, and Sakel was aware of these. The rate of relapse was high, although Sakel himself seems not to have carried out follow-up studies.

The importance of this therapy was not only in raising an awareness of treatment possibilities for schizophrenia but also in drawing attention to the convulsions that were noted in many patients. Sakel wrote, "The provocation of a convulsion in the third or fourth hour of hypoglycaemia is meant in this case to be an *'early'* convulsion (midbrain convulsion), resembling a spontaneous insulin convulsive shock only for the intended *purpose of raising the emotional level.* In unresponsive, difficult cases, deliberate *'late convulsions'* (cortical convulsions) sometimes prove beneficial." By "late convulsions," Sakel was referring to those provoked by stimulants such as metrazole and later by electricity.

There was a celebrated confrontation at the First World Congress of Psychiatry in Paris in 1950 between Sakel and both Ladislas von Meduna (1896–1964) and Ugo Cerletti (1877–1963). Sakel for a long time refused to acknowledge the relevance of the convulsive seizure for his results. The previous quote is from a time after the seizures had attracted more interest as a therapeutic factor. He continued, "Such a therapeutic variation (i.e. deliberately inducing a convulsion) represents the so called *'intensive shock'* where the attempt is made to add a convulsion to the coma." This, he opined, could "force the dormant reserve cells, whose intensity of biologic action is increased by the long drawn-out treatment through insulin-forced glycogen reserve accumulation to awaken and take over." The insulin coma was still for Sakel the important agent for any treatment effect, based on his biochemical theories. He later acknowledged his mistake, and he then claimed priority for introducing seizure therapy for schizophrenia. But it was too late, as others by then were already emphasizing the importance of the seizure as the therapeutic agent. It may be of some interest that in postmortem analysis the brains of those patients who died after irreversible coma showed loss of cells especially in the frontal and temporal lobes.[20]

Antagonisms

Von Meduna was working in Budapest in the laboratory of Karl Schaffer (1864–1939), who was interested in the brain pathology of schizophrenia. Von Meduna made the remarkable observation that histologically there were differences between the brains of patients with schizophrenia and those with epilepsy. In schizophrenia there was loss of neurons and no glial reaction, while in epilepsy he observed slight loss of neurons and massive glial reactions. He reasoned from these findings that the cause of epilepsy stimulated the glia, while that of schizophrenia did the opposite, and he decided to try inducing seizures as a treatment for schizophrenia based on clinical observations of an inverse relationship between epilepsy and schizophrenia.

Several investigators had recorded improvements of psychoses if patients had seizures, and both Maudsley and Savage had commented on this a generation before. Dr. Glaus in 1931 examined over 6,000 cases of schizophrenia from the Zurich clinic and noted that epileptic seizures were rare. These observations were rather in keeping with the unitary psychosis model, since for Glaus and Eugene Bleuler, if seizures were present, the diagnosis was still assumed to be schizophrenia; it was not possible to have two forms of endogenous psychosis present in the same patient.[21]

Two other Hungarians, Nyírő and Jablonsky, actually tried injecting the blood of schizophrenics into people with epilepsy to improve their seizures. This was based on their observations, published in 1930, that in their hospital only 1.05 percent of epileptic patients without schizophrenia were discharged as cured, while in epileptic patients with schizophrenia the figure was 16.5 percent. Von Meduna wrote, "From his clinical data Professor Nyiro concluded that there was an antagonism between epilepsy and schizophrenia, meaning that the schizophrenic process would work against the epileptic process. . . . I accepted the concept of a biological antagonism between the two diseases, but I thought it worked in the opposite direction." It seems likely from the figures he produced and his subsequent discussion that the antagonism he was referring to was not between two diseases, epilepsy and schizophrenia, but between signs and symptoms, namely, seizures and psychosis.[22]

Von Meduna first induced a seizure in a patient with schizophrenia in 1934. He used intramuscular camphor, publishing favorable results in 10 out of 26 cases a year later. He also gave this treatment to patients with depressive disorders, with even better results. Camphor did not reliably bring on a seizure, so he then used intravenous metrazole, which would produce a convulsion within seconds. His publications led to physicians trying combinations of insulin and metrazole, but this was overtaken by the observations of Cerletti. He had worked with Kraepelin, Alzheimer, and Nissl, and with Lucio Bini (1908–1964) he induced seizures with electricity in dogs, as well as in hogs in a slaughterhouse. A problem was that 50 percent of the experimental animals died, related in part to the positioning of the electrodes, passing current through the heart. They found out that a bitemporal location was safer. Bini built the first electroconvulsive therapy (ECT) machine, eliminating most of the dangers of the therapy. Their first patient received an 80-volt shock for 1.5 seconds, which failed, so they gave another at 110 volts, which provoked a convulsion; Cerletti christened the treatment

"*L'elettro schock*" (electroshock) in April 1938. He showed that the convulsion itself rather than the electric current was important for the therapeutic effect, and subsequently associated muscular contractions were shown to be irrelevant; in 1940 curare was introduced to prevent physical injury.

Sakel was not a man to endear others to him; he expressed himself poorly in his writings, and he left no "school" of followers. He claimed that he was a direct descendant of the twelfth-century physician Maimonides. He believed that there was a conspiracy between von Meduna and the Nazis to steal the Nobel Prize from him, a mental state that became delusional. He did not take opposition lightly, and as the use of his method waned, his megalomania increased. He refused to shake hands with von Meduna at the 1950 congress. He died, in obscurity, of heart failure, by which time insulin coma therapy had virtually disappeared. What there was left of his method had been overtaken by the developments of convulsive therapy and then later by the advent of neuroleptics.

Von Meduna immigrated to America just before the Second World War, and after 1943 he was at Illinois with Gibbs, where the neurosurgeon/psychiatrist Percival Bailey (1882–1973) was performing temporal lobectomies on patients with epilepsy. Von Meduna was polite and sensitive enough to withdraw his controversial discussion with Sakel from the publication of the Paris meeting's proceedings. He commented as follows on the view of some of his colleagues regarding these treatments: "Many psychiatrists, when they first saw patients recover after a few convulsions artificially induced, were shaken in their beliefs regarding the psychogenic origin of the psychosis. Some of these psychiatrists recovered from their shock and tried—and are still trying—to coin psychological theories to explain the physiological effects of the convulsions upon the psychotic brain. . . . I shall sympathise with those who are baffled, although I feel I shall be unable to accept their theories."[23]

Digging Deeper

The advancement in physical therapies for major psychiatric disorders progressed further with the work of António Egas Moniz (1874–1955). He was a wealthy Portuguese neurologist and a member of parliament, who studied psychoanalysis and hypnotism, and was a good friend of Babinski. He developed the technique of cerebral angiography, experimenting with intracarotid injections of various radio-opaque substances in dogs and cadavers, with success coming in 1927 with the use of sodium iodide.

This opened up more possibilities for exploration of the brain without opening the skull; Walter Dandy (1886–1946) had replaced ventricular cerebrospinal fluid with air in 1918. Moniz developed the leucotome and initiated trials of leucotomy by cutting fibers between the frontal cortex and the thalamus, making him the founder of modern-day neurosurgery for psychiatric disorder, which he referred to as psychosurgery.

Moniz's concepts of psychiatry were very neurological, with obsessive ideas being the result of abnormal activity of certain intracerebral connections, which needed to be interrupted. He implicated the most recently developed frontal lobes in the theory. He commented, "You psychiatrists always get lost in the dark undergrowth of dialectic, and psychopathology, and here and there you rub shoulders with the domain of metaphysics. . . . There is only Kleist . . . who takes a clear stand in psychiatry as I understand it: he starts out from an accurate knowledge of the organ and takes a precise study of the brain as his basis."[24] Moniz was shot seven times by a paranoid patient when he was aged 65, one bullet lodging near his spine, another in his leg. He thought he was going to die, but he survived, although for the rest of his life he required use of a wheelchair. Ten years later, in 1949, he received the Nobel Prize, in part because of his work on the treatment of psychosis, an honor he shared with Walter Hess (1881–1973).

Hess was a Swiss physician, with a considerable clinical and private practice interest in ophthalmology, but renowned for his research in physiology. His aim was to look for correlations between psychically motivated behaviors and the functional integrated organization of the brain. On his travels he had visited England and befriended Sherrington. He was impressed by studies of others of free-living animal behavior, either in zoological gardens or in the laboratory. These included the learning experiments of Pavlov and the studies of innate and instinctual behavior in various animals by ethologist-zoologists such as Konrad Lorenz (1903–1989) and Nikolaas Tinbergen (1907–1988). Hess wanted to understand the basic endogenous elementary drives of behavior via comparative psychophysiology.

Much of Hess's work was related to the diencephalon, especially the thalamus, subthalamus, and hypothalamus. He used awake cats whose brains were stimulated by direct current pulses passed through small steel electrodes fastened to a frame attached to the skull. The site of the placements was recorded by creating lesions with higher-frequency stimulations, which were afterward checked histologically. The animal's behavior was carefully recorded. This was a considerable departure from the methods of British and American physiologists, who concerned themselves with fixed preparations of isolated segments of the nervous system. By exploring the brain below the cortex in live animals, Hess followed physiological links between behavior, the diencephalon, and the brain stem. His work emphasized the hypothalamus as an integrative center of autonomic and somatic functions, yet he could not discover isolated "centers" of activity, stressing, à la Sherrington, integrative activity.

His studies led him to discuss psychosomatic theories and the transition from physiological to pathological states when "abnormal tension in the vegetative system becomes conscious." He won a Nobel Prize for his work on the central representations of the vegetative system in 1949.[25]

Yet this work was quite neglected at the time, as was that of Walter Cannon (1871–1945). Cannon had also stimulated the hypothalamus, and he famously coined the phrase "flight or fight" to describe an animal's response to threats. In complete contrast to the James–Lange theory of emotion, the Cannon–Bard theory emphasized the hypothalamic–thalamic necessity for emotional expression; with thalamic discharge, simultaneous bodily changes occur along with the emotion. But the emotion was independent from the cortex and bodily expression, so physiological arousal does not have to precede the emotional experience. The idea that a small electrical stimulus to a part of the diencephalon could dramatically elicit emotions was an anathema to psychiatry and of little interest to neurology. But it was the start of a renewed paradigm of exploring brain–behavior relationships which opened up a neuroanatomical focus upon which current neuropsychiatry is founded.[26]

Electricity versus Chemistry

The reticular theory of neuron connections waned, Sherrington had defined the synapse, and the synaptic cleft was observed with developments in microscopy, ultimately leading to electron microscopy. The discovery of chemicals that altered activity between nerves and muscles and then between nerves themselves echoes back to the French physiologist Claude Bernard (1817–1878). He puzzled over the mechanism whereby curare blocked muscle action, since stimulation of the muscle evoked a contraction, but there was no such contraction when the nerve was excited. He postulated some event between the nerve and the corresponding muscle, raising doubts about the purely electrical nature of neural transmission. Otto Loewi (1873–1961), a German pharmacologist who was interested in studying the autonomic nervous system, was much preoccupied by these findings. In a state between sleep and wakefulness he found a solution, wrote down the answer, and went back to sleep, but when he awoke the next morning he could not read what he had written. The same dream-thought came to him the next night, and he immediately went to his laboratory to do an experiment (1921). Two frogs' hearts were immersed in Ringer's solution, one of which had the parasympathetic nerves severed. He stimulated the vagus nerve of the latter and transferred the solution to the other, and he found that the fluid now had an inhibitory action on the denervated heart. He repeated the experiment, severing the sympathetic supply to the heart, and the transferred fluid led to acceleration. These studies rather undermined the electrical transmission theory, but they revealed how a nerve impulse might lead not only to excitation but also to inhibition. He referred to the two substances involved as *Vagusstoff* (vagus material) and *Acceleranstoff* (acceleration material).

On his travels, Loewi had visited England and met Henry Dale (1875–1968). Dale had discovered noradrenaline and experimented with the effects of acetylcholine, so the results of Loewi were of interest to him. In 1929 he extracted acetylcholine from the spleens of horses killed at a slaughterhouse, and after further work, he concluded that the two compounds Loewi had identified were acetylcholine and noradrenaline, respectively. The first neurotransmitters were thus nominated, and nerves became categorized as "cholinergic" or "adrenergic" (by Dale), even if initially without direct evidence of their presence in the CNS, but with profound importance for the neurosciences.

Whether or not because his fame had apparently emerged via a dream, Loewi met with Freud in Vienna and in London and was able to talk with him about the unconscious nature of discovery.[27] In 1936, Loewi and Dale shared the Noble Prize, Loewi being forced to give the prize money to the Nazis.

Destructive Progress

In 1939, the outbreak of the Second World War had several important consequences for neuropsychiatry, not the least being a renewed interest in war-related neuroses, but also considerable concern with the consequences of head injury in wounded soldiers. Technological developments necessitated by the war effort gave newer and better instruments for investigation of and damage to the CNS, and biological treatments for neuropsychiatric disorders were explored. Military neuropsychiatry was a recognized discipline, in part necessitated by a paradigm shift in evaluating and treating the "war neuroses." There were several neuropsychiatry units in America and England,

and a book called the *Manual of Military Psychiatry* was edited in 1944 by Harry Solomon (1889–1982) and Paul Yakovlev (1894–1983). Forces of destruction were being unleashed throughout the world. After the explosions of the atomic bombs in Japan in 1945, abstract expressionist art portrayed raw energy in the paintings of such artists as Jackson Pollock (1912–1956) and Willem de Kooning (1904–1997). We had the emergence of pop art, the Theatre of the Absurd (Eugène Ionesco, 1909–1994; Samuel Beckett, 1906–1989), and minimalist and aleatoric (or "chance") music. Some of those involved in neuroscience, such as Hess, concerned themselves with forces within: "psychic forces that are responsible for the excitation and coordination of muscle forces."[28]

NOTES

Epigraph: The opera *Victory over the Sun* was performed in St. Petersburg in 1913. The libretto was written by the poet Aleksei Kruchenykh, the music written by Mikhail Matyushin, and the staging and costumes designed by Kazimir Malevich.

1. Proust M, 1913/2002, 49–50.

2. Malevich inaugurated a movement he called *suprematism*—referring to a new, nonobjective culture, most perfect. There were several versions of *Black Square*, and later he also painted *White on White* (1918).

3. *"Zum Raum wird hier die Zeit"*—*Parsifal*, act 1, scene 1.

4. The enchanted loom—where millions of flashing shuttles weave a dissolving pattern, always a meaningful pattern though never an abiding one—was a phrase of Sherrington's from *Man on His Nature*. Grey Walter W, 1953, 35.

5. Hill JDN, Parr G, 1950, 319–364.

6. Dennis Williams was one of the neurologists at the National Hospital for Nervous Diseases who was interested in the personalities of his patients and in what are now called comorbid psychiatric disorders associated with neurological disease. Several of his important observations were published in *Brain*, for which he served as editor for many years, which may not have seen the light of day under a different editor. Included are his important observations on links between obsessional personality and neurological symptoms and the affective disorders associated with the periictal period in epilepsy. He also opened up an interest in the problems of living with epilepsy, the stigma, and the importance of involving family and friends in management. See, for example, Williams D, 1981, 49–59.

It has perhaps become apparent that the name of the National Hospital, affectionately known as "Queen Square," has changed at various times in this book. The first patients were admitted in 1860, and it was the first hospital devoted to the care of "nervous diseases." The official name was the National Hospital for the Relief and Cure of the Paralysed and Epileptic. The name changed to the National Hospital for Nervous Diseases in 1948. Reflecting on the pejorative connotations of "nervous" and the divisions between neurology and psychiatry that developed, the name was changed again to the National Hospital for Neurology and Neurosurgery in 1988, although some of the medical committee at the time protested at this last revision.

7. Cobb S, 1944, 1–4.

8. Cobb S, 83, 84.

9. White BV, 1984.

10. For a complete description of many of Penfield's cases, see Penfield W, Erickson TC, 1941. The described case is on p. 132.

11. For example, see Penfield W, Jasper H, 1954. This book was dedicated to Hughlings Jackson and Sherrington.

12. The story is recounted in Hughes JR, Stone JL, 1990.

13. Personal communication. John Hughes, who took over Gibbs's laboratory in Chicago, told me that Gibbs was initially interested in neuropathology, his father having died of a brain tumor. He became much more interested in the EEG, but he apparently never had a license to practice medicine and so never examined patients personally.

14. Gibbs FA, Stamps, FW 1953, 78.

15. This led to pressures from other departments, which led to his resignation in 1943. The influx of analysts was prodigious. Some 250 analysts or analytically trained psychologists immigrated to America in the 1930s, many of whom became leaders of the analytic movement in the States, which became the world center for psychoanalysis. Kubie was clinical professor of psychiatry at Yale University from 1947 to 1959.

16. Cohen J, Brown Clark S, 2010, 75.

17. Engel GL, 1959. This paper is essential reading for anyone managing patients with chronic pain.

18. Note that the biopsychosocial model referred to by Engel is not that of today's use of the term. "There are other options (to today's biopsychosocial model) besides dehumanised biological reductionism." See Ghaemi SN, 2010, 213.

19. I recall going to Rochester to visit him and choosing such a day. However, two days later he cordially gave me time in his office to explore ideas with him.

20. For the whole episode and quotes, see Freeman W, 1968, 35–37.

21. Glaus actually described 8 cases out of his 6,000, with either simultaneous or successive associations, the latter hinting at some form of antagonism. Bleuler was also in Zurich at the Burghölzli Hospital. He suggested the term *schizophrenia* instead of Kraepelin's *dementia praecox*. The *Einheitspsychose*, or "unitary psychosis," asserted, as noted before, that there was only one psychosis, which may alter presentation over time, or the presentation could be influenced by pathoplastic effects of life events or personality. This view was supported by a number of eminent researchers, including Reil, Griesinger, and Bonhoeffer. It coincides in some ways with degeneration theory, and it was challenged by Kraepelin. It has been resurrected in recent times by Tim Crow.

22. Von Meduna L, 1985, 54. The situation was complicated since although he was suggesting an antagonism, in his series there was a good percentage of patients with a combination of epilepsy and psychosis (46%), of whom only 16.5% were cured. Wolf and Trimble (1985) tried to resolve this paradox by pointing out that von Meduna was a Hungarian writing in German and was not always very precise with his terminology. He used the terms *Krankheit* (disease) and *Krankheistbild* (disorder) interchangeably and confused seizures with epilepsy (still a common problem).

23. In Freeman W, 1968, 44.

24. Pichot P, 1983, 129.

25. Hess WR, 1964, 5. Tinbergen and Lorenz also received Nobel Prizes for their work in 1973.

26. Philip Bard (1898–1977) was Cannon's student. They also experimented with stimulation of decorticated cats, which still elicited emotions (so-called sham rage), and noted how the latter disappeared with thalamic destruction.

27. The other celebrated scientific discovery allegedly to appear in a dream was Friedrich August Kekulé's (1829–1896) discovery of benzene after dreaming of a snake catching its own tail.

28. Hess WR, 1964, 161.

REFERENCES

Cobb S. *Foundations of Neuropsychiatry*. 3rd ed. Williams and Wilkins, Baltimore; 1944.

Cohen J, Brown Clark S. *John Romano and George Engel: Their Lives and Work*. Meliora Press, Rochester, NY; 2010.

Engel GL. "Psychogenic" pain and the pain-prone patient. *American Journal of Medicine* 1959;26:899–918.

Freeman W. *The Psychiatrist: Personalities and Patterns*. Grune and Stratton, New York; 1968.

Ghaemi SN. *The Rise and Fall of the Biopsychosocial Model: Reconciling Art and Science in Psychiatry*. Johns Hopkins University Press, Baltimore; 2010.

Gibbs FA, Stamps, FW. *Epilepsy Handbook*. Charles C. Thomas, Springfield, IL; 1953.

Grey Walter W. *The Living Brain*. Gerald Duckworth, London; 1953.

Hess WR. *The Biology of Mind*. Trans. Von Bonin G. University of Chicago Press, Chicago; 1964.

Hill JDN, Parr G. *Electroencephalography*. Macdonald, London; 1950.

Hughes JR, Stone JL. An interview with Frederic A. Gibbs and Erna L. Gibbs. *Clinical Electroencephalography* 1990;21:175–187.

Penfield W, Erickson TC. *Epilepsy and Cerebral Localisation*. Ballière, Tindall and Cox, London; 1941.

Penfield W, Jasper H. *Epilepsy and the Functional Anatomy of the Human Brain*. Little, Brown, Boston; 1954.

Pichot P. *A Century of Psychiatry*. Roger Dacosta, Paris; 1983.

Proust M. *The Way by Swann's*. Trans. Davis L. Penguin, London; 1913/2002.

Von Meduna L. Autobiography of L. von Meduna. *Convulsive Therapy* 1985;1:43–57.

White BV. *Stanley Cobb, a Builder of the Modern Neurosciences*. Francis A. Countway Library of Medicine, Boston. University Press of Virginia, Charlottesville; 1984.

Williams D. The emotions and epilepsy. In: Reynolds E, Trimble M, eds. *Epilepsy and Psychiatry*. Churchill Livingstone, Edinburgh; 1981.

Wolf P, Trimble MR. Biological antagonism and epileptic psychosis. *British Journal of Psychiatry* 1985;146:272–276.

CHAPTER TWELVE

After the War

Back to the Front

The boredom of living is replaced by the suffering of being.

Samuel Beckett, 1931

"One fine summer night in June 1933 I was sitting on a lawn after dinner with three colleagues, two women and one man. We liked each other well enough but we were certainly not intimate friends, nor had any one of us a sexual interest in another. Incidentally, we had not drunk any alcohol. We were talking casually about everyday matters when, quite suddenly and unexpectedly, something happened. I felt myself invaded by a power which, though I consented to it, was irresistible and certainly not mine. For the first time in my life I knew exactly—because, thanks to the power, I was doing it—what it means to love one's neighbor as oneself. I was also certain, though the conversation continued to be perfectly ordinary, that my three colleagues were having the same experience. (In the case of one of them, I was later able to confirm this.) . . . I felt their existence as themselves to be of infinite value and rejoiced in it."[1]

William James was extremely interested in mysticism, explored thoroughly in his *Varieties of Religious Experience* (1902), subtitled *A Study in Human Nature.* He considered it absurd that science should exclude from investigation or even consideration all feelings that involved the spiritual and those associated with "the invisible pinch of destiny," such as those of Wystan Auden (1907–1973) above.[2] Auden's experience lasted at its full intensity for about two hours, but it did not vanish completely for some two days. Walt Whitman, celebrating himself and his body electric, and while, like Auden, lying on the grass, put it thus:

I believe in you my soul . . .
Loafe with me on the grass, loose the stop from your throat . . .
Only the lull I like, the hum of your valvèd voice.

I mind how once we lay such a transparent summer morning . . .
Swiftly arose and spread around me the peace and knowledge that pass all
    the argument of the earth,
And I know that the hand of God is the promise of my own,
And I know that the spirit of God is the brother of my own,
And that all the men ever born are also my brothers, and the women my
    sisters and lovers,
And that a kelson of the creation is love.[3]

James's perspective was to study such experiences as he would study other psychological phenomena, accepting not only their reality but also their vulnerability to scientific inquiry. He was concerned with immediate personal experiences, which he considered to be universal to mankind and which could be placed at the forefront of the psychology of religion. For many these may involve distinct religious sentiments, in others they are best described as mystical, but they are as convincing to those who have them as any direct sensible experience, and more "convincing than the results established by mere logic. . . . The unreasoned and immediate assurance is the deep thing in us, the reasoned argument is but a surface exhibition."[4]

These experiences, he opined, should have natural antecedents. The title of his first chapter was "Religion and Neurology," in which he explored the potential psychophysical associations with religious feelings. But he never developed a neurology of religious experiences, in part because of the rudimentary development of neurology at the time, and perhaps because he was not a neurologist. In addition, his own neurophysiology of emotions, as we have seen, considered the experiences derived from widespread bodily changes and their reception by the cortex.

Epilepsy Again

The discovery of the EEG and its continuous modulations linked, at least crudely, some features of the mental state to neurophysiology. But it was the links between psychopathology and epilepsy that were central to the reunification of neurology and psychiatry and are a cornerstone of neuropsychiatry. A background summary is shown in table 12.1.[5]

This actually embodies a history of thinking leading up to today's neuropsychiatry. Period 1 refers to the era of degeneration, the ideas discussed in chapter 6. The second period was a part of the psychoanalytic era, when the personality of the patient preceded the onset of epilepsy and was linked to

TABLE 12.1
*Summary of the Relationship between Epilepsy
and Personality Disorders*

Period	Years
1. Of epileptic deterioration	Circa 1900
2. Of the epileptic character	1900–1930
3. Of normality	1930–present time?
4. Of psychomotor peculiarity	1950–present time

it. Perhaps a quote from Aldren Turner (1864–1945), physician at the National Hospital for the Paralysed and Epileptic with a special interest in epilepsy, may help: "In early days, the convulsion or fit, was regarded as the sole element of importance in the clinical study of epilepsy; but in more recent years the psychical factor has come to be looked upon as of almost equal importance, and both are regarded as manifestations of a predisposition associated with inheritance . . . *the mental condition is not solely a consequence of seizures but is an expression of the same nervous constitution which gives rise to the convulsion.*"[6]

These theories considered epilepsy a constitutional disorder, others implying even more direct psychoanalytic views, such that the seizure itself was simply a psychological regression, protecting an overstressed ego. The seizures were viewed as arising out of the personality, and the "epileptic character" was lined up with various other "psychosomatic" disorders, such as the diabetic or the cancer personality.

These ideas were firmly rejected by, among others, the epileptologist William Lennox (1884–1960), whose two-volume book *Epilepsy and Related Disorders* (cowritten with his wife Margaret) became, for a long time, the revered text. The view was that while patients could have psychological disorders, these were secondary to factors other than the seizures themselves, such as head injury, effects of sedative medications, and social stigma. The personality of people with epilepsy was considered quite ordinary, hence the "normality" referred to in the table. This view is still widely held, especially with lay societies and psychologists, but has fallen more and more into disfavor on account of neuroscientific advances and comorbidity studies.

The term "period of psychomotor peculiarity" summed up everything. It was the EEG and the findings of Gibbs and colleagues that heralded a re-volution (re-turning) within clinical neuroscience. The French epileptologist Henri Gastaut (1915–1995) reported similar findings to those of the

Gibbses, emphasizing the underlying neuropathology of "psychomotor" seizures (Ammon's horn sclerosis, as it was then) and the personality differences
between patients with psychomotor epilepsy and those with generalized
epilepsy. He pointed out the high frequency of people with epilepsy in
mental hospitals. He noted that many of those with temporal lobe epilepsy
presented "bizarre" behaviors interictally, but "psychical troubles in epileptics result, not from the repetition of the epileptic fits, but from the lesion
which causes them." With regard to temporal lobe epilepsy, he noted that
the "lesions are known to be in the rhinencephalon, a region long ago shown
by Herrick and by Papez to be concerned with affect."[7]

What Is It about Smell?

Proust's powerful somatic sensations linked to autobiographical memory
brought about by smell and taste bring us to a discussion of the rhinencephalon, the "nose brain," and the great limbic lobe of Broca. The term *rhinencephalon* was probably first used by the biologist and comparative anatomist
Richard Owen (1804–1892) to refer to the olfactory bulb and peduncle. Albert von Kölliker (1817–1905) linked it to a group of cerebral structures that,
although apparently cortical, could be easily differentiated from the rest of
the cerebral mantle. Meynert also had distinguished between cortex with a
gray surface and cortex with a white surface, the latter constituting Kölliker's "rhinencephalon" and including the "olfactory lobe," hippocampus,
and septal region. The neuroanatomist Grafton Elliot Smith (1871–1937) suggested three basic levels of cerebral organization: the purely olfactory paleo-
pallium of fishes, the archipallium of reptiles and early mammals, and the
neopallium that had expanded through mammalian development. More relevant was his suggestion that olfactory structures, unlike other exteroceptive
areas, gave "'affective tone' linking together anticipation and consummation
into one experience and, in so doing, provided a germ for memory." Broca
himself thought that the rhinencephalon had to do with "the brute within."[8]

The definition of the rhinencephalon was thus a matter of considerable
debate, and the term is now confined to the historical past. The arguments were largely between comparative neuroanatomists, but they diminished clinical interest in the limbic lobe and its relevance for mammalian
evolution.

During the development of the mammalian brain, increased expansion of
the surrounding gray matter to the nonolfactory inputs to the telencephalon
(essentially the cortex and basal ganglia) crowded out but coalesced with

the olfactory afferents. The "rhinencephalic" area of the brain expanded and secondary connections increased, leading to polysensory integration and increased memory and emotional abilities, essential for evolutionary progression. Thus, while cognitive functions of the neocortex are mainly orientated to the external world, the older areas of cortex dealt with what MacLean referred to as "paleomentation," regulating master routines and subroutines and expressive displays linked to nonverbal communication. This and emotional mentation (subjective feelings) were to be distinguished from rational mentation.

Critchley berated our lack of clinical interest in the sense of smell, which he referred to as the "citadel of the senses." For our macrosomatic subhominid mammal predecessors, with a nose close to the ground, smell was vital to survival. As our line became bipedal and the nose moved farther and farther away from the floor, visual and then auditory signals dominated sensory input over that from the first cranial nerve: neurology was simply uninterested in the rhinencephalon. This was strange. Hughlings Jackson referred to smell as the most suggestive of the senses and described a form of aura characterized by smell, the uncinate seizure. From perfumes to Proustian evocations, hardly anyone is born without a sense of smell, and those who develop anosmia will tell of their loss of a quality of life. But it is very difficult to classify smell, and olfactory perception quickly fades. It is a sense quite unlike the others, distinguished neuroanatomically by not first relaying through the thalamus. In view of its evolutionary significance, it seems very unlikely that the sense of smell is trivial in guiding human behavior.[9] The rhinencephalon was renamed by MacLean as the "visceral brain" and then the limbic system, echoing the original designation of Broca, *le grande lobe limbique*.

At the Border—the Limbus

Although great strides in neuroanatomy and neurophysiology were made in the nineteenth century, little progress was made in understanding how emotions were represented in the brain. At the end of the century the James–Lange hypothesis was popular, implying that emotions were derived from peripheral and proprioceptive sensory inputs, but with no obvious cerebral location.

The Jamesian hypothesis was soon tested and shown to be wrong. The animal studies of Cannon and others had shown that removal of the cortices of the brain on both sides did not abolish the expression of emotion, and stimulation of subcortical structures could lead to the release of emotions. These observations formed the basis for a neuroscience revolution: the concept that certain brain structures could form the foundation of an "emotional brain system" was a stunning departure for neurology. This became fundamental to the disciplines of behavioral neurology—that branch of neurology which tries to understand how brain lesions alter behavior—and a related but distinguishably different area of expertise, namely, today's neuropsychiatry. The importance of understanding the distinction between these two specialties has been discussed in my preface.

Broca described *le grande lobe limbique* in 1878, which included the cingulate superiorly, the hippocampal lobe inferiorly, and the olfactory lobe anteriorly. He had found the edge of the cerebral cortex. Six-layered isocortex (also referred to as neocortex) was distinguished from cortex with fewer layers, allocortex having three layers; in between are transitional areas. In 1937 James Papez (1883–1958; pronounced "papes," like "grapes") published a paper entitled "A Proposed Mechanism of Emotion," in which he identified the hypothalamus, the anterior thalamic nuclei, the cingulate gyri, the hippocampi, and their connecting fibers as forming a "harmonious mechanism which may elaborate the functions of central emotion as well as participate in emotional expression."[10] It is somewhat unclear where Papez derived his ideas from, although comparative anatomical studies had revealed that the medial surface of the hemisphere was connected anatomically and physiologically with the hypothalamus (see fig. 12.1). This paper, which implied a corticothalamic mechanism for emotion, seemed to come out of the blue, and it was a mind-breaking intuition.

Beyond the Papez Circuit

MacLean graduated from Yale and then became an intern at the Johns Hopkins Hospital, before returning to Yale, where at one point he directed the EEG laboratory after Cobb. From 1971 to 1975 he was chief of the Laboratory of Brain Evolution and Behavior at the National Institutes of Health. He was fascinated by epilepsy: "Psychomotor epilepsy provides evidence that the limbic system is involved in self realization . . . it is of great consequence apropos of a sense of being that the phenomenology of psychomotor epilepsy reveals that even the least obtrusive feelings generated by limbic activity are tinged with some degree of affect."[11] Thus, some of MacLean's original ideas came from his observations in the clinic, and he was aware, not only from seeing patients with temporal lobe epilepsy but also from some case reports

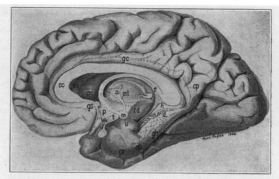

Medial view of the right cerebral hemisphere, showing the hippocampus and its connection with the mamillary body through the fornix and also the connections of the mamillary body to the anterior thalamic nuclei and thence to the cortex of the gyrus cinguli. In this specimen an unusually large exposed (nude) hippocampus is seen.

Figure 12.1. The Papez circuit for emotion. A number of key structures are shown: a = anterior nucleus of thalamus; gc = cingulate gyrus; gh = hippocampal gyrus; h = hippocampus; m = mammillary body; mt = mammillothalamic tract. From Papez JW, 1937, 727

of people with cortical lesions, especially frontal lobe lesions, that markedly altered personality and behavior were consequences of damage to such brain regions.

The concept that certain brain structures, referred to as limbic, could form the foundation of an emotional brain system was a revelation as profound to human science as the discoveries of Copernicus, Darwin, or Freud. The key pioneers in unraveling this neuroanatomy after Papez, one of the most exciting neurological expeditions of all time, were Yakovlev, Walle Nauta (1916–1994), MacLean, and Lennart Heimer (1930–2007), as well as their numerous collaborators.[12]

Yakovlev was a Russian who worked with Flechsig, was influenced by Pavlov, was an assistant in Paris to both Marie and Babinski, and was hired by Cobb after immigrating to America to work at Boston City Hospital and then at Harvard. He was particularly interested in epilepsy, and Lennox asked him to find out about the state of epilepsy at the Monson State Hospital for Epileptics, which had some 1,600 patients. Yakovlev was a skilled neuropathologist, and he developed a microtome large enough to hold a human brain. At the Connecticut State Hospital, and later back at Harvard, he enlarged his collection of brains, which contained many of those referred to as "mentally retarded" and from people who had had frontal lobotomies.

The neuroanatomical organization of the limbic cortex and thalamocortical relationships were of particular interest to him, and he put forward a theory of a tripartite division of the brain, hence his influence on MacLean.[13]

MacLean added to the limbic construct the amygdala and some nuclei in the brain stem which had connections to the structures given by Broca. MacLean used strychnine neuronography to reveal association pathways from sensory cortical areas to the hippocampus, and he made electrical recordings from and stimulations to the brains of freely moving animals. Also importantly, he revealed that "the thalamocingulate division of the limbic system is intimately linked to the prefrontal cortex which reaches its greatest development in human beings and which, among its other functions appears to underlie altruistic sentiments."[14] MacLean and others thus challenged the belief that cortical and subcortical systems were distinctly separated, noting the strong connectivity between limbic structures, the basal ganglia, and the neocortex. This overturned entirely the idea that the limbic system was a discrete "system" regulating only emotion, unable to influence motor systems or the neocortex (and thus speaking [*phasis*], doing [*praxis*], and knowing [*gnosis*]).

These researches also revealed the extended influence of the limbic system to the midbrain (referred to as limbic midbrain) and brain stem, including connecting with the cell structures that we now know are the origin of the ascending neurochemical systems, especially of those neurotransmitters called monoamines. The inputs to the limbic structures were thus both interoceptive (from within the body) and exteroceptive (conveying information about the immediate environment).

MacLean tied an evolutionary neuroanatomy to our most human behaviors. In arguing that the human brain was composed of three subdivisions, two of which were similar to those of other animals, he used the expression the "triune brain." There was the protoreptilian (the striatal complex), the paleomammalian (limbic), and the neomammalian (neocortex) (see fig. 12.2). By *triune* he was referring to "three-in-one," the whole being greater than the independent parts, but this view caused some criticism by those who considered it a simplistic representation.

The Triune Brain of MacLean

MacLean suggested three forms of behavior which distinguished mammals from reptiles, namely, nursing and maternal care, audiovocal communication for maintaining maternal–offspring contact, and play. He gave the summary

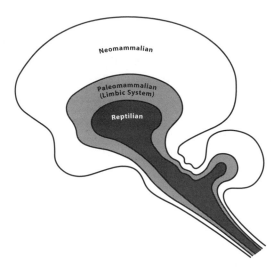

Figure 12.2. MacLean's trilogy. This image has been reproduced many times and was helpful in recognizing that an evolutionary approach to neuroanatomy was essential in order to understand the behavior of *Homo sapiens*. It has been criticized for its simplicity, and it certainly has been misapplied by many people with no grasp of neuroanatomy.

already quoted that the history of the evolution of the limbic system provides the kernel to understanding our history as compassionate social mammals.[15] MacLean's point was that the history of the evolution of mammals, and hence the family, came with infant–mother attachment and bonding. These became more complex as the phylogeny advanced through to *Homo sapiens*, possessing the most elaborate limbic system and most complex social structures.

Going further, he discussed how the integration of limbic and neocortical activity was involved in the sense of self, and how the limbic cortex generates free-floating affective feelings, conveying a sense of what is real and true for the individual, but which cannot be expressed vocally. Feelings were "visceral," using his early-preferred term.

The initial anatomy associated with the concept of the rhinencephalon rather implied that there were very few direct cortical projections from the cortex to the hypothalamus, and that the hypothalamus was to be regarded as the principal subcortical projection of the limbic system. This led to an interesting but rather damaging conclusion, which had implications not only for understanding brain–behavior relationships but also for the developing fields of behavioral neurology and neuropsychiatry. Thus, those neurologists who, a half century ago, might reluctantly concede that there was an underlying neurology of behavioral expression could say that the limbic system–hypothalamic axis was explanatory enough to understand how there might be a neuroanatomical basis for some emotions, and hence of relevance for psychiatry, but this was very different from accepting a more complex neuroanatomy of emotional disorders. Clinical neurology at this point was essentially concerned with the neocortex and its main outputs, especially the basal ganglia and the pyramidal motor system. Psychiatry and neurology had little in common from a brain-orientated perspective.

Sympathy

The autonomic system as we now understand it is closely tied to our emotional state. Sympathy has been noted along the way as a term to describe a "consensus" or an affinity between bodily organs that are not connected anatomically. While mystical associations were sometimes invoked, it became linked to the developing concept of the reflex via Descartes, Whytt, and Willis.

The chain of sympathetic thoracolumbar ganglia was described by Galen based on his animal dissections, and Willis mistakenly considered the *nervous intercostalis* to have an origin in the brain like cranial nerves, allowing for a "sympathy of the parts." Willis gave us the term *solar plexus*, the celiac plexus being like the sun, with rays emerging from it. The term *sympathetic response* was used well into the nineteenth century. Reil used *vegetative Nervensystem* (1807), James Johnstone (1730–1802) introduced *visceral* in 1795, and Ferriar gave us *conversion*, using the term to explain how, in the course of an illness, apparently unconnected symptoms could occur in different parts of the body. Walter Gaskell (1847–1914) in 1898 suggested the term *autonomic*, emphasizing a degree of independent action, which by this time included the activity of the vagus nerve.[16]

Further anatomical studies continued to undermine the belief that cortical and subcortical systems were so distinctly separated, and revealed how nearly all of the neocortex is connected to the basal ganglia by extensive efferent connections, and that there was serial linkage between limbic structures, the basal ganglia, and the neocortex. It then became appreciated that the rostral parts of the basal ganglia, far from being exclusively motor in function, were actually innervated by the limbic system, the main outflows from the medial temporal limbic structures being to the striatum. Heimer and colleagues showed that, contrary to the held belief that the limbic output from medial temporal cortex innervated mainly the hypothalamus, the input also heavily influenced the ventral striatum (nucleus accumbens—a basal ganglia structure), completely altering views on the way the basal forebrain was organized. Emotion was united anatomically with motion, as it is in many languages and in clinical practice. This radical shift in our understanding of this neuroanatomy forms a basis for the conceptual understanding of modern neuropsychiatry.[17]

It was the contributions of neuroanatomists such as Rudolf Nieuwenhuys and Nauta that united the limbic structures elaborated by MacLean and Heimer with the brain stem autonomic ganglia. Nieuwenhuys described the "greater limbic system," including pontine and medullary nuclear groups, and Nauta the "limbic midbrain area."

Giuseppe Moruzzi (1910–1986) and Horace Magoun (1907–1991) introduced the term *ascending reticular activating system* in 1949, emphasizing the importance of this area for attention and arousal, but its ascending and descending projections were quite unclear. Using modified silver staining techniques, Nauta identified the medial forebrain bundle as a "limbic traffic artery" for descending and ascending (including monoamine) pathways, and limbic structures innervating brain stem visceral afferents and efferents (vagus nerve) and other autonomic areas. As Andrew Lautin summarized this work, "Anatomic integration of this degree, a long sought after goal of such pioneers of the limbic brain as Papez and MacLean, points the way to a resolution of old disputes between such views as the James–Lange theory (peripherally weighted theory of emotion) with more centrencephalic determined theories."[18]

Remarkable Russians

Pavlov, like Freud, has had his name attached to an adjective, widely used in common parlance. He was a pioneer of Russian neurophysiology, yet Pavlov and his predecessor Ivan Sechenov (1829–1905), referred to by Pavlov as the father of Russian physiology, remained of the view that the reflex arc was the important basis for explaining mental phenomena, and they largely excluded psychology from their work.[19]

Vladimir Bekhterev (1857–1927) is best known for his work on ankylosing spondylitis, which disease is eponymously named. However, he was a renowned neurologist, who was literally written out of history, at least for a number of years. He traveled widely in Europe, visiting such notables as

Meynert, Flechsig, and Charcot, and was influenced by the psychology of Wilhelm Wundt. His knowledge of the brain was such that one person claimed that "only two know the mysteries of the brain: God and Bekhterev."[20] He established a neurological school in Kazan, and later the Psychoneurological Institute in St. Petersburg, where Pavlov worked. The two men were quite bitter enemies, always disputing each other's ideas, even insulting each other in the street.

In December 1927, Bekhterev received a telegram from the Kremlin ordering him to come and examine Stalin. On the twenty-third he left a congress he was attending and returned some hours later. He was reported to have said to a colleague that he had just examined a paranoiac with a short, dry hand. That evening he went to the theater, and in the interval he had drinks and cakes with unknown men. When he returned to his apartment, he became ill with vomiting, which was worse by the next morning. This was diagnosed as gastroenteritis. Two uninvited doctors visited him, but he became unconscious, could not be resuscitated, and died. In such circumstances it would be expected that a postmortem would have been carried out. But this did not happen, and he was cremated against the will of his relatives, but after his brain had been removed, apparently the latter being a request from Bekhterev himself. The two doctors were later identified as Klimenkov and Konstantinovski, members of the Secret Service. Later Bekhterev's children were arrested and died in a concentration camp.

It is conjectured that the suggestion that Stalin was paranoid was reported to the authorities, and as a consequence Bekhterev was poisoned. Apparently, the remark about the hand referred to the fact that Stalin was known to have some atrophy of the left arm and hand. In 1928, the Moscow Brain Research Institute was founded, where there was to be a large collection of brains. Bekhterev's brain was one of them, weighing 1,720 grams (huge by any standards!). But this might be something to do with Russian brains. The famous author Ivan Turgenev (1818–1883) is reported to have had the heaviest brain ever recorded at 2,021 grams (Einstein's weighed only 1,230 grams).

Alexander Luria (1902–1977) was born in Kazan and was much influenced by the work of Bekhterev. The son of a physician interested in psychosomatic medicine, and well versed in the German romantic tradition, Luria entered the world of psychology at a time when Wundt in Leipzig had founded the first experimental laboratory for a scientific approach to psychology and William James was doing his experiments at Harvard. Luria argued against associationism and was influenced by the *reale Psychologie* of the psychologist and sociologist Wilhelm Dilthey (1833–1911). The latter discipline implied that the human being should be studied as a unified dynamic system, referred to by Dilthey as *Geisteswissenschaften* (human sciences), contrasting with *Naturwissenschaften* (natural sciences), at once revealing tensions between the humanities and natural sciences which required some resolution.

Luria's legacy was substantial, and he is perhaps best known in neuropsychiatry for his motor tests, especially the serial hand-position tasks, used to assess integration between basal ganglia and frontal cortical functions. He argued against the localizationist ideas of Broca and was influenced by Hughlings Jackson and Head. His ideas implied that a functional system required a series of afferent or adjusting and efferent or effector impulses. He accepted that there were unobservable states of mind but founded a neuropsychology based on physiological and neuroanatomical considerations. He insisted on objective data and a holistic vision such that understanding behavior could not emerge from a study of isolated parts.

His books *The Mind of a Mnemonist* and *The Man with a Shattered World* described memory disorders of extreme kinds, but more important were his observations of the effects of memory and its disturbances on personality. The second book described the effects on a patient of a bombshell wound to the parietal cortex of the left hemisphere, after which he lost all memories and the meaning of words. Luria followed him up for 30 years.

A follower of Tolstoy, Luria was influenced by Marxism as it permeated Soviet psychology in the 1920s, at which time all theories of psychology came in for scrutiny and criticism, including his own. He was dismissed from his post at the Institute of Neurosurgery, and Pavlov's theories became the dominant model to be adhered to. He was even obliged to state that his work on aphasia and restoration of brain function after injury was flawed, but the latter anticipated concepts of brain plasticity and reorganization of the brain, now such a popular area of research. He was later reaccepted to the Institute of Neurosurgery, and the discipline of neuropsychology was considerably advanced by his endeavors.[21]

The Frontal Assault

War injuries in the Second World War gave those interested in neuropsychiatry a plethora of material to explore the associations between brain damage, especially focal injuries, and behavior. They provided Luria with much neurological material, which gave him further insights into the relevance of the frontal lobes for behavior. He observed that patients with frontal lobe damage presented with relatively preserved mental processes but were unable to use them effectively since the active flow of their complex psychological processes was disturbed.

We have seen how, especially since the days of the phrenologists, the frontal predominance of the human skull has assumed relevance for certain human attributes. The case history of Phineas Gage had been described in the middle of the nineteenth century, although not arousing the considerable interest it has done in more recent times.[22]

Becky and Lucy became for a while famous chimpanzees, after they were presented at the Second International Neurology Congress in London in 1935. They were both laboratory animals, well known to their keepers. Becky was described as "affectionate," while Lucy was "somewhat crotchety," and both had been trained to press levers to get food rewards. Their "neurosis" was induced by getting the animals to respond to different stimuli requiring opposite responses, and then forcing a response by showing both of the original stimuli at the same time. Carlyle Jacobsen and John Fulton (1899–1960), neurophysiologists at Yale, extirpated the frontal lobes of the chimpanzees, after which the latter made numerous mistakes, their performance on a delayed matching to sample memory task falling to chance levels. Their personalities changed. Becky became less agitated, Lucy more violent, but they seemed impervious to frustration and were described as if they had joined "the happiness cult of Elder Michaux and had placed its burdens on the Lord."[23]

Moniz and Walter Freeman (1895–1972) were at this conference. Moniz led a discussion on the possibilities of using such a procedure to alleviate mental illness, and while Fulton thought that this was hardly possible, Moniz followed this up. He suggested that in mental illness connecting groups of cells lose flexibility and become "more or less fixed," the arrangements leading to persistent ideas and delusions. In order to cure patients, "it is necessary to destroy the arrangements of cellular connections, more or less fixed, that must exist in the brain and particularly those that are linked with the frontal lobes."[24]

Within a month of the congress, Moniz performed his first operation. Encouraged by Cobb and Fulton, Freeman and the neurosurgeon James Watts (1904–1994), both at George Washington University, carried out a frontal lobotomy in 1936, and by 1942 they had performed 200 cases. Originally called *leucotomy*, the name was changed to *lobotomy* since the operation involved cutting the lobe rather than just the white matter fibers of the fron-tothalamic radiations. Variants of the procedure were developed, from the subtotal resection by Penfield, to radical bilateral removal, to that used by Freeman. This was an orbital approach, briefly anesthetizing the patient with electroshock and inserting a stilette above the inner canthus of the eyeball, forcing it through the roof of the orbit for a measured distance with a mallet. He then swept the instrument from one side to another, severing frontal areas from the rest of the cortex. This procedure could be done in a matter of minutes, allowing several patients to be operated on in a day.[25] No less a colossus than the 70-year-old Meyer, on hearing the early results of Freeman and Watts, considered that there were "more possibilities to this operation than appear on the surface."[26]

Watts did not approve of this procedure and severed his associations with Freeman. The subsequent history of neurosurgery for mental disorder is beyond the scope of this book; however, these animal studies and human operations revealed much information on the importance of the frontal cortex for social and cognitive abilities. After damage, perceptual power dilapidated, important situations passed unobserved, and memory became unreliable: delayed reaction responses were impaired.[27]

One of the authors whose work had persuaded Moniz to target the frontal lobes was the German neuropsychiatrist Karl Kleist. Kleist studied with Kraepelin in Heidelberg and with Alzheimer in Munich, but he was much influenced by the work of Wernicke and Gabriel Anton (1858–1933), who succeeded Wernicke at the University of Halle. Anton is known for the Anton–Babinski syndrome of visual anosognosia, in which patients after occipital injury develop "cortical blindness," confabulating and denying their lack of vision.

This was a heritage of identification of neuropsychiatric symptoms, such as aphasia and apraxia with cerebral lesions. Kleist did major work in these areas, but he also studied schizophrenia, disorders of the brain stem, con-sciousness, and the structure and function of the cerebral cortex. He stud-ied war-related brain injuries, encephalitis lethargica, and several forms of catatonia. Kleist attempted a psychopathological localizational basis for the symptoms and signs of psychoses, by analogy with neurological diseases. He was particularly concerned with the frontal areas, especially the orbital frontal cortex, the cingulum, and their associations with the striatum, the thalamus, the diencephalon, and the left temporal lobe. In 1948 in Freiburg, during one of Freeman's visits to Europe, Kleist watched him perform a lobotomy but quite disapproved.[28]

Gestalt Theory

Kurt Goldstein (1878–1965) was a German neuropsychiatrist, much influenced by *Gestalt* psychology. In part a reaction to the growing trends of strict localization theories and the continuation of associationist psychology, in which the mind was viewed as made up of independent sensory elements, the *Gestalt* school took as their basis the earlier theories of the likes of von Monakow and Lashley, who discussed mental processes as the activity of the brain as a whole, seeking what was united rather than divided. There was in the *Gestalt* perspective a dynamic interaction of parts into a unity, but the latter was more than simply the sum of the parts.

They also took their lead from the German philosopher-psychologist Franz Brentano. They wanted an "act psychology" as opposed to a "content psychology." They wished to develop a scientifically based psychology, the starting point of which was that mental processes were always aimed at performing some function, which implied an active and creative agency. A mental act is always directed at or refers to something. Brentano was against Cartesian dualism, took an Aristotelian perspective, and was interested in the spatial dimension that characterized our external world, but he wanted to understand "inner" primary experiences and their relationship to the external. This was not akin to introspection, as such reflection can only be focused on past mental phenomena. He wanted to explore how we know about things in the world (which Kant, with his phenomenon/noumenon split, considered unknowable)—how is the subject connected to the object; the "now" starts with intentionality—the intentional "in-existence" of an object exclusively characterizing mental phenomena. Such inner perception is immediate, ineffable, and self-evident. Goldstein studied over 2,000 people with head injuries in the First World War and made pertinent observations on acute, but also longer-lasting, alterations of behavior. Like others, he documented the changes after frontal injuries, especially prefrontal ones, which he referred to as resulting in defects in the capacity for "abstraction." Influenced by the theories of Hughlings Jackson, he considered the resulting clinical picture from a *Gestalt* perspective. Cerebral injury affected the patient's personality as a whole and led to struggles to cope with demands that could no longer be met. Thus, while focal lesions leave a particular signature, a group of symptoms emerge that are not due to localized injury. There are two facets to these: the struggle itself, and the escape from the struggle. A third consequence is an impairment of "special attitudes," namely, a "defect of attitude" which decreases the capacity for learning with concretization of thoughts. In this way Goldstein explained fatigue, variable cognitive performance, catastrophic outbursts of anger, self-exclusion from the world, the onset of seizures, excessive orderliness, lack of insight, and alteration of volitional attitude, commonly observed after such injuries.

"Closer examination shows that in order to readjust itself to the world, the injured organism has withdrawn from numerous points of contact with it and has thus attained a readaption to a shrunken environment." The individual reaches a new equilibrium. Goldstein pointed out that these readaptations are not consciously effected; will is uninvolved. He drew attention to the negative and positive symptoms, the former being direct, the latter indirect, impairment of parts putting the whole out of function. As with Hughlings Jackson, he appealed that "any assumption of localisation of a function is always a theory, not a statement of fact." Symptoms of "systematic disintegration" result from damage at any level of the CNS, altering thresholds of excitation, with increased responses to external factors. This leads to flitting attention, with abnormal fixations and blurring of the normally sharp boundaries between "figure" and "ground"—a direct reference to *Gestalt* psychology.

Goldstein's view was that every process in the nervous system had the character of "a figure-ground" process. While this was often taken to refer to visual assimilations, it also referred not only to words, which are only meaningful in context, but also to the body and its postures (e.g., a raised arm [figure] and the rest of the body [background]). A change of background will alter the figure, and "figure-ground performances have their counterpart in processes of the nervous system." Brain damage alters this relation, leading to difficulties in grasping the essentials of a given whole.[29]

Goldstein developed many psychomotor tests for evaluating the effects of brain injury and based his treatments on the various alterations of function after damage. This was linked to his observations on what was the most relevant for the individual, from the localized lesion to the more general effects, although he considered that the latter were always in evidence with even minor injuries.

Neuropsychiatry United—Not Quite

The importance of neuroanatomical findings in the mid-twentieth century for unifying neurology and psychiatry would seem obvious. But even today, many relevant textbooks fail to emphasize or even describe them. Epilepsy opened up close associations, but in recent classifications of epilepsy and seizures, descriptive terms such as *psychomotor seizures* or anatomical ones such as *temporal lobe epilepsy* have vanished.[30] It is of further interest that at midcentury the frontal lobes were still considered by many to be of little interest to neuroscience or clinical neurology. These neuroanatomical underpinnings of human behavior were likewise ignored by mainstream psychiatry (still arguing about schizophrenogenic mothers) and psychology (behavioralist in outlook). Something else was about to happen.

Doing the Splits

The metaphorical contest between Apollo and Dionysus resonates through Western culture.

Nietzsche's most famous book, *The Birth of Tragedy*, begins with the two gods and their fundamental importance to the cultural development of Western art: "art derives its continuous development from the duality of the *Apolline* and *Dionysian*."[31] Apollo represents the rational, the world of reason, the slayer of chaos. Dionysus, founder of wine, god of ecstasy, rapture, and rupture of boundaries, claims, in contrast, melody, lyric poetry, and the dance. However, it is not so much their apparent opposition but their synthesis, as well as their *psychological* and *physiological* as opposed to any classical or theological meaning, that is important. These are metaphors for "artistic powers which spring from nature itself."[32]

Nietzsche was interested in science, and although he lived through an age when *The Origin of the Species* was published (1859) and discussed, evolution of a Darwinian kind does not feature in his works. There was much contemporary scientific discourse on the nature of nature's forces and the relationship between force and matter. Nietzsche's Dionysus was representative of creative forces, and later of power, which became central to his evolving philosophies. Apollo and Dionysus are personified as drives (*Kunsttriebe*), physiological phenomena underpinning the creative brain. He even referred to the creative act as associated with "the cerebral system bursting with sexual energy."[33] More interestingly, at another time, he referred to "two chambers of the brain, as it were, one to experience science and the other non-science: lying juxtaposed, without confusion, divisible, able to be sealed off; this is necessary to preserve health. The source of power is located in the one region: the regulator in the other. Illusion, partialities, and passions must provide the heat."[34] These drives are embedded in nature and in us, are not generally considered conscious, and yet "they weave the web of our character and our destiny."[35] They are part of our physiology. But our cognitive structures, as discussed in the next chapter in more detail, are also embodied, as we are embedded in our social world.

As noted, even before Freud, Nietzsche was aware of the large subterranean edifice of the human unconscious. Nietzsche's views implied a new way of viewing the relationship between perception and emotion, in which the emotions dominate, and outlined psychology based not on the dualism of the Cartesian cataract but on Apollo and Dionysus. Consciousness for Nietzsche was but "a fanciful commentary on an unknown, perhaps unknowable, but felt text."[36]

Giving little credit to Nietzsche, the critic Camille Paglia contrasted Apollo, the law giver, definer of boundaries, and representative of sculptural integrity, with Dionysus, the god of fluids, of *sparagmos* and of pleasure/pain—objectification and identification, respectively. Her thesis was that the Apollonian and the Dionysian were two great principles governing the sexual personae in art, but also in life, and she noted the developments in neurobiology:

> Apollo, is the hard cold separatism of Western personality and categorical thought. Dionysus, is energy, ecstasy, hysteria, promiscuity, emotionalism— heedless indiscriminateness of idea or practice. Apollo is obsessiveness, voyeurism, idolatry, fascism—frigidity and aggression of the eye, petrifaction of objects. Human imagination rolls through the world seeking cathexis. Here, there, everywhere, it invests itself in perishable things of flesh, silk, marble, and metal, materialisations of desire. Words themselves the West makes into objects. Complete harmony is impossible. Our brains are split, and brain is split from body. The quarrel between Apollo and Dionysus is the quarrel between the higher cortex and the older limbic and reptilian brains.[37]

NOTES

Epigraph: Beckett continues: "The laws of memory are subject to the more general laws of habit. Habit is a compromise effected between the individual and his environment, or between the individual and his own organic eccentricities, the guarantee of a dull inviolability, the lightning-conductor of his existence. Habit is the ballast that chains the dog to his vomit. Breathing is habit. Life is habit. Or rather life is a succession of habits." Beckett S, 1931, 8.

1. Mendelson E, 1999, 160–161.

2. James W, 1902/1982, 500. Published just over 100 years ago, William James's collection of the Gifford Lectures remains the most revealing investigation into the psychology of religion ever attempted. The book was variously greeted as "unique" and "one of the great books of our time."

3. Whitman W, "Song of Myself," pt. 5, lines 82, 84–86, 91–95, 710.

4. James W, 1902/1982, 72, 74.

5. Table taken from Guerrant J, et al., 1962.

6. Turner A, 1907, 2; italics added.

7. Gastaut H, 1954, 62.

8. MacLean PD, 1990, 264; Heimer L, et al., 2008, 2. Anyone who wants to understand the development of our knowledge of the limbic system, from virtually all angles, but most definitely the comparative anatomy, should read the masterful book by MacLean (1990). Anyone who wants to see the dissection of the human brain and limbic system in all its glory by a master of the techniques should see the DVD that accompanies Heimer L, et al., 2008. Paul MacLean and Lennart Heimer were close friends and collaborators of mine, and I remain forever grateful for treasured memories of times spent with them, and for their profound neuroanatomical insights they shared with me.

9. Critchley M, 1986, 1–14. The olfactory system is involved in several neurological and psychiatric disorders. In some, such as depression, the sense of smell can be diminished, along with a decrement of other sensations such as taste or touch. In Alzheimer's disease and in Parkinson's disease, especially the Lewy body variant, diminished smell sensation has been reported. In schizophrenia, decreased sense of smell is found early in the course of the disorder and may be associated with smaller perirhinal cortices, as measured with magnetic resonance imaging. Disturbances of smell in such disorders are likely interlinked with underlying neuroanatomical deficits, rather than being simply a manifestation of a psychosis or deteriorating intellect.

10. Papez JW, 1937, 743.

11. MacLean PD, 1990, 578.

12. There were, of course, many others, as there are with any scientific journey. The author apologizes to anyone who feels that their name should be among these, but his own ideas have been influenced most by these neuroscientists and his personal recollections of either working with them or discussing ideas with them during his career.

13. Kemper T, 1984. The collection of brains must be one of the largest and best documented ever.

14. MacLean PD, 1985, 221.

15. MacLean PD, 1990, 247.

16. Hunter R, Macalpine I, 1963, 546. For an excellent review of the history, see Clarke E, Jacyna LS, 1987, 308–370.

17. It was the development of new techniques in neuroanatomy, such as the use of horseradish peroxidase, histofluorescence, autoradiography, and improved silver staining methods, that led to a renaissance of interest and understanding of the structure, connectivity, and functions of the limbic lobe. Especially important were cortical–subcortical interactions, as well as the relevance of them for an understanding of the neuroanatomy of emotion and behavior. See Heimer L, et al., 2008 for more details and the anatomico-clinical relevance of these studies. Six-sevenths of the word *emotion* is *motion*, and many languages have double meanings of equivalents to the English "moved" and "being moved." For example, the German term *Bewegung* translates as both "motion" and "emotion."

18. Fluorescent microscopy (histofluorescence) mapping individual nuclei and neural connectivity for individual neurotransmitters soon confirmed these findings, leading to an explosion of information about neurotransmitter pathways in the brain. Quote from Lautin A, 2001, 111.

There is now good agreement as to the areas of the brain involved in the generation of emotional states, and this has received verification from recent brain imaging studies. Damasio and his group have examined in detail the cerebral events that follow emotional arousal. Arguing from observations of the changed behaviors of patients with ventral and medial prefrontal brain lesions, Damasio's somatic marker hypothesis proposes that the stimuli that induce an emotion activate preexisting neuronal circuits (initially in the frontal cortex), triggering activity in both body and brain. These somatic-marker associations relate to the present state of the body, even if they do not arise in the actual body, but are in the representations of the body in the brain. They serve to bias responses, either consciously or unconsciously, to the environmental situation.

19. Pavlov never used the term *conditioned reflex*, which was a translation from the Russian but has become widely used in relation to his ideas. He was awarded a Nobel Prize in 1904.

20. This quotation is occasionally attributed to Friedrich Wilhelm Theodor Kopsch (1868–1955), the eminent anatomist. Bekhterev described several reflexes, and the medial raphe nucleus and superior vestibular nucleus are named after him.

21. Luria AR, 1979.

22. See Malcolm B, 2008; Damasio H, et al., 1994. There remains controversy about the actual interpretation of the behavioral changes that were reported on Gage, as well as on the amount of recovery that he achieved.

23. *Elder Michaux* was a religious TV show on the DuMont Television Network, hosted by evangelist Lightfoot Solomon Michaux. Freeman F, 1968, 52.

24. Freeman W, 53.

25. For good accounts of the development of our understanding of the frontal lobes and the early work with neurosurgery, see Fulton J, 1949, 1951.

26. El-Hai J, 2005, 117.

27. Moniz, in his justification for the surgery, suggested that for some years before the first procedure he had been considering the operation, having read of the effects of frontal lesions and removal in the neurological case literature. He was a careful scientist, and not the knife-happy neurosurgeon he is sometimes made out to be. He was criticized for the short length of his follow-up of patients and for never demonstrating that his theories of fixed neural connections were correct.

The case for or against the operations of Freeman has been argued variously. Watts did not consider him a proper neurosurgeon and considered him a showman, even if he acknowledged that Freeman knew more neuroanatomy than most neurosurgeons. Until the rift they were great friends. Freeman was well versed in neuropathology and convinced of the organic nature of mental illness, at a time when treatments for severe psychiatric illnesses were quite ineffective. In 1933 he published *Neuropathology: The Anatomic Foundation of Nervous Diseases*. Unconvinced by psychoanalysis, Freeman was interested to work with Watts, having read of his work on the effects of frontal lesions on the gastrointestinal system of animals—at a time when the frontal cortex was not known to influence autonomic function. Freeman wanted some method of helping patients before they became so incurably ill and thus developed his technique, which allowed him to operate away from a traditional surgical theater and on a large number of patients. He refined his method such that in seven minutes he could accomplish the same as the Freeman–Watts standard lobotomy, and his treatment was sought by mental hospitals around the United States. Unlike many, Freeman followed up with his patients carefully, often visiting them at home, and was keenly interested in their adaptation to life after the operation. The title of his book *Frontal Lobotomy 1936–1956: A Follow-up Study of 3,000 Patients from One to Twenty Years* says it all. For a superb biography, see El-Hai J, 2005.

28. For a comprehensive biography of Kleist's life and work, see Neumärker K-J, Bartsch J, 2003.

29. Goldstein K, 1942, 77, 84, 88. Anyone who has an interest in head injuries who has not read the observations of Goldstein will fail to understand the magnitude of the problem and the potential or not for recovery of various defects.

30. Simple partial and complex partial seizures, as introduced into the official classification in 1981, lost all anatomical or neuropsychiatric relevance, and even newer proposals to drop these terms fail to emphasize the importance of anatomy.

31. Nietzsche F, 1993, 14.

32. Nietzsche F, 18.

33. Moore G, 2002, 108.

34. Safranski R, 2002, 200.

35. Richardson J, 2004, 36.

36. Richardson J, 208.

37. Paglia C, 1990, 96.

REFERENCES

Beckett S. *Proust*. Chatto and Windus, London; 1931.

Clarke E, Jacyna, LS. *Nineteenth-Century Origins of Neuroscientific Concepts*. University of California Press, Berkeley; 1987.

Critchley M. *The Citadel of the Senses*. Raven Press, New York; 1986.

Damasio H, Grabowski T, Frank R, Galaburda AM, Damasio AR. The return of Phineas Gage: Clues about the brain from the skull of a famous patient. *Science* 1994;264:1102–1105.

El-Hai J. *The Lobotomist: A Maverick Medical Genius and His Tragic Quest to Rid the World of Mental Illness*. John Wiley and Sons, Hoboken, NJ; 2005.

Freeman F. *The Psychiatrist: Personalities and Patterns*. Grune and Stratton, New York; 1968.

Fulton J. *Functional Localisation in the Frontal Lobes and Cerebellum*. Clarendon Press, Oxford; 1949.

Fulton J. *Frontal Lobotomy and Affective Behaviour: A Neurophysiological Analysis*. Chapman and Hall, London; 1951.

Gastaut, H. *The Epilepsies: Electro-clinical Correlations*. Trans. Brazier M. Charles C. Thomas, Springfield, IL; 1954.

Goldstein K. *Aftereffects of Brain Injuries in War: Their Evaluation and Treatment*. William Heinemann, London; 1942.

Guerrant J, Anderson WW, Fischer A, Weinstein MR, Jaros RM, Deskins A. *Personality in Epilepsy*. Charles C. Thomas, Springfield, IL; 1962.

Heimer L, Van Hoesen G, Trimble MR, Zahm, DS. *Anatomy of Neuropsychiatry: The New Neuroanatomy of the Basal Forebrain and Its Implications for Neuropsychiatric Illness*. Academic Press, London; 2008.

Hunter R, Macalpine I. *Three Hundred Years of Psychiatry, 1535–1860*. Oxford University Press, London; 1963.

James W. *Varieties of Religious Experience: A Study in Human Nature*. Penguin, London; 1902/1982.

Kemper T. Paul Ivan Yakovlev 1894–1983. *Archives of Neurology* 1984;41:536–540.

Lautin A. *The Limbic Brain*. Kluwer Academic, New York; 2001.

Luria AR. *The Making of Mind: A Personal Account of Soviet Psychology*. Ed. Cole M, Cole S. Harvard University Press, Cambridge, MA; 1979.

MacLean PD. Evolutionary psychiatry and the triune brain. *Psychological Medicine* 1985; 15:219–221.

MacLean PD. *The Triune Brain in Evolution: Role in Paleocerebral Functions*. Plenum, New York; 1990.

Malcolm B. Phineas Gage—Unravelling the myth. *Psychologist* 2008;21:828–831.

Mendelson E. *Early Auden*. Faber and Faber, London; 1999.

Moore G. *Nietzsche, Biology and Metaphor*. Cambridge University Press, Cambridge; 2002.

Neumärker K-J, Bartsch J. Karl Kleist (1879–1960)—a pioneer of neuropsychiatry. *History of Psychiatry* 2003;14:411–458.

Nietzsche F. *The Birth of Tragedy out of the Spirit of Music*. Trans. Whiteside S. Penguin Classics, New York; 1993.

Paglia C. *Sexual Personae: Art and Decadence from Nefertiti to Emily Dickinson*. Yale University Press, New Haven, CT; 1990.

Papez JW. A proposed mechanism of emotion. *Archives of Neurology and Psychiatry* 1937;38:725–743.

Richardson J. *Nietzsche's New Darwinism*. Oxford University Press, Oxford; 2004.

Safranski R. *Nietzsche—a Philosophical Biography*. Trans. Frisch S. W. W. Norton, London; 2002.

Turner A. *Epilepsy*. Macmillan, London; 1907.

Whitman W. *Leaves of Grass and Other Writings*. Ed. Moon M. W. W. Norton, New York: 2002.

CHAPTER THIRTEEN

Coda

In the conscience I have, which every one of us has—you see—we think we are "one" with "one" conscience, but it is not true: it is "many," sir, "many" according to all the possibilities of being that are in us: "one" with this, "one" with that—all very different! So we have the illusion of always being at the same time "one for everyone" and always "this one" that we believe we are in everything we do. It's not true! It's not true!

Luigi Pirandello, 1995, 26, act 1

"From what I had read of the mescaline experience I was convinced in advance that the drug would admit me, at least for a few hours, into the kind of inner world described by Blake. . . . Half an hour after swallowing the drug I became aware of a slow dance of golden lights. A little later there were sumptuous red surfaces swelling and expanding from bright nodes of energy that vibrated with a continuously changing, patterned life. At another time the closing of my eyes revealed a complex of gray structures, within which pale bluish spheres kept emerging into intense solidity and, having emerged, would slide noiselessly upwards, out of sight. The other world to which mescaline admitted me was not the world of visions; it existed out there, in what I could see with my eyes open. The great change was in the realm of objective fact. . . . I was seeing what Adam had seen on the morning of his creation—the miracle, moment by moment, of naked existence."[1]

Aldous Huxley (1894–1963) was encouraged to take mescaline by the neurophilosopher-psychiatrist John Smythies. After the discovery of synapses and the first neurotransmitters, there was a growing interest in the chemical composition of the brain. In the 1950s several monoamines were found in the CNS, although proving they had function as transmitters took some time to unravel. Serotonin and noradrenaline were initially the most intriguing, but others followed, including gamma amino butyric acid (GABA) and dopamine. Huxley thought that giving LSD to people with terminal cancer would make dying a more spiritual process, and on his deathbed he had his wife inject him with it. The psychologist Timothy Leary (1920–1996), who, along with Huxley, set up the Harvard Psilocybin Project, famously and utterly prophetically said, "The game is about to change. . . . Drugs are the religion of the twenty-first century. . . . Turn on. Tune in. Drop out."[2]

The psychiatrist Humphry Osmond (1971–2004), who worked with Smythies, first proposed the term *psychedelic* in 1957, and together they identified that small amounts of lysergic acid diethylamide (LSD) could block the action of serotonin, speculating that if this occurred in the brain, marked behavioral effects were likely. Their "transmethylation" hypothesis was the first specific biochemical theory of schizophrenia to be developed. Reserpine, a sedative drug, was shown to deplete neurons of serotonin and be associated with clinical depression. Since depression was linked also to Parkinson's disease, the biochemist Oleh Hornykiewicz suggested that decreased serotonin was the chemical basis of Parkinson's disease. Other neurotransmitters were soon discovered, first noradrenaline and then dopamine.[3]

In 1950, the surgeon Dr. Morel-Fatio presented a paper to the French Anesthetic Society about an operation he had done on Madame X, who wanted her nose straightened. On the day of the operation, she was anxious and so agitated that the operation could not proceed, especially as no nose mask for inhaled anesthesia was possible. She was given an injection referred to as Dip-Dol. She instantly calmed, but she was not asleep and responded to questions asked. She reported that it was as if the operation was being carried out on someone else's nose. One of the assistants suggested that this was like a pharmacological lobotomy. The substance used was diethazine, a phenothiazine. Seeking even better anesthetic agents, in December 1950 Paul Charpentier synthesized chlorpromazine (4560 RP—for Rhone-Pôulenc), which altered the face of psychiatry and the fate of psychiatric patients thereafter. The psychiatrists Jean Delay (1907–1987) and Pierre Deniker (1917–1998) at Sainte-Anne Hospital in Paris realized the potential use of these compounds for psychiatry and obtained samples from Rhone-Pôulenc. They described the effects as an apparent emotional detachment of patients, slower reaction to external stimuli, and a reduction of initiative and anxiety but without loss of alertness or intellectual functioning. They presented their data in 1952, and shortly afterward Largactil was made commercially available.

One of the side effects of chlorpromazine was the development of Parkinsonism. Since the drug was shown to quell the psychotogenic properties of LSD, attention turned to its effects on monoamines. Further analysis of postmortem tissue from brains of those with Parkinson's disease clarified the prime role of dopamine in that disorder.[4] Neurology and psychiatry had a common language in the neurotransmitters and began to use a similar range of drugs to influence them in the brain.

The use of insulin-related treatments and forced restraint declined, the number of patients in long-term asylums fell, and psychoanalysis began its final drift away from mainstream psychiatry. The discovery of lithium by John Cade (1912–1980) in 1949 and the antidepressant effects of impramine and two antituberculous drugs isoniazid and iproniazid soon followed (1952–1957).

Awakenings

Enter Oliver Sacks (1933–2015). Having left England to practice neurology in America, he found himself looking after a large number of patients with chronic encephalitis lethargica at Mount Carmel, New York. His book *Awakenings* (1973) describes in some detail the remarkable effects of giving what was then a new and experimental drug, L-dopa, to 20 severely affected patients. He started the therapy in 1969, and with wonder and amazement he observed that the patients not only were awoken from their organic slumber but developed marked alterations of posture, especially dystonias, and many features of psychopathology. This all harked back to the clinical descriptions given by von Economo. Siding with those who prefer a holistic understanding of brain function, Sacks explored the way that illness and these "awakenings" changed the patients' total relationship between the self and the world.

After a period of "perfection of being," the "happy state . . . starts to crack, slip, break down and crumble." Pathological dysphorias with restless, choreic, akathisic movements and tics with "urges . . . flaming into manias, passions and greeds, led into climactic voracities, surges, and frenzies . . . until the crash comes at last." Relying heavily on Freud, Schopenhauer, and the poet John Donne (1572–1631) in his descriptions, even if criticized by some as being anecdotal and overly romantic, his observations were another milestone in the reunification of neurology and psychiatry, but as neuropsychiatry. The close association between movement disorders and emotional expression was now supported clinically and underpinned by neuroanatomy and neurochemistry. Dopamine was a chemical necessity for understanding both neuro- and psychopathology. No dopamine, no movement, too much dopamine, excess movement: decrease dopamine and lyse a psychosis, increase dopamine and provoke one.[5]

Full Circle Back to Epilepsy

Of even greater importance for understanding the development of today's neuropsychiatry is the continued study of the links between epilepsy and psychopathology, a theme that has reemerged in the past few chapters and echoes back to the insights of Hippocrates discussed in chapter 1.

First, there was the continuing debate over whether or not there was an increased frequency of psychosis in epilepsy, or whether there was an inverse relationship. There were also discussions as to the frequency of epilepsy in people with schizophrenia. Some of the confusion was based on the inadequacy of the populations investigated (different hospital settings and limited community studies), and in part on conceptual difficulties. Seizures and epilepsy were confused, as were acute psychoses (now recognized as postictal) versus chronic psychoses (interictal); the clinical features of the psychoses were poorly documented. As already discussed, there were arguments as to whether the relationship was simultaneous or successive, whether one disorder had a pathoplastic effect on the other, and the specter of the *Einheitspsychose* made things worse. The spectrum of conclusions reached around the mid-twentieth century was summarized by Berrios as follows:

1. The link as a chance combination
   a. As a result of defective statistics
   b. As the result of a chance genetic combination
2. The link as real
   a. Epilepsy producing psychosis
   b. Psychosis producing epilepsy
   c. Both as the result of a common organic factor.[6]

The burden of the evidence favored 2a, although Hughlings Jackson would have approved of 2c.

There was also the continuing problem of epileptic equivalents, noted in chapter 8. The term *psychomotor* was introduced by Griesinger to refer to patients with epileptic seizures who in the course of their history develop a "seizure" in which they become very aggressive—an apparent substitute for the "epileptic" seizure, sometimes called an "epileptoid state." Mood swings seen in epilepsy were discussed by Gustav Aschaffenburg (1866–1944). He wrote about a sudden onset and termination of an affective disorder with no clear relationship to a seizure.[7] It was then but a short step to the introduction of *affect epilepsy*, attacks of rage with loss of consciousness precipitated by psychological factors. *Épilepsie larvée* (larval epilepsy—Morel), *transformierte Epilepsie* ("transformed epilepsy"—Samt), and epileptic rudiments were other expressions. Much of this literature was published in France and Germany and has always been ignored or received poor attention outside those countries. However, these expressions have permeated through to the present day, "episodic dyscontrol syndrome" and "intermittent explosive disorder" being other epithets hinting at the epileptic equivalent.

Forced Normalization

Heinrich Landolt (1917–1971) worked at the Swiss Epilepsy Centre in Zürich and had psychiatric experience working with Manfred Bleuler (1903–1994). He published largely in German, and his discovery of one of the most important neuropsychiatric syndromes of the previous century went unnoticed in English-speaking countries by two generations of neurologists.[8]

Landolt was one of the first epileptologists to use the EEG, and he published several papers on the relationship between EEG rhythms and the mental state in epilepsy. He was given a new antiepileptic drug to try on patients with intractable seizures resident at the center, and he noted in many cases that with suppression of seizures their mental state deteriorated. This was not the first clinical observation of this association: Friedrich Hoffmann had described the first case of a possible drug-related alternative psychosis using an antiepileptic herbal remedy, and there were a few cases described with bromide treatment (see discussion in chap. 8).

The link to the EEG findings was startling. Thus, in epilepsy, when the mental state deteriorates, the EEG reveals alterations either with signs of an encephalopathy (slowing) or with a signature pattern of a seizure or status epilepticus. Landolt had described *Forcierte Normalisierung* (forced normalization). The "Landolt phenomenon" refers to the fact that during the behavior disturbance, the EEG normalizes, and as the behavior problems resolve, the EEG resorts to its abnormal configuration. "Forced Normalisation is the phenomenon characterised by the fact that, with the occurrence of psychotic states, the EEG becomes more normal or entirely normal as compared with previous and subsequent EEG findings."[9] The seizure suppression was usually, but not always, the result of a new drug (he at first noted it with ethosuccimide), and while originally the effect was thought to involve only psychoses, other behaviors became documented, including depression, anxiety, and paranoid agitation. In these situations, the EEG is paradoxically normalized when the behavior is abnormal, leading to a later rechristening of the phenomenon by Peter Wolf as "paradoxical normalization."

Forced normalization is essentially an EEG diagnosis, but the clinical counterpart referred to by Hubert Tellenbach (1914–1994) as "alternative psychosis" was also frequently reported. These were patients whose behavior alternated between deterioration with control of seizures and improvement with return of seizures over a period of time.

Other observations were in keeping with the Landolt phenomenon. Sakel and others had helped psychoses by bringing on seizures, and there were clinical descriptions of patients with epilepsy who, over years, gradually lost their seizures but developed a slowly evolving psychosis, which then became the main clinical problem.

These findings had and still have profound implications for neuropsychiatry. The clinical dilemma, as Tellenbach put it, was between the Scylla of seizures and the Charybdis of psychosis, which in some patients required careful adjustments of medications against seizure activity. Theoretically, it raised complex problems about the "meaning" of seizures—a propensity to have them, which every brain possesses, and the biological links between neuroanatomy, neurochemistry, and psychopathological states. Landolt attributed his phenomenon to a "supra-normal breaking action" in the brain achieving an electrobiological homeostasis, a new level of inhibition, and at first he linked it with temporal lobe epilepsy, a theme taken up by Janice Stevens.[10]

All That Spikes Is Not Fits

Recall that the anatomy of the temporal lobes, especially the medially situated structures and their connectivity with the basal ganglia and cerebral cortex, was well understood by the mid-1970s. The era of "psychomotor peculiarity" had been ushered in by Gibbs and others with the identification of temporal lobe epilepsy. A discrete behavioral disorder had also been described following extirpation of both temporal lobes by Heinrich Klüver (1897–1979) and Paul Bucy (1904–1992), the Klüver–Bucy syndrome being one of the first limbic syndromes to be recognized (1939). The main features were an alteration of personality with loss of aggressivity and taming, a tendency to repetitively explore the environment orally (hypermetamorphosis), hypersexuality, and visual agnosia.

Stevens was a resident at Yale, spent much of her professional life as a neurologist and psychiatrist at Portland, and was involved in the early days of EEG monitoring and the developing neuroanatomy of the basal fore-brain. She was influenced by the work of Gibbs, although she was somewhat critical of his findings on epidemiological grounds, the selected nature of the patients he tested, and the fact that he did not make the psychiatric diagnoses himself.

Stevens noted similarities of symptoms between some epilepsies and schizophrenia and the temporal lobe EEG abnormalities that were common to both. Anatomically, she described how the dopaminergic input to the (ventral) striatum may serve a gating function for limbic impulses, altering the threshold of transmission to neocortical activity, giving rise to psychotic symptoms and altering the seizure threshold. She went on to suggest that "while excessive excitatory neuronal discharge causes seizures, excessive inhibition—maintained by increased dopaminergic or other inhibitory neu-rotransmitters (norepinephrine, GABA)—contributes to psychosis." She not only provided a synthesis of the Landolt phenomena with neurotransmitter activity but also brought together the relevant neuroanatomy of the limbic forebrain for a fuller understanding of neuropsychiatric disorders.

One of her most remarkable papers was entitled "All That Spikes Is Not Fits," in which she argued that spike activity in the CNS was "a special message used by the nervous system for communication of imperative messages." She suggested that spikes are physiological, related to pulsatile neuroendocrine release, to sleep activity, and to "reward" systems, and that the monoamine pathways regulated their activity, restricting unwanted spread. She drew at-tention to the low threshold of limbic structures for epileptic discharge, which she considered had evolutionary neurobiological implications.[11]

In support of her findings were the imaginative studies of Robert Heath (1915–1999). Professor of neurology, psychiatry, and neurosurgery and chairman of the department of neurology and psychiatry at Tulane for over 30 years, Heath had a commanding position to explore brain-related behav-ior disorders. In 1950 he used stereotactic techniques to implant electrodes in cortical and subcortical sites in intractably ill psychiatric patients. With skull fixation methods, EEG recordings could be continued over many months. He attempted to correlate changes of activity with patients' subjec-tive accounts of their feelings, but via stimulation to improve symptomatol-ogy. Most of his subjects had epilepsy, schizophrenia, or pain, the latter in part providing comparison data.

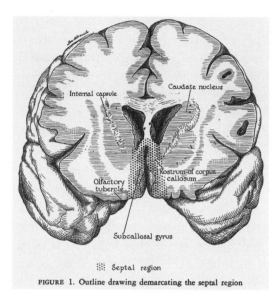

FIGURE 1. Outline drawing demarcating the septal region

Figure 13.1. The septal area as defined by Heath. From Heath RG, Tulane University Department of Psychiatry and Neurology, 1954, 2

In a series of 66 implanted patients he reported spiking in the amygdala and hippocampus interictally in epilepsy (but no cortical signals), spreading to the area of the brain he referred to as the septal region (see fig. 13.1).[12]

Injections of acetylcholine at the latter site inhibited hippocampal discharges, while stimulation of the hippocampus with acetylcholine increased activity in the septal area and was associated with feelings of pleasure. He described two cases monitored during intense sexual arousal and orgasm, which showed spike and large slow-wave activity with superimposed fast rhythms in the septal region, similar to the electrical activity of the cortex in patients with epilepsy. Recalling painful emotions led to synchronous activation of the hippocampus, cerebellum, cingulate gyrus, and sensory thalamus. With regard to psychoses, these were also linked to activity in the septal area which correlated with symptoms, and in those who had leads in the cerebellum, there was activity in the vermis. While most of the psychotic patients had schizophrenia, those with epilepsy who had psychotic episodes likewise had similar changes in the septal area.

Heath was interested in the possibilities of brain stimulation either electrically or chemically as therapy. In animals he had anatomically discovered links between the vermis of the cerebellum and the septal area, so he studied the effects of stimulation of the cerebellum on psychoses and in aggressive patients. He reported several successes. These were quite remarkable findings, the results of which should never be forgotten and will never be repeated.

Heath did experiments on the intracerebral effects of LSD and mescaline, and he attempted to find neurochemical toxins in schizophrenia. He identified a protein antibody taraxein, which was reported to cause behavioral abnormalities in monkeys. He started looking for CNS antibodies years before it became the fashionable quest that it is today.[13]

Heath was not alone in using drugs and electrical stimulation to influence behavior at this time. José Delgado (1915–2011), a neurosurgeon at Yale, developed the stimoceiver, a miniature intracerebral electrode capable of receiving and transmitting electronic signals controlled from a distance. He used this in patients with epilepsy and schizophrenia, but he became famous with his recordings of himself in a bullring suddenly stopping a charging bull before it actually hit him. Vernon Mark and Frank Ervin, neurosurgeon and psychiatrist, respectively, at Harvard, were attempting to treat violent patients with brain stimulation or ablation, especially of the amygdala.

Heath ran into problems over funding from the MKUltra project, a secret CIA-backed initiative to investigate "mind control." The Tulane department had received funds from the army for research into the effects of LSD and mescaline. Heath was then approached in 1956 by the CIA to test another compound called bulbocapnine (an anticholinesterase inhibitor), which he gave to several monkeys and one human; the compound led to symptoms similar to alcohol intoxication. This work was motivated by the concern that hostile governments, such as the Chinese and Russians, were also investigating the possibilities of mind control. However, these secret investigations were revealed only later when MKUltra was officially halted in 1973 and investigated in 1975 by a congressional commission. By this time many of the documents relating to it had been deliberately destroyed. Heath was not incriminated, but after the investigation, and given the adverse publicity that biological psychiatry generally was receiving, especially in America as an emerging counterfoil to psychoanalysis, he was unable to obtain any more government grants.

Mark and Ervin also came to grief. The author Michael Crichton (1942–2008), who was also at Harvard and studied with Ervin, published *The Terminal Man* in 1972. A patient who had been treated by them, known as

"Thomas R," claimed that the story in the book and the character Harry Benson were based on him. Benson had epilepsy and violent episodes, but after electrodes were implanted, he would turn them on and stimulate himself for pleasurable effects. Legal suits followed, their grants were revoked, and death threats were received. These events called a virtual halt to neurosurgery for mental illness in America for the next 25 years.[14]

Heath had been alerted to some of the possibilities of brain stimulation by the work at McGill University by James Olds (1922–1976) and Peter Milner. They had reported that rats with electrodes implanted in the brain's septal area would press levers at rates up to 2,000 times per hour to receive stimulation, even to the exclusion of food. They had identified "pleasure centres," linked to the medial forebrain bundle and dopamine. Heath had identified these in the human brain.

The Interictal Behavior Disorder Ordered

Stevens, although acknowledging that there were associations between temporal lobe epilepsy and psychoses, never accepted a strict causal localization hypothesis, insisting that multiple areas of brain pathology were involved. In this she had some formidable rivals.

Norman Geschwind, a man who is said to have had a theory about almost everything, revived interest in the neuropsychiatric observations of the late nineteenth century. His paper in *Brain* entitled "Disconnection Syndromes in Animals and Man" (1965) was devoted to such syndromes as the aphasias, apraxias, agnosias, and alexias and their cortical and subcortical associations. His interest was stimulated by Fred Quadfasel (1902–1981) at Boston's Veterans Administration Hospital, a neurologist who had trained in Germany under Bonhoeffer and Goldstein and whom Geschwind replaced upon Quadfasel's retirement as chief of service in 1963. The trail of influences included Wernicke and the writings of Meynert, Dejerine (especially on alexia without agraphia), and Liepmann (apraxias). Geschwind's perspective was that every behavior had an underlying neuroanatomy. He gave us the title *behavioral neurology.*

Geschwind took over the chair of neurology at Harvard after Derek Denny-Brown (1901–1981). At the time of his paper on disconnection syndromes, neurology was poised between the more holistic views of Head, von Monakow, and Goldstein and the focalizationalists such as Kleist and the earlier German school. Central to Geschwind's views was the Flechsig doctrine that in the human brain primary sensory areas connected only to second-order areas, and then via those to polymodal association cortices, especially in the parietal cortex, which was important for human language. In lower mammals this rule does not apply, primary sensory areas being connected directly to one another and the limbic cortex. In primates, primary cortices connect only to their association cortices with limbic intermediaries, but in man, the majority of intermodality connections are mediated by higher-order association cortices. As explained by Marco Catani and Dominic ffytche, "In man, intermodality connections are freed from the limbic system through the development of the inferior parietal lobe (the angular and supramarginal gyri), an area connecting visual, auditory and somatosensory association areas. For Geschwind, the area and its multisensory connections played a particular role in language development: it is only in man that associations between two nonlimbic stimuli are readily formed and it is this ability which underlies the learning of names of objects" (see fig. 13.2).[15] His theories concerned syndromes that were secondary to lesions of white matter and association cortices, and his implications of a loosening of the human cortex from limbic influences were surely incorrect.[16]

Most controversial was his work on the interictal behavior syndrome in temporal lobe epilepsy. Geschwind was less interested in the acute behavioral manifestations of the seizures than in the slowly developing personality changes that were associated with temporal lobe epilepsy.

The key features of this interictal syndrome were changes in sexual behavior, excessive religious preoccupations and activities (hyperreligiosity), hypergraphia (a tendency toward extensive and often compulsive writing), and viscosity (described below). Some of these features were described in the nineteenth century by authors such as Samt, but they became better acknowledged with the work of Gastaut and were clearly documented by Geschwind—the term *hypergraphia* being coined by him. The clinical picture could be contrasted both with frontal lobe syndromes and with the Klüver–Bucy syndrome, having some characteristics almost opposite to the latter.[17]

The hyperreligiosity included an excessive interest in the cosmic and supernatural, a conviction that the person had some special significance in the world or some messianic mission. Viscosity, another feature of this syndrome, refers to a stickiness of thought processes, a bradyphrenia, but also to an "interpersonal adhesiveness." The latter refers to increased social adhesion: patients simply will not leave the consulting room and tend to prolong interpersonal encounters, for example, with physicians, beyond that indicated by social cues. Some display circumstantiality of thought

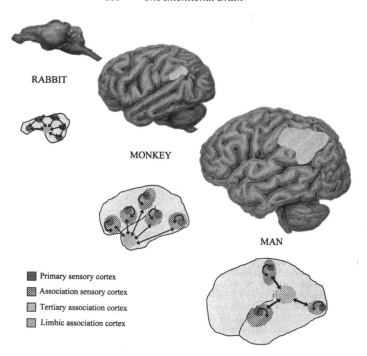

Figure 13.2. Geschwind's scheme of the different connections between primary and association cortices and the limbic system in rabbits, monkeys, and humans. The growing freedom in the human brain from limbic influence is suggested. From Catani M, ffytche DH, 2005, 2230.

with slow rambling speech, having difficulty terminating conversations. The pathogenesis of the syndrome was suggested to be ongoing interictal abnormal electrical activity in the limbic system—a hyperconnection as opposed to a disconnection.

There was a poor reception to the description of the interictal behavior syndrome by contemporary epileptologists (greeted with rehearsed hisses at one conference), and its existence is still challenged by some who are unable to embrace a neuropsychiatric approach to their discipline. Contemporary psychiatric diagnostic manuals such as the *DSM V* are of no help in understanding these syndromes, as they are simply not included within them. It is now christened the Gastaut–Geschwind syndrome and is a neuropsychiatric cipher.

Geschwind's work on disconnection syndromes has been somewhat eclipsed by findings of modern brain imaging, but his attempt to develop a perspective in between the extremes of localization and integrative aspects of brain function was a move that has now progressed to models of distributed networks and connectionist theories. He inspired a generation of behavioral neurologists, including Elliott Ross, Marsel Mesulam, Ken Heilman, Albert Galaburda, Bruce Price, Frank Benson (1928–1996), and Damasio; the neuropsychologists Howard Gardner and Nelson Butters (1937–1995); and the epileptologists Donald Schomer, Steven Schachter, and Orrin Devinsky.

Phenotypes and Genotypes

Geschwind's interest in epilepsy had been in part stimulated by the work of Eliot Slater at the National Hospital for Nervous Diseases, where he was appointed in 1946. Slater and William Sargent (1907–1988) were two of the most significant neuropsychiatrists of postwar England. Their textbook *Physical Methods of Treatment in Psychiatry* (1944) stamped their approach to mental illness, recalling as they did the helplessness of so many chronic patients in the prewar eras, when no effective treatments were available. Mosquitoes were colonized in London for malarial therapy, and there was insulin coma, but they sought alternatives. They introduced amytal narcosis and abreaction therapy for the war neuroses, continuous sleep treatment for severe manic states or anxiety, and amphetamine for depression, and in 1941 they began work with ECT.

Their book was dedicated to Mapother, who had little time for either psychoanalysis or purely psychologically based therapies. At this time England's most prominent psychiatric research hospital, the Maudsley, which had been under the leadership of Mapother, was in the hands of the formidable Aubrey Lewis, who was against the organic approaches of his predecessor. Slater and Sargent were thus compelled to carry out their work away from the Maudsley, but they surely saved British psychiatry from the onslaught of psychoanalysis that came to dominate much European and American psychiatry at the time.

Slater is well known for his work on genetics, and his *The Genetics of Mental Disorders* was the first authoritative text on the subject. Although twins were noted even by Hippocrates to have similar illnesses, Slater's genetic studies with twins and his statistical prowess brought much methodological rigor to the subject. For neuropsychiatry, his work on the schizophrenia-like psychoses of epilepsy and hysteria is still much quoted.

Slater's study of 69 cases of psychoses and epilepsy documented the insidious onset of the psychosis, with many characteristics typical for schizophrenia, such as auditory hallucinations, Schneiderian first-rank symptoms, thought blocking, and neologisms. He also commented on the frequency of religious delusions, a thought disorder with circumstantiality and a tendency to ramble. He pointed out that in the absence of epilepsy, the psychoses in their study group would have been diagnosed as schizophrenia. However, there were distinct differences, such as a better prognosis, fewer motor signs, better maintenance of affect, and episodes of affective disturbance shown by all of the patients, hence the term *schizophrenia-like*. Most important was the absence of a family history of psychosis; this was then a clinical picture of a genetic disorder with an organic basis—so much for schizophrenogenic mothers.

Hysteria Again

Slater's 1965 paper on the prognosis of hysteria documented the poor prognosis and the frequently discovered organic illness at follow-up. Famously, he considered the label *hysteria* to be a "disguise for ignorance and a fertile source of clinical error. It is in fact not only a delusion but also a snare."[18]

Known for his intellectual rigor and clear thinking, Slater could be provocative. He resigned from the National Hospital for Nervous Diseases after the medical committee turned down an academic unit of neuropsychiatry for which he had raised money. He took his money to the Maudsley Hospital and got on with his genetic studies. He gave his lecture on hysteria at the hospital as a final farewell. His observations on the high frequency of organic findings in patients diagnosed with hysteria by the neurologists implied that they could not diagnose organic diseases. This was soon vigorously contested by Sir Francis Walshe (1885–1973) and Sir Charles Symonds (1890–1978).[19]

Gifted in other areas, Slater painted well, wrote poetry, and was married to Lydia Pasternak, sister of the Russian poet Boris Pasternak. After retirement, he was awarded a PhD for his work seeking to clarify the authorship of the play *Edward III* by statistical word analysis. Rather aloof from colleagues and patients, his first question to a new patient was reputed to be, "Are you a twin?"[20]

Moving Sideways

A further significant contribution to neuropsychiatry from Maudsley Hospital came from the work of the neurosurgeon Murray Falconer (1910–1977) on temporal lobectomy and the observations of Pierre Flor-Henry. Flor-Henry's paper "Psychosis and Temporal Lobe Epilepsy" (1969) revived a forgotten jewel of older neuropsychiatric thinking awaiting rediscovery, namely, laterality.

Wigan, whom we met in chapter 8, wrote *The Duality of the Mind* (1844) with the belief that the two sides of the brain were separate organs, with independent mental functions—two different organs of thought. Thinking could be carried out by both hemispheres independently, and mental illness was then caused by a disruption of the normal harmonious balance between the two sides. The book was dedicated to Sir Henry Holland (1788–1873), physician to the royal family, who had published *On the Brain as a Double Organ* in 1839. Wigan took the idea much further than Holland. Regarding his own unusual experiences he referred to a double identity, sometimes finding it difficult to hold himself together and to keep a mental equilibrium. His description of one of his episodes is clearly a déjà fait or déjà vu experience he had at the funeral of Princess Charlotte when he was standing by the side of her coffin.[21]

Flor-Henry published on a group of patients who had undergone unilateral temporal lobectomy and had psychopathology. They had all been worked up as far as possible in those days with regard to the side of the epileptic focus, with electroencephalography and electrocorticography, and their psychiatric features were well documented. He reported that schizophrenia was associated with left-sided (dominant) foci and manic depressive illness with right-sided (nondominant) foci. This was the first reporting of psychological associations with laterality since the work of Broca.

Laterality

In the course of evolution our brains have become lopsided.[22] Potentialities to lateralization have been shown in other species, but there are good evolutionary explanations as to why this occurred in *Homo sapiens*. By differentiating the activities of the two hemispheres, abilities that did not require bilateral representation in the brain, such as speech (as opposed to sight), became preferentially localized in one or the other hemisphere. This minimized brain size and energy consumption needed for an increase in cognitive capacity, and it allowed for continued safety of the birth process. Fetal head size was crucial, since with a bipedal gait, the hips of females required a special anatomical structure with size limitation. The infant head could not increase too much, as it would not be able to travel through the birth canal, dangerous for the mother as well as the infant. Further, lateralization of function allowed neurons with the same type of computational activity to be placed together, aiding brain development and minimizing the need for long-distance connecting axons.

In neurology most attention has been given to the activities of the left hemisphere, which, in most right-handed people, regulates syntax, propositional language, and the motor abilities of the right side of the body. It is referred to as dominant and contrasted with its nondominant counterpart, the right hemisphere. Wigan's book was approved of by Maudsley, but it was Hughlings Jackson who preferred the term "leading hemisphere" for the left cortex, and who almost alone sang the praises of the other side. Since the left side of the brain was so closely aligned with language, it was metaphorically "civilised," in contrast to the more "primitive" right side. In literature, interest in dual personalities is found in Robert Louis Stevenson's (1850–1894) *Strange Case of Dr Jekyll and Mr Hyde* (1866), in which the moral Jekyll gives way to the evil instincts of Dr. Hyde. Poor Jekyll despaired, "I have been doomed to such a dreadful shipwreck: . . . of the two natures that contended in my field of consciousness . . . I was radically both. . . . It was the curse of mankind that these incongruous faggots thus were bound together— that in the agonised womb of consciousness, these polar twins should be continually struggling."[23]

Anatomical investigations had shown that the left anterior and the right posterior cortices were greater in dimension than their opposites, and the phrenologists had signaled differences between the activities of the hemispheres. Brown-Séquard considered that the two hemispheres were separate, the left one being "predominant" and the right one subserving emotions. He considered that the latter could be educated to raise it up to the level of the other, and various popular movements were started to try to bring this about (ambidextrous cultural societies). The Scottish psychiatrist Lewis Campbell Bruce (1866–1945) published a paper on dual brain action in 1895, and in 1897 one on dual action in relation to epilepsy. In the latter he suggested that unilateral seizures disrupted the activity of one or the other hemisphere, altering the personality. He even gave the names Welsh and English for two distinct consciousnesses, the former relating to the right hemisphere and the latter (superior!) to the left.

This is not the place to relate the belated growth of interest in the functions of the right hemisphere, but there are some very important concepts to be considered. Neuropsychologists were particularly interested in the operation of section of the corpus callosum for severe epilepsy, as this allowed for psychological testing of one hemisphere at a time. Roger Sperry (1913–1994),

Michael Gazzaniga, and their coworkers used sophisticated tachistoscopic techniques for this purpose. The right hemisphere has been known for some time to be associated with visuospatial functions, attention, and integration of perceptual processes capturing the *Gestalt*, but it is only recently that a much wider perspective of the actions of that hemisphere has become appreciated and investigated.[24] The nondominant hemisphere has recently had its champions, including the psychiatrist and philosopher John Cutting, and more recently Iain McGilchrist in his brilliant analysis of the interactions between the two hemispheres, *The Master and His Emissary: The Divided Brain and the Making of the Western World.*

The book deals with the relationship between the hemispheres from not only an evolutionary but also a historical-sociological perspective. McGilchrist analyzes asymmetrical biases between the functions of the hemispheres and the way that the so-called dominant hemisphere, that which is language rich for propositions, has created for the human mind a "sort of self reflective virtual world"; this is essentially parasitical on the master, the right hemisphere. But it is the right hemisphere that tolerates metaphor, that is creative, that is insightful into problem solving: it is deductive, not suppressive. Emotion precedes cognition from an evolutionary perspective, and within us *feeling is at the heart of being*. While language, logic, and linearity are under the control of the left hemisphere, knowledge gained by it is *knowing that (wissen Ger)*, as opposed to personal knowledge and knowing myself *(kennen Ger)*.[25]

McGilchrist's book does not stop there, however. Not accepting the philosophical paradoxes of the language of the left hemisphere and a philosophy that concerns itself only with language (the linguistic turn of the twentieth century), McGilchrist takes the reader into philosophical realms that many Anglo-Saxon philosophers avoid, and which are not the currency of most neuroscientists. This is not an empirical or Cartesian worldview; in the book he refers to the works of such philosophers as Nietzsche, Husserl, Martin Heidegger (1889–1976), Merleau-Ponty (1908–1961), and others of a "continental" philosophical orientation. For McGilchrist, philosophy begins and ends with the right hemisphere: traditional philosophers and neurologists examine the life of the right hemisphere only from the standpoint of the left.

Thus, in contrast to the way that the left hemisphere deals with the world, the right hemisphere seems dominant for attending to bodily space, for integrating cerebral circuits for spatial representation of the body, and importantly

for regulation of mood and certain aspects of speech, the emotion in pros-
ody. The right hemisphere utilizes holistic, global representations (*Gestalts*),
rather than the discrete, particularized, boundaried computations of the
left hemisphere. According to McGilchrist, the right hemisphere has virtually all of the attributes that make us human. It is attracted to what is new in
the world and is attentive to the world around us before the left hemisphere.
It is the repository of our emotional biographical memories, and it is where
our embodied sense of self is located. The syntax-bound left hemisphere con-
stricts meanings that are not relevant for its own achievement. The creative
right hemisphere speaks not the language of most philosophers, but that of
poets and musicians, and it responds to the minor key.[26]

Music and the Romantic Brain

Exploration of the romantic brain cannot avoid the exclusive ability of
Homo sapiens to make and shake to music. It will forever remain a mystery
as to exactly when and how in our long evolutionary trail these remarkable
capacities emerged. There are persistent arguments revolving around the
relation of music to language. These include whether music itself can be
referred to as a language, whether music or spoken language evolved first,
or whether they both had a precursor from which they branched off in their
own separate ways.

It is important to deal briefly with some of these issues of relevance to
the active, intentional, and romantic brain. It makes no sense to think that
spoken languages, the Tower of Babel, emerged before some form of musical
intonation. While some arguments hinge hopelessly on what the definition
of *music* is, this sad semantic confuses the most obvious. We need to go back
to the days of our hunter-gatherer ancestors and what they did with the
great gift of Prometheus. To even get a feeling for this, it is important to realize two things. First, this was a time when there was no predicated language;
surely, communication between individuals must have been one of high expressive emotional content. Secondly, although for Darwin music remained
somewhat of a mystery, he noted its emotional power and the love theme of
so many songs: "musical tones and rhythm were used by the half-human
progenitors of man during the season of courtship . . . they aroused each
other's ardent passions, during their mutual courtship and rivalry . . . women
are generally thought to possess sweeter voices than men, and as far as
any guide we may infer that they first acquired musical powers to attract the
other sex."[27] After all, the book in which he discussed this is titled *The*

Descent of Man, and Selection in Relation to Sex. There was more to natural selection than the "struggle for life." Sexual selection is anything but a passive environmental influence on survival; it is an active, intentional seeking *Trieb*, driven from and by individuals. As Denis Dutton puts it in *The Art Instinct*, "This Darwinian process of self domestication introduces a place for consciousness and purpose which is absent in principle from natural selection. To that extent *Homo sapiens* is a self designed species."[28]

The central features of music making, including emotional intonations, dancing, and precise vocal and rhythmic control, emphasize fitness and health, and musical creativity implies special cognitive abilities. Our ancestors' aesthetic sensitivity and perceptual auditory bias allowed for a combination of the emotive power of music and rhapsody, from which arose beautiful and provocative courtship displays, which are still ours to enjoy. It is difficult not to conclude that music was an important factor allowing the development of complex social structures and spoken languages. Musical rituals, with gesture, dance, and some kind of incantations involving group interaction, must have had relevance for the emergence of more sophisticated social encounters, not only at the group level but also between individuals. Dancing to the rhythm of life must have enhanced life, as the fittest, attracted to each other, copulated and contributed to group survival.

Making music is an intentional act. Acoustic novelty requires a creativity allowing the blending of structural combinations of essential musical components such as repetition, variation, prediction of synchronization, and rhythmic modulation, all giving rise to a sense of expectancy and an urge to move. Our reactions to music are not in the form of a stimulus and response since the response and stimulus coincide—rhythm and music embodied.[29]

Full Circle

I have suggested that the first case vignettes described in this book were related to epilepsy and an associated psychotic disorder. Epilepsy was an important disorder for Hippocrates and has been one cornerstone for the development of twentieth-century neuropsychiatry. This in part has been brought about by the development of technologies, especially the EEG, but other developments in the past 70 or so years have served to remind us how the rupture between neurology and psychiatry unfolded and skewed clinical practice. This has been quite a short era over our 2,500-year journey, and the emergence of today's neuropsychiatry from unforetold beginnings may come to be seen as inevitable by future generations.

At this juncture we have seen several branches that have grown from the "neuroscience" historical tree, with its clinical trunk composed of what became acknowledged as neurology and psychiatry in the twentieth century. Offshoots are behavioral neurology, biological psychiatry, organic psychiatry, neuropsychology, and neuropsychiatry. There are those who still view the invasion of neuroscience into psychiatry with considerable skepticism, and many who consider that the penetration will soon be so great that all psychiatry will crumple like a house of cards, just leaving a refashioned neurology. This neglects an understanding as to how and why there has been a reconstitution of neuropsychiatry. The dramatic decline of therapies based on psychoanalytic and Pavlovian principles and the burgeoning of new treatments exploiting discoveries in neuroanatomy and neurochemistry certainly constructed bridges between neurology and psychiatry. There was and is no way of going back, but there is more to this story of neuropsychiatry and the romantic brain.[30]

Sacks, in describing the responses of his patients to L-dopa, referred to a "continuous proliferation of 'new' excitations (*de novo*), until the brain is *lit up* with innumerable excitations . . . flares being lit in the cerebral city, until finally it goes up in flames."[31] Prometheus re-membered.

NOTES

Epigraph: In this play within a play, the characters seek immutability, fixed forms, as they become part of a play written by someone else.

1. Huxley A, 1954, 4.

2. Quoted in Watson P, 2014, 428. Leary used this apothegm at the Human Be-In, a gathering of 30,000 hippies in Golden Gate Park in San Francisco in 1967.

3. Smythies studied neurology at the National Hospital for Nervous Diseases, studied psychiatry at St. George's and Maudsley Hospitals, and had an acute interest in philosophy and psychical research. For many years he was a neuropsychiatrist in Birmingham, Alabama. His associations with Huxley are described in Smythies J, Smythies V, 2005.

4. The story of the discovery of the neuroleptics is well told by Thullier J, 1999, 103–153.

5. Sacks O, 1973, 212, 217. Dopamine was essential for sensory perception in the olfactory bulb, the latter of which considerably increased in size in early mammals. However, it was also present in the rudimentary olfactory bulbs of earlier vertebrates, thus being present as an evolutionary force for half a billion years. Dopamine is a precursor of adrenaline and noradrenaline.

6. The papers and background are all reviewed in Trimble MR, 1991.

7. Aschaffenberg G, 1906.

8. Landolt published only one paper in English on forced normalization. His life and work are written up in Trimble MR, Schmitz B, 1998.

9. Trimble MR, Schmitz B, 38.

10. The observations of Landolt were ignored for several reasons. Firstly, he was writing in German; secondly, he meddled with psychiatry; and thirdly, he shifted the axis of management of epilepsy away from seizure control as the only management strategy. His center was next to the Burghölzli Hospital, and he could relieve the psychoses with a course of ECT if necessary. His observations sat uncomfortably with especially American neurologists, who by now were divorcing anything to do with psychiatry from neurology and epilepsy.

11. See Stevens J, 1986; quotes from p. 97.

12. Heath's definition of the septal region was as follows. This comprised several structures. Caudally there was the anterior commissure, rostrally was the tip of the anterior horn of the lateral ventricle, medially was the midline space separating the hemispheres, and the dorsal extent was the septum pellucidum proper and the base of the lateral ventricles. It extended centrally to the base of the brain and laterally about 5 millimeters from the midline. He noted that the area was a part of the olfactory system and interposed between higher neocortical structures and the diencephalon and midbrain structures. This embraced the nucleus accumbens and the area referred to as the ventral striatum. Heath RG, Tulane University Department of Psychiatry and Neurology, 1954, 3.

13. Heath RG, Tulane University Department of Psychiatry and Neurology, 1954; Heath RG, 1986, 126–138. He had had psychoanalysis with Sandor Rado (1890–1972). On one occasion when I visited Heath's department, discussing the use of stimulation of the cerebellum in persistently aggressive behavior, he recounted to me how in some patients the effects were dramatic, and that stopping the stimulation would bring back the symptoms. He recounted the history of one patient whose behavior had been much improved for several years but who suddenly deteriorated. When the parents brought him back to the clinic, the electrodes were found to have broken. When they were repaired and the stimulation resumed, he again became calmer. I think that Heath didn't believe that I was convinced, so he picked up the phone and called the boy's father, and the latter and his son were at Heath's office within half an hour. I was able to interview father and son, who confirmed the events as described.

14. See Faria MA, 2013, for an excellent account of these events.

15. Catani M, ffytche DH, 2005, 2230.

16. For a tribute to and compendium of Geschwind's life and work, see Schachter SC, Devinsky O, 1997; Benson F, 1997. Also for a recent appraisal, see Catani M, ffytche DH, 2005.

17. The hypergraphia, with detailed accounts of daily events being recorded, or with the elaboration of texts often with a moral or religious theme, was an important feature. Geschwind told me that he considered this one of the most missed neurological signs in clinical practice. However, if patients were not asked questions about their religious activities or their writing habits, then the syndrome would be missed. The sign is the prodigious writing, in diaries or texts that patients will bring with them to the clinic or will produce if asked for. There is also a proclivity to write poetry, which again will be missed if not inquired about.

18. Slater E, 1965, 1399.

19. Sir Charles Symonds had considerable experience in assessing and managing military personnel with war neuroses, and he had an extensive private practice, many of his patients having psychiatric disorders—not uncommon in neurological practice then and now, except they so often go unrecognized.

20. Slater never left behind him a "school," but he bequeathed a milieu that, like that of Hughlings Jackson, can still be felt. His direct successor was Lishman, followed by Merskey, and then myself. When I was appointed, the other neuropsychiatrist was Richard Pratt. He joined the staff in 1962 when Slater resigned. His original plans were to do neurology, but he

had abandoned these on account of his diabetes and a mistaken diagnosis of lymphoma given to his wife.

Pratt's MD thesis was on the genetics of multiple sclerosis, and to compliment Slater's work on the genetics of mental illness, he wrote *The Genetics of Neurological Disorders* (1967). He was unimpressed with the psychosomatic theories at that time, especially the idea of predisposition to illness through personality. But he was quite dedicated to his patients, who remained with him for years. He did his best to treat those referred to him from neurologists, and he always assessed the patients as if they had the degree of psychopathology that the neurologists thought they had (even if knowing that this was incorrect). He used much ECT, and with the neuropsychologist Elizabeth Warrington he tried to find the most effective way of treating patients with minimal neuropsychological impairment. He was a strong advocate of nondominant hemisphere unilateral application. He proofread texts by reading the lines from right to left for better accuracy, and nothing pleased him more than reading some medieval text in its Latin original—such was the scholarship on the staff of the hospital at that time.

21. Charlotte was the eldest daughter of King George III.

22. See Corballis MC, 1993.

23. Stevenson RL, 1886, 113–114.

24. The properties and functions of the nondominant side of the brain, as well as their importance for regulating so many aspects of human behavior, are discussed in some detail in Trimble MR, 2007.

25. In English there is only one verb for "know," but in other languages there are multiple, with clear distinctions. In German the difference between *wissen*, "to know that (a fact)," and *kennen*, "to know personally," is central to much philosophy, but it is also representative of our different ways of knowing the world we live in.

26. See Cutting J, 1990; McGilchrist I, 2009. McGilchrist's brilliant exposure of the quite extensive but diverse and well-neglected literature on the functions of the right hemisphere proclaims a warning about the dangers for human civilization posed by the now overriding dominance of the dominant hemisphere in Western culture. He reveals how the functions of the two hemispheres of our bicameral brain have shifted, intermingled, conflicted, competed, and settled for us, in the twenty-first century, in such a skewed relationship between them that the imbalance puts our very future existence as *Homo sapiens* in considerable danger. His thesis is that at one historical time or another, the tendencies and proclivities of one hemisphere have been the more dominant, shaping the immediate culture from a historical perspective, and he acknowledges the fact that the left hemisphere has "helped us achieve nothing less than civilisation itself" (93). But he thinks that we are coming close to forfeiting that very civilization. The left hemisphere (the emissary) has taken over and dominated the right (the master), the emissary now acting like a Faust, with no insight into what is going on and who cannot see the destruction that is coming. Toward the end of the book, McGilchrist acknowledges the limits not only of his own understanding but also of any human understanding. In view of the Orwellian world that reflects the left hemisphere's language-dominant rules, making a goad to our existence, a grip that appears to be tightening almost daily, he is not as pessimistic as others would consider his book should be. He is cautious to undermine reason and is content to take a scientific approach.

27. Darwin, 1871, quoted in Wilson EO, 2006, 1209.

28. Dutton D, 2009, 166.

29. The octave is a fact of nature, and it must be intriguing that nearly every scaled form of music involves seven or less pitches to the octave. Further, most rhythmic patterns are based in two or three time. Rising sound patterns lead to different emotional effects from descending ones, and faster tempi lead to different expressions compared to slower ones. All

these combine to influence the emotional resonance of sounds. Repetition is essential in so many human activities; the fact that it is central to musical composition above all the other arts again reveals the early biological imperative of our love of music.

30. The Society for Biological Psychiatry was founded in 1954, and the World Federation of the Societies of Biological Psychiatry in 1974. There has never been a society devoted to organic psychiatry.

31. Sacks O, 1973, 217.

REFERENCES

Aschaffenberg G. *Über die Stimmungs-Schwankungen der Epileptiker.* Marhold, Halle; 1906.

Benson F. Disconnection syndromes revisited. In: Trimble MR, Cummings J, eds. *Contemporary Behavioural Neurology.* Butterworth Heinemann, Oxford; 1997.

Catani M, ffytche DH. The rises and falls of disconnection syndromes. *Brain* 2005;128: 2224–2239.

Corballis MC. *The Lopsided Ape: Evolution of the Generative Mind.* Oxford University Press, Oxford; 1993.

Cutting J. *The Right Cerebral Hemisphere and Psychiatric Disorders.* Oxford Medical Publications, Oxford; 1990.

Dutton D. *The Art Instinct: Beauty, Pleasure, and Human Evolution.* Bloomsbury, London; 2009.

Faria MA. Violence, mental illness, and the brain—a brief history of psychosurgery: Part 3—from deep brain stimulation to amygdalotomy for violent behaviour, seizures, and pathological aggression in humans. *Surgical Neurology International* 2013;91. doi: 10.4103/2152-7806.115162. eCollection.

Heath RG. Studies with deep electrodes in patients with intractable epilepsy and other disorders. In: Trimble MR, Reynolds EH, eds. *What Is Epilepsy? The Clinical and Scientific Basis of Epilepsy.* Churchill Livingstone, Edinburgh; 1986.

Heath RG, and the Tulane University Department of Psychiatry and Neurology. *Studies in Schizophrenia.* Harvard University Press, Cambridge, MA; 1954.

Huxley A. *The Doors of Perception.* Chatto and Windus, London; 1954.

McGilchrist I. *The Master and His Emissary: The Divided Brain and the Making of the Western World.* Yale University Press, London; 2009.

Pirandello L. *Six Characters in Search of an Author.* Trans. Musa M. Penguin, London; 1995.

Sacks O. *Awakenings.* Duckworth, London; 1973.

Schachter SC, Devinsky O. *Behavioural Neurology and the Legacy of Norman Geschwind.* Lippencott-Raven, New York; 1997.

Slater E. Diagnosis of "hysteria." *British Medical Journal* 1965;1:1395–1399.

Smythies J, Smythies V. *Two Coins in the Fountain: A Love Story.* Privately printed; 2005.

Stevens J. All that spikes is not fits. In: Trimble MR, Reynolds EH, eds. *What Is Epilepsy? The Clinical and Scientific Basis of Epilepsy.* Churchill Livingstone, Edinburgh; 1986.

Stevenson RL. *Strange Case of Dr Jekyll and Mr Hyde.* Longmans, Green, London; 1886.

Thullier J. *Ten Years That Changed the Face of Mental Illness.* Taylor and Francis, London; 1999.

Trimble MR. *The Psychoses of Epilepsy.* Raven Press, New York; 1991.

Trimble MR. *The Soul in the Brain: The Cerebral Basis of Language Art and Belief.* Johns Hopkins University Press, Baltimore; 2007.

Trimble MR, Schmitz, B. *Forced Normalisation and Alternative Psychoses of Epilepsy.* Wrightson Biomedical, Petersfield, UK; 1998.

Watson P. *The Age of Nothing: How We Have Sought to Live since the Death of God.* Weidenfeld and Nicholson, London; 2014.

Wilson EO. *From So Simple a Beginning: The Four Great Books of Charles Darwin.* W. W. Norton, New York; 2006.

CHAPTER FOURTEEN

Neuropsychiatry,
Then and Now

If God had not created woman's flesh I would not have become a painter.

Renoir, quoted in Roberto Calasso, 2012, 204

BRADY. Now, Howard, how did man come out of this slimy mess of bugs
and serpents, according to your—"Professor"?
HOWARD. Man was sort of evoluted. From the "Old World Monkeys."
BRADY. Did you hear that my friends? "Old World Monkeys"! According to
Mr Cates, you and I aren't even descended from good American monkeys!

Jerome Lawrence, Robert Edwin Lee, 1955, 69

Moving Forward

Movement is vital to life. Movement implies change; movement implies
evolution. In order to fully appreciate the romantic brain, the organ intimately connected to the development of *Homo sapiens* and our stunning
intellectual achievements, we must recognize that none of the latter would
have been possible without movement.

The premodern movement for movement began with Hughlings Jackson. He viewed human activities in terms of motion, the function of the
brain co-ordinating and integrating the body through physiological
representations of sensory-motor activities. He envisaged the higher centers as re-representations or re-re-representations of sensory-motor functions and integrated reflex arcs. For Hughlings Jackson, thinking of a
movement was comparable to thinking of an object. The accompanying
neural activity was faint, but there was an idea of movement, and thinking
of a movement or remembering it involved the highest centers: this was a
physical process; there was no pure sensory event, and before the act there is
a faint image.

Sherrington noted how in evolution, motor acts preceded recognizable minds. The brain is an organ that coordinates motor acts; it integrates the "motor individual." The nervous system is never at rest, and sensory "trigger-organs" impressed change in an activity already in process, a means of realigning a motor act. "Mind, recognisable mind, seems to have arisen in connection with the motor act. . . . Evolution brought it; natural selection sanctioned it; it had survival value."[1]

The German word *Gestalt* originally referred to the shape and structure of the human body, although it later came to refer to configurations of the outside world. For Schilder there was a combination, the *Gestaltung* being a constructive perceptual experience. For him, the body image was never static.

Neuroscientists such as Cannon opened up explorations of the brain's interior structures as drivers of emotion; for Hess, "an important factor in the development of goal-directed psychomotory action was also the driving curiosity which, when we are awake, keeps the body in motion by exploring."[2]

In 1964, Grey Walter discovered the *Bereitschaftspotential* (the readiness potential)—a negative electrical potential appearing in the brain before making a movement. Benjamin Libet and colleagues, in investigations that have been replicated several times, found that although the readiness potential appeared some 550 milliseconds before the actual movement took place, the moment of conscious intention to move was noted at 350 milliseconds. In other words, the brain's processing of the motor action occurs unconsciously, some 350 milliseconds *before* conscious awareness of the intent to action.

Notwithstanding the controversies that such findings have caused, especially for arguments about free will and the nature of voluntary acts, they reveal that we live in the past, and even our treasured, thoughtfully self-willed actions have unconscious beginnings. As Gazzaniga puts it, "The brain finishes the work half a second before the information it processes reaches our consciousness."[3] This preattentive emotional processing is just another example of the way the human brain processes our sensations and behaviors, leaving us consciously unaware of the whys and wherefores of our predispositions and actions, which we then have to justify by conformity to our own personal narratives. The neuroscientist Marc Jeannerod pithily stated, "Consciousness . . . reads behaviour rather than starting it."[4]

Throughout this book, and starting with the Greeks, associations between philosophy and neuroscience have been continually emphasized. Three twentieth-century philosophers who also have made interesting and important contributions to neuropsychiatry, but whose names are rarely discussed in neuroscience circles, are now noted.

Active Perception

We have noted how Brentano, who recognized the empirical possibilities of psychology, reintroduced the concept of intentionality. His anti-Cartesian perspective dethroned the objective necessity of empirical philosophy, appealing for an understanding of our "inner" world and its relation to the external. This was not avowing introspection, but a quest to understand how the subject was connected to the object. Inner perception is immediate, ineffable, and self-evident; we are always conscious of something.

Henri Bergson (1859–1941), related to Proust through marriage, familiar with the writings of Spencer and Charles Darwin, and friend of William James, brought the biological sciences into his theories within a Heraclitan thrust. *Matter and Memory* was published in 1903, and his concept of the *élan vital* first appeared in *Creative Evolution* in 1907.

He reasoned that existence for a conscious human being implied change, time being central to the construction of reality. Life was a continuum of the evolutionary impetus, and there was an inner directing principal to life, pace Darwin: "it springs from the very effort of the living being to adapt itself to the circumstances of its existence." Action in the world implies anticipation of several possible actions, these being marked out before action itself. Moving beyond the moment of existence is the link to the future. Our actions are the outcome of a preceding series of anticipatory potentials, and action itself is involved with what he referred to as the body image of our conscious states.[5] The present is charged with the past but has a foot in the future: memory, as Bergson implied, is the past pushing into the present and not vice versa: what is found in the effect is already in the cause. We need help from time and discontinuity, and "our memory solidifies into sensible qualities the continuous flow of things."[6]

Bergson in his writings quotes clinical cases, for example, of Charcot and Heinrich Lissauer,[7] bringing in apraxia, aphasia, word blindness, echolalia, and loss of topographical orientation. He was highly critical of the ever more complicated box and wire diagrams that were produced to explain simply the complicated to the simple, and he refuted the idea of "centers" in the brain.

In his writings, he frequently referred to the brain, which he called "an instrument of action and not representation." "To know how to use a thing . . . is to take a certain attitude, to have a tendency to do so through what the Germans call motor impulses (*Bewegungsantriebe*)." It is these motor tendencies that give us feelings of recognition, such that perception is no simple photographic reproduction.[8] "All of animal and vegetable life in its essence seems like an effort to accumulate energy and then to release it via flexible vessels, vessels that can be reshaped, at whose extremity it will accomplish infinitely varied tasks. That is what the *élan vital*, traversing matter, seeks to obtain at one stroke.[9]

Maurice Merleau-Ponty (1908–1961) collaborated with Jean-Paul Sartre (1905–1980) in an intellectual resistance group and in 1949 became professor of child psychology at the Sorbonne, where he collaborated with Jean Piaget (1896–1980). He took a scientific approach to philosophy, and his most celebrated work, *The Phenomenology of Perception* (1945; English translation, 1962), was grounded in a philosophy that emphasized the body and its engagement with the world: he was after the nature of the body–world dialogue. The body was for him the intermediary between mind and matter, the "body subject" being anchored in a pre-cognitive, pre-objective, pre-reflexive world.

Merleau-Ponty was influenced by the infant development theories of Piaget, the anthropology of Claude Lévi-Strauss (1908–2009), the psychology of the *Gestalt* school, and the philosophies of Kant, Husserl, Heidegger, and Sartre. He rejected the Cartesian perspective and considered the *Lebenswelt*, the lived world that our bodies move around in, as most relevant to interpreting the intersubjective world of individual consciousness. Bergson had stated that "there is no perception which is not full of memories," and that "the essential process of recognition is not centripetal, but centrifugal."[10] In the same vein, perception as envisaged by Merleau-Ponty was not just the passive receiving of sensory stimuli, but an active process of exploration of the environment: the "body subject" had a "grip" on objects, which was nevertheless ambiguous and time bound. Things, so to speak, "speak" for themselves: "The object which presents itself to the gaze or touch arouses a certain motor intention . . . [there is] a coition . . . of our body with things."[11]

Although interested in visual perception, unlike so many philosophers, especially of the empiricist schools, he was concerned with our other senses, particularly the tactile—that of touching and being touched. The empiricists, who would conceive of each of the senses as being separate, lacked a framework for the binding problem, as it is now called—that is, how do we experience the combined unified sensation of, say, viewing an apple, as an object, complete, as an apple with its sensory and metaphorical surround? The physicist and mathematician Hermann Weyl (1885–1955), as quoted by the Nobel Prize–winning physicist Frank Wilczek, put it thus: "The objective world simply *is*, it does not *happen*. Only to the gaze of my consciousness, crawling along the lifeline of my body, does a section of this world come to life as a fleeting image in space which continuously changes in time."[12] The whole is more than the sum of the parts.[13] Further, for Merleau-Ponty, the importance of perception was to be found not in the perceived object but in our experience of it. He would not adopt an idealistic position, that consciousness was somehow outside of the world and different from it, but concerned himself with the phenomenal field, that which is presented to the perceiver, which is ontological and not ontic. His complete rejection of the Cartesian view, his understanding of the importance of the *Gestalt* school of psychology, and his use of medical case histories, like Bergson, informed his insistence on the embodied nature of perceptual experiences and hence consciousness.[14]

Merleau-Ponty accepts that perception has certain rules, his "perceptual syntax," but these are not grounded in the causality of logic (derived by an idealism that claims that the world is interpreted by a consciousness amenable to reason), since sensations are synthesized in noncausal relationships.

But—and here is a crucial twist—included in those sensory perceptions are the proprioceptive ones, those that give us information about our own bodies, coming from our limbs and interior organs. We are visual creatures, but also especially tactile—we feel our way in the world. Further, perceptual experience gives us access to things that have value for us; we have meaningful interactions with them. Where that meaning comes from will be suggested below, but this philosophy gives perception a binding with movement and a bonding with emotion.[15] Nicolai Rostrup, in relation to the field of neuroaesthetics, put the following emphasis on this idea: "The evaluation of future rewards seems to be an automatic feature of the brain continually guiding behavior . . . phenomenally we recognize this process as an emotion."[16]

One of the key ideas in *The Phenomenology of Perception* is *motor intentionality*. This for Merleau-Ponty is the way in which the body directs itself toward and "grasps" objects in a pre-cognitive manner. This is something that is an anticipation of or arrival at the objective and is ensured by the body itself as a motor power, a "motor project," a (pre-cognitive) "motor intentionality." Perceiving is a *motor skill*. This is a fundamental change of perspective for philosophy, psychology, and neurology. The main line of thinking has always been the Cartesian reflex, that stimulus leads to response, perception leads to action. But he inverts this relationship, putting the active, seeking brain behind our perceptions. There is a background to perception, based on motor possibilities. The environment draws the subject to action, the body "being-in-the-world" is offered the world: "motility as basic intentionality. Consciousness is in the first place not a matter of 'I think,' but of 'I can.'" To quote George Steiner, "It is verbs, particularly verbs of motion, which enunciate the otherwise inexpressible nature of being. The verb 'to be,' and the assertion 'is' have determined the destiny of man."[17]

Feelings infuse with and are inseparable from perception, linking to the concept of familiarity. They alter the way things look and hence the resulting behavior associated with an object—Freud's *Besetzung* or cathexis. This of necessity is associated with one's social upbringing and the individual's past and future, but since most human bodies have the same basic structure and encounter similar objects in the world, there is a common core of human experience and knowledge.

Merleau-Ponty died of a heart attack at the young age of 53, and Sartrean existentialism rapidly came to dominate the chatter of the Parisian cafés.[18] Yet Merleau-Ponty's ideas are now rebounding in neuroscience, which has much interest in "the social brain," theory of mind (that other people have minds like one's own), empathy, mirror neurons, and mental time travel (how we see the possibility of future events based on past experiences). Most significantly, there is considerable interest in the concept of embodiment.[19]

The development of these ideas emerges in the nineteenth century, "movement images" or an equivalent being discussed by the neurologist Henry Charlton Bastian, William James, Liepmann (*Bewegungsformel*—movement formula), and Head. In effect, the brain anticipates the immediate *Umwelt*; actions are "projacent," as coined by the philosopher John Searle; and we use intentions in action, not reactions. Jeannerod refers to this as "motor cognition": "the motor system (now) stands as a probe that explores the external world, for interacting with other people and gathering new knowledge."[20] Merleau-Ponty put it thus: "Sensations, 'sensible qualities' are then far away from being reducible to a certain indescribable state or *quale*; they present themselves with a motor physiognomy, and are enveloped in a living significance."[21]

The mainstream of Western philosophy, as summed up by George Lakoff and Mark Johnson, has for so long clung to a view that rational thought is conscious, logical, transcendent, and dispassionate.[22] However, the above ideas place the body's physiological processes and anatomical structure as fundamental, not only for the rise of consciousness but also for knowledge, reasoning, and creativity. Here we are after pre-science as prescience. The implications of the arguments of Jackoff and Johnson are profound. Since abstract thoughts are largely metaphorical, and since metaphor is shaped by our bodily interactions with the world, our lived experience, rationality itself is embodied. Rational thought is not disembodied, found in some ether-floating Cartesian ego; concept formation is embedded through the body during ontogeny. As Damasio's theories imply, emotions are integral to this. As he comments, "Our brains receive signals from deep in the living flesh and thus provide local as well as global maps of the intimate anatomy and intimate functional state of the living flesh."[23]

The last philosopher to present is Jaspers. As a young man he studied medicine and worked with Nissl at Heidelberg, the latter a student of Kraepelin, who was also at Heidelberg before going to Munich. Recognized as a major twentieth-century philosopher, his book of concern to neuropsychiatry, *Allgemeine Psychopathogie* (General Psychopathology), was written when he was recuperating from illness, and first published in German in 1913, when he was 30 years old. The seventh edition was translated into English in 1963 and has had a considerable impact on Anglo-Saxon psychiatry.[24]

His name is closely linked to phenomenology, a philosophical term much abused by psychiatrists. The Greek *phainómenon* ("things appearing") is combined with *logia*, hence the science of appearing. The philosophical background of phenomenology links to Husserl, but the theories of Kant and Dilthey all played into Jaspers's thoughts. Recall that Dilthey distinguished between natural and human sciences, separating the law-based explanations of the former from those required to have an understanding of the human person. His *verstehende Psychologie* (understanding) was much concerned with the latter and emphasized attempting to put oneself in the shoes of the other. This was not an ideal science, but an examination of the living connections of reality as experienced in the mind, revealed through mental connections. The dichotomy can be summed up as follows: "We explain nature, but we understand mental life."[25]

Jaspers wanted to capture the essences of mental states, attempting an "objective" descriptive psychology of patients' inner experiences. He wanted freedom from outmoded constructs and biases of psychology (including Freudian theories), examining individual components of mental states as descriptive as those of the histologist: "Whoever has no eyes to see cannot practice histology: whoever is unwilling or incapable of actualising psychic events and representing them vividly cannot acquire an understanding of phenomenology."[26] Understanding required empathy, to immerse oneself in the patient's mental life with knowledge of our own experience. Through this method we come to realize that there are some experiences that are beyond such understanding, essentially the psychotic. Seeking meaningful connections, how one psychic event merges with and emerges from another, what Jaspers called genetic understanding, required a longitudinal dissection of the psyche and was central to this process. Causal explanation is different, hence his distinction between understanding (*Verstehen*) and the latter, referred to as explanation (*Erklären*). Psychic processes alter psychic life without destroying it, while organic processes despoil it.

Jaspers's text is full of descriptions of mental phenomena that are variously classified, and which still form the basis of many of our current psychopathological terms, feeding through to diagnostic manuals such as the later ICD and DSM publications. Sadly, the concept of phenomenology and the elegant analysis of Jaspers's psychopathology, seeking not just causes but meaning, of such relevance for neuropsychiatry, have become reduced to descriptions of signs and symptoms allocated to usually unvalidated ratings scales of psychopathology, and degraded further by requiring only a computer to print out the patient's diagnosis.

Envoi

For many, invading neurology with philosophical speculation seems irrelevant and likely to muddy clear waters, but we have seen how philosophical ideas have been deeply embedded in neuroscience for over 2,500 years. Hughlings Jackson, the founder of modern neuropsychiatry, considered philosophy essential to understanding the organization and functions of the brain. The Pre-Socratics gave us air, water, earth, and fire, the last especially important for Heraclitus—fire animating the soul but also having destructive properties. The four humors, blood, yellow and black bile, and phlegm, gave us four temperaments, sanguine, choleric, melancholic, and phlegmatic, respectively. We have seen how ideas of the constitution of the body via humors and spirits have lingered on. These were vital spirits for Galen, the animal spirits of Descartes, Sydenham's nervous energy, the vital force of the early romantics, and the *élan vital* of Bergson. Four, a magic number, so often underpins such conceptions, aided and abetted by trilogies, another favorite human construct.[27]

The Platonic vision keeps returning and is a thread that runs through the story of the neurosciences and the romantic brain. It is encountered in modern physics: "The Real is more compelling for being the Ideal, and the Ideal is more compelling for being Real."[28] The skillful capacity of the human mind to try to dissect nature at the joints is overwhelming, but as a method for understanding the human brain and its relation to the individual life it would seem to lead to profound paradoxes and be literally and metaphorically mind bending.

Dichotomies are everywhere: Adam and Eve, right and left, black and white, back and front, up and down, and so on; two is the number of division, and the human mind cannot avoid splitting. But philosophical ideas relating to subjective and objective, inner and outer, mind and brain, and body and soul have found the liveliest of thinkers deficient. Historical explanations of how the brain functions fall back on metaphors that are related to the culture of the time. A list of these includes a society and a river (Egypt), signs of the zodiac, horses and chariots, a cathedral, an alembic, a clock, a storage battery, a radio receiver, and the latest contrivance—a computer. As Luria expressed it, the "tendency to introduce naïve concepts to explain the nervous system on the basis of analogies with artificial things is more common in the study of behaviour than anywhere else."[29] The list is lamentable, and we await the next brain-wave.

Other models over time have emphasized hierarchies within the brain and sympathy within and between parts of the body and the brain. These are relevant for the romantic brain and my perspective on significant differences between the neurology that emerged from the late nineteenth century, coalescing into that of the twentieth century, and today's neuropsychiatry. I emphasize not only a scientific interest in the active intellect and creative brain but also a phenomenological approach to clinical practice and an embrace of the embodied nature of human personality and knowledge. The neuropsychiatry that developed sometime in the mid-twentieth century was cast like a high relief on a long historical backdrop but born out of clinical necessity. Too many patients fell between a neurology that had lost touch with the drives, motivations, and personalities that encase symptoms and often signs and a psychiatry that was simply uninterested in the mental states and management of those with disorders such as epilepsy, movement disorders, dementias, the effects of trauma and brain injury, hysteria (now renamed), and the other neuropsychiatric conditions discussed in this book.

Aquinas accepted an active intellect and mind's *potentia*. Freud and Hughlings Jackson used terms such as *energy*. Nietzsche gave us *Rausch*, driving "forces" of emotion behind motion; Schilder used *drive intendings*. William James emphasized the *continuous continuity of consciousness;* the *integrative action of the body and brain* was central to the ideas of Sherrington; and Hess focused on the notion of *zest*. The close ties between thought and motion noted by those as far apart as Coleridge, Hughlings Jackson, Freud, and Schilder, the intrusion of the unconscious into the molding of clinical signs and symptoms, and an emphasis on intentionality and embodiment have eroded the fallacy of the passive receptive brain and the empirical associationist and behavioralist views of the past.

Descartes bypassed all that goes to make up human nature. For Spinoza, "the object of our mind is the body as it exists, and nothing else."[30] We live in three-dimensional space but a four-dimensional universe, and we relate to our world with our bodies; there is no disembodied Cartesian empire that occupies some space that is no place. We are bridled by the drives that have brought *Homo sapiens* to where we now find ourselves; Apollo and Dionysus are metaphorical molding forces of human evolution, the very underpinnings of life and art. Yet this is not simply another split, since, as Nietzsche so strongly emphasized, form and force must coincide, must harmonize, and therein lies creative power. Prometheus is an image of human nature itself; endowed with an unblessed foresight and riveted to a narrow existence, he had nothing with which to oppose the combined and inexorable powers of nature but an unshaken will and the desire to be free.

In the preface I cautioned that while I am able to envisage, as others have done, many different perspectives of historical exegesis, the approach I have taken in this book is that perhaps best referred to as creating a narrative, interweaving past events and personalities, which is what most of us do most of the time. I have tried to relate a story with two separate themes. The first is that of the changing perspectives over time on the way the brain, especially our human brain, interacts with the world it encounters, and the slow but evident and highly significant shift from being viewed as a passive receptacle of sensations (predominantly visual) to an active, creative organ. This perspective, although hinted at in earlier times, first crystallized in the romantic era, and it has been influenced by and has in turn affected associated artistic and literary representations.

Secondly, and most importantly, is the view I have taken that in order to understand the origin of anything that has somehow evolved in the eons-long trajectory of the development of the human brain, there is no advantage—and indeed a considerable intellectual difficulty—in starting out a discovery by taking things as they are today and working backward. A better approach is working from the past to the present. To start out trying to understand, say, our musical or linguistic capacities by taking such skills as we now have them and trying to discern their origins cannot work simply by tracking backward. This will not allow for an illumination of the proto-musical and proto-linguistic elements that developed within the hominid line from a neurological and behavioral, incomplete and nonlinear, database extending back millions of years. This reflection surely also applies to the development of neuroscientific discovery and the evolution of ideas, the latter influenced by cultural and social factors, themselves intertwined with developing and developed constructs.

Neuropsychiatry as a clinical discipline has coalesced out of a lineage that has its heroes and helpers, its decriers and detractors, and its patients with their special problems, in need of special clinical skills. The narrative of my text has carefully avoided such limiting epithets for the discipline as the "bridge" between neurology and psychiatry, "psychiatry is neurology without signs," "neurology is psychiatry—and vice versa," and others. Further, I have avoided the derivative idea that it forms an indiscrete narrow space shaded in between neurology and psychiatry in a Venn diagram. The current status of an independent neuropsychiatry seems secure, and its importance as a recognized discipline within the clinical neurosciences is increasingly accepted. The future, as with all such constructs, is quite open to question. The historian Edward Shorter has discussed how in the course of just over a century, psychiatry has had two tsunamis that have shattered its citadel. The overwhelming embrace of psychoanalytic theory and practice of the first two-thirds of the twentieth century wiped out the psychopathologists' contributions of the previous 100 years. In its turn, the imposition of politically driven and committee-based but apparently nontheoretical structures of ideology and practice from the 1980s, with the introduction of DSM 3 and its successors, toppled the house that Freud built. There is every reason to believe that this house of cards may tumble in the future, and where the residence of neuropsychiatry will be then, no one can tell.[31]

All good stories have a beginning and an ending, but the narrative of our evolution from the campfire to the present has left us with an open, insecure ending. The longest case history in psychiatry must be that of Don Quixote, whose monomania took him on a quest to find and honor the elusive Dulcinea del Toboso, the embodiment of purity and beauty. Ill and dying, the don developed a fever; the illusions that had carried him onward to so many adventures for so many years went away, and he became depressed and despondent. As T. S. Eliot said, "Humankind cannot bear very much reality." We are part being but always becoming, and magnetized by our illusions, we are drawn forward to our futures. But after all, are we not akin to that other Promethean hero, Goethe's Faust, whose fabulous tale was also that of *Das Ewig-Weibliche / Zieht uns hinan* (the eternal feminine draws us onward)?[32]

NOTES

Epigraph: Howard is a 13-year-old boy, being interrogated by the attorney Matthew Harrison Brady, in the play about the Scopes trial that took place in Dayton, Tennessee, in July 1925. Bertram Cates, his teacher, was accused of teaching about evolution and the writings of Charles Darwin.

1. Sherrington C, 1940, 213–214.

2. Hess WR, 1964, 6.

3. Gazzaniga MS, 1998, 63. The motor preintention is but half of the story. Thus, after a sensory stimulus, peripheral sensations reach the cortex of our brains rapidly, in some 20 milliseconds, but they become conscious to us rather later, at a time lapse of around 500 milliseconds. We then refer backward in time when considering the timing of the onset of the stimulus.

4. Jeannerod M, 2006, 65.

5. Bergson H, 1911, 76.

6. Bergson H, 1908/2004, 279.

7. Lissauer was an assistant at one time to Wernicke and is especially known for his descriptions of visual agnosias.

8. Bergson H, 1908/2004, 83, 111. Bergson does identify two kinds of memory, one habit and the other true memory, the latter moving "in the past and not [like habit memory] in an ever renewed present" (195).

9. Translation from Steiner G, 2011, 221.

10. Bergson H, 1908/2004, 24, 168.

11. Merleau-Ponty M, 2002, 370, 373.

12. Wilczek F, 2015, 116.

13. There are several binding problems. The other of significance is how the brain "binds" together information that reaches different areas of the brain at different times to form a complete sensation for us, the phenomenon of our experience. The mind–body problem is still *the* binding problem that exercises so many, especially of a naïve Cartesian outlook.

14. Merleau-Ponty considered that sensations are intentional; one sees green things, not just greenness; sensations are more than sense qualities. Further, they have a figure/background

construction. Merleau-Ponty did not accept total perceptual clarification, since perceptual horizons are always moving. Sense experience is thus not to be confused with sensation, a fundamental philosophical mistake. Ontic: relating to the real, not the phenomenal.

15. Merleau-Ponty M, 2002, 146. He took the neurological case history of a patient of Goldstein's, Herr Schneider. After a stroke, the patient had visual agnosia and apraxia. Merleau-Ponty called this state "psychological blindness" and opined that Schneider's basic problem was not one of vision, but one of tactile sense. Sight and touch are not independent, but provide an integrated experience, which had been lost to Schneider.

Schneider lost "his power of apprehending simultaneous wholes, . . . of taking a bird's eye view of movement and projecting it outside himself. . . . It is never our objective body that we move, but our phenomenal body, and there is no mystery in that, since our body, as the potentiality of this or that part of the world, surges towards objects to be grasped and perceives them" (121).

16. Rostrup N, 2014, 202.

17. Merleau-Ponty M, 2002, 164, 159, 169. Steiner G, 2011, 204. Again, the German language is most helpful. There is an etymological link between the verbs *kennen* (to know) and *können*, the latter of which means "can do," or "to be able," but also has a further meaning, "to know."

18. Sartre was three years older than Merleau-Ponty and collaborated with him as political editor of *Les Temps Modern*. Merleau-Ponty was unhappy with the Sartrean dichotomies of the "in-itself" (unconscious being, unaware and unable to change) and the "for-itself" (conscious being, indeterminate with no predetermined essence, which therefore creates itself out of nothingness—existence precedes essence), as outlined in *Being and Nothingness*. He reconfigured Sartre's "we are condemned to freedom" to "we are condemned to meaning." He mounted several critiques of Sartre's works. He disliked his politics and his theories about the experience of others. The self was not in primordial conflict with the other, selves forever negating each other, but self and other encroached on each other, like a hinge, the subject and the surrounding world viewed as a chiasmus. Merleau-Ponty also had a distaste for the Cartesian aspects of *Being and Nothingness*.

19. This is explored in some detail in my book *Why Humans Like to Cry*. There is a growing interest in what has been referred to as motor pre-presentation. There is growing evidence in neuroscience that there is a motor pre-presentation, a readiness or protension that guides not only what it is that we perceive but even what it is that we might want to perceive.

20. Jeannerod M, 2006, vi. He uses the concept of action representations, a state that represents future rather than present events, which anticipate the action to be performed, and the "state of the world that will be created by the action." Action representations "encode goals" (2, 5). Much early interest emerged from observations on patients with apraxias.

21. Merleau-Ponty M, 2002, 243.

22. Lakoff G, Johnson M, 1999, 513.

23. Damasio A, 2003, 128.

24. Wilhelm Mayer-Gross (1889–1961) was central to introducing Jaspers's ideas to England. Mayer-Gross was also at Heidelberg, but he had to leave Germany. With the help of Mapother, he obtained a position at Maudsley Hospital. The views of Jaspers were not in keeping with Meyerian psychobiology, nor psychoanalysis, hence the relative neglect of Jaspers's ideas in America, until later, when his views, in part echoed by Kurt Schneider, found their way into diagnostic manuals. Schneider's *Klinische Psychopathologie*, published in 1950, contained his delineation of first-rank symptoms.

25. Dilthey W, 1894/1976, 88–97.

26. Jaspers K, 1912/2012, 94.

27. Aside from the religious connotations of the trilogy, this division is in the neuroanatomy of Hughlings Jackson, Elliott Smith, Yakovlev, and MacLean. Four also abounds: four rivers in the garden of Eden, four testaments, four seasons, four compass directions, four horsemen of the Apocalypse, four bases in DNA and RNA, four blood groups, the valence of carbon (fundamental to life), four-letter words, the number of movements in a traditional symphony, and many more.

28. Wilczek F, 2015, 114. Wilczek's book extols the beauty of nature's deep design, found not only in art but in our understanding of the universe. From Pythagoras and Plato to Einstein and quantum mechanics, he finds symmetry and emphasizes it as the basis for comprehending beauty. "The world, in so far as we speak of the world of chemistry, biology, astrophysics, engineering, and everyday life, *does* embody beautiful ideas. . . . Symmetry really does determine structure. A pure and perfect Music of the Spheres really does animate the soul of reality. Plato and Pythagoras: We salute you!" Energy is there to: *"almost all of the nucleon's mass, and therefore almost all of the mass of ordinary matter in the Universe, arises from pure energy."* Wilczek F, 276, 259.

29. Luria A, 1979, 10.

30. Spinoza B, 1883/1955, 92, prop. 13.

31. Shorter E, paper presented at the American Psychiatric Association annual meeting, Toronto, May 20, 2015.

32. "Burnt Norton," pt. 1, lines 42–43, in Eliot TS, 1944. The final quote from Goethe's *Faust* sounds better in German. *Faust*, pt. 2, act 4, lines 12110–12111.

REFERENCES

Bergson H. *Matter and Memory*. Trans. Paul NM, Scott Palmer W. Dover Philosophical Classics, New York; 1908/2004.

Bergson H. *Creative Evolution*. MacMillan, London; 1911.

Calasso R. *La Folie Baudelaire*. Penguin Books, London; 2012.

Damasio A. *The Feeling of What Happens*. Heinemann, London; 2003.

Dilthey W. Ideas about a descriptive and analytical psychology. In: Dilthey W, *Selected Writings*. Ed. Rickman HP. Cambridge University Press, Cambridge; 1894/1976.

Eliot TS. *Four Quartets*. Faber and Faber, London; 1944.

Gazzaniga MS. *The Mind's Past*. University of California Press, Los Angeles; 1998.

Hess WR. *The Biology of Mind*. Trans. Von Bonin G. University of Chicago Press, London; 1964.

Jaspers K. The Phenomenological approach in psychopathology. Trans. Curran JN. Reproduced in: *The Maudsley Reader in Phenomenological Psychiatry*. Ed. Broome MR, Harland R, Owen GS, Stringaris A. Cambridge University Press, Cambridge; 1912/2012.

Jeannerod M. *Motor Cognition: What Actions Tell the Self*. Oxford University Press, Oxford; 2006.

Lakoff G, Johnson M. *Philosophy in the Flesh*. Basic Books, New York; 1999.

Lawrence J, and Lee RE. *Inherit the Wind*. Ballantine Books, New York; 1955.

Luria A. *The Making of Mind: A Personal Account of Soviet Psychology*. Ed. Cole M, Cole S. Harvard University Press, Cambridge, MA; 1979.

Merleau-Ponty M. *The Phenomenology of Perception*. Trans. Smith C. Routledge, London; 2002.

Rostrup N. Environmental neuroaesthetics. In: Lauring JO, ed. *An Introduction to Neuroaesthetics: The Neuroscientific Approach to Aesthetic Experience, Artistic Creativity, and*

Arts Appreciation. Museum Tusculanum Press, University of Copenhagen, Copenhagen; 2014.

Sherrington C. *Man on His Nature*. Cambridge University Press, Cambridge; 1940.

Spinoza B. *On the Improvement of the Understanding; The Ethics; Correspondence*. Dover, Mineola, NY; 1883/1955.

Steiner G. *The Poetry of Thought: From Hellenism to Celan*. New Directions Books, New York; 2011.

Wilczek F. *A Beautiful Question: Finding Nature's Deep Design*. Allen Lane, London; 2015.

ACKNOWLEDGMENTS

In writing this book I am aware of all those colleagues and students who have furnished me with such a rich and stimulating academic and social environment over nearly 40 years of studying, from a clinical and neuroscience perspective, normal and pathological human behavior, and the borderland between the two. The importance of trying to understand the historical backdrop to any discipline was imparted to me by my first consultant colleagues, Dick Pratt and Richard Hunter. Their encouragement kindled my instinct to collect things, especially antiquarian books, hence the ready availability on my own bookshelves of many of the texts that find their way into this book. My interests in comparative neuroanatomy were kindled at medical school, especially by Joe Herbert. Exploration of the subject beyond medical school was continually revived, especially by Paul MacLean and Lennart Heimer; the importance of neuroanatomy for the topic at hand can never be underestimated.

I am continually aware that my clinical life and the all-embracing area of the clinical neurosciences would not exist if it were not for the many patients who present with disorders for which they seek specialist advice. They arouse our curiosity as to the links between the mind, the brain, and human suffering, which compels us to seek an understanding of what it is that has led to such clinical signs and symptoms, with hopeful therapeutic possibilities. I am grateful to them all, many of whom, over years of treatment, became friends.

Many people in one way or another have contributed to the ideas in this book, and several people have commented on various sections. I thank Ted Reynolds, for whom the history of neurology and psychiatry has been an enduring interest. Thanks also to Steve Schachter, Jeff Cummings, Antonio Damasio, Hugh Rickards, Gary Price, Michael Graubart, and especially Dale Hesdorffer for reviewing sections of the text. I thank Nancy Heim for working on some of the images.

Jacqueline Wehmueller at Johns Hopkins University Press has been encouraging and supportive of my endeavors to tackle the difficult task of writing a story of the development of the discipline that today is referred to as neuropsychiatry, but which has absorbed my intellectual fascination from the beginning. Thanks also to Jeremy Horsefield, Courtney Bond, and Hilary Jacqmin for careful attention to the manuscript through to publication. Jackie Ashmenall has helped in the preparation and organization of the book from its outset to completion of the final version.

Page numbers followed by *f* and *t* indicate figures and tables, respectively.

Page numbers followed by *f* and *t* indicate figures and tables, respectively.